Information Dynamics and Open Systems

T0205605

Fundamental Theories of Physics

An International Book Series on The Fundamental Theories of Physics:
Their Clarification, Development and Application

Editor: ALWYN VAN DER MERWE
University of Denver, U.S.A.

Volume 86

Information Dynamics and Open Systems

Classical and Quantum Approach

by

R. S. Ingarden
Institute of Physics,
N. Copernicus University,
Toruń, Poland

A. Kossakowski
Institute of Physics,
N. Copernicus University,
Toruń, Poland

and

M. Ohya
Department of Information Sciences,
Science University of Tokyo,
Tokyo-Noda City, Japan

KLUWER ACADEMIC PUBLISHERS
DORDRECHT / BOSTON / LONDON

A C.I.P. Catalogue record for this book is available from the Library of Congress.

ISBN 978-90-481-4819-6

Published by Kluwer Academic Publishers,
P.O. Box 17, 3300 AA Dordrecht, The Netherlands.

Kluwer Academic Publishers incorporates
the publishing programmes of
D. Reidel, Martinus Nijhoff, Dr W. Junk and MTP Press.

Sold and distributed in the U.S.A. and Canada
by Kluwer Academic Publishers,
101 Philip Drive, Norwell, MA 02061, U.S.A.

In all other countries, sold and distributed
by Kluwer Academic Publishers,
P.O. Box 322, 3300 AH Dordrecht, The Netherlands.

Printed on acid-free paper

Table of Contents

PREFACE

This book has a long history of more than 20 years. The first attempt to write a monograph on information-theoretic approach to thermodynamics was done by one of the authors (RSI) in 1974 when he published, in the preprint form, two volumes of the book "Information Theory and Thermodynamics" concerning classical and quantum information theory, [153] (220 pp.), [154] (185 pp.). In spite of the encouraging remarks by some of the readers, the physical part of this book was never written except for the first chapter. Now this material is written completely anew and in much greater extent.

A few years earlier, in 1970, second author of the present book, (AK), a doctoral student and collaborator of RSI in Toruń, published in Polish, also as a preprint, his habilitation dissertation "Information-theoretical decision scheme in quantum statistical mechanics" [196] (96 pp.). This small monograph presented his original results in the physical part of the theory developed in the Toruń school. Unfortunately, this preprint was never published in English. The present book contains all these results in a much more modern and developed form.

One of the reasons of such a delay in the preparation and publication of the book was the relatively fresh and unfinished state of research in this field and the fast pace of its development. Monographs can be written only when some field is sufficiently developed and relatively stable and ripe. We hope that now such a state is more or less reached, actually only if we strictly delimit our subject from the newest, quickly developing neighbouring fields, such as algebraic and algorithmic information theory, quantum stochastic processes, new trends in phenomenological thermodynamics, and such differential-geometrical methods of thermodynamics, as Finsler geometry to which a separate book, with the co-authorship of RSI, has been devoted recently, published by Kluwer in 1993 [24] (308 pp., in particular Sec. 4.5, pp. 185–199). Going also into these fascinating new fields would cause a substantial increse in the size of our book and a new delay.

A new impulse for writing the present book was given by the invitation of RSI for a visiting professorship and lectures at the Philipps University of

Marburg by Professors G. Ludwig and O. Mehlsheimer (1985, 1987), and then by a series of invitations to Science University of Tokyo by Professor M. Ohya (1986, 1989, 1994). During the latter visits the idea to write the book in collaboration with MO and AK took shape. Thus we came to the project of a book of three authors: two from Toruń and one from Tokyo.

Of course, the book is concentrated mainly, though not only, on the interests and results of the authors. We consider mainly equilibrium thermodynamics (RSI), but generalized to systems with higher-order temperatures (those with strong interactions or relatively small number of particles). In the non-stationary case, elaborated by AK in Chaps. 6 and 7 on isolated and open systems, we confine ourselves mainly, but not exlusively, to relatively slow processes which can be described as Markovian ones. We discuss adiabatic (iso-entropic) processes by means of the generalized Hartree-Fock and Pruski methods. As a general background of the theory, or rather as the second main subject of the book, we take quantum dynamical semigroups, i.e., a formalism of open systems (containing so-called dissipators) in the form initiated and developed by AK and other authors. It is presented here in a new, much more extended and generalized, abstract formulation. Unfortunately, there was no more place and time for elaborating many special examples and applications. We hope that they will be presented soon is some original publications by AK in the newly created international journal "Open Systems and Information Dynamics" (OSID) which will be published by Kluwer Academic Publishers beginning 1997 (since 1992 it was published by Nicholas Copernicus University Press).

In the title of our book we use, except for open systems, the term "information dynamics" and not "information thermodynamics", as used previously, because we have in mind not only the thermal physical application of this formalism. E.g., it may be used in other parts of physics, and also in biology, economy, theory of shapes (pattern recognition), statistical linguistics, etc. In general, it is a formalism of non-complete or modal description (estimation), as it is explained in the Introduction (Chap 1).

MO introduced the concept of the compound state expressing the correlation between the input (initial) state and the output (final) state through a channel transformation. This compound state plays a similar role as joint probability in classical systems, by which he could formulate the mutual entropy (mutual information) in fully quantum systems. Various operator-theoretic and algebraic approaches to quantum entropy are discussed in the book "Quantum entropy and its use" [256] (335 pp.) written by MO and D. Petz, and published by Springer in 1993. In the present book, we discuss quantum entropy, first by discussing more elementary notions and results with full proofs, and then going to more advanced formulations, so that this book is somehow complementary with the one just mentioned.

The term "information dynamics" was first proposed by MO in 1989, presenting yet another point of view. Namely, it is a synthesis of the dynamics of state change and the theory of complexity, which provides a common framework to treat both physical and nonphysical systems together. Quantum mutual entropy and the complexity of information dynamics enabled to define fractal dimensions of a state giving new measures of chaos, different from the entropy, and to define Kolmogorov-Sinai type entropies, improving the original definition of Kolmogorov-Sinai entropy making it more general and easier to compute. Information dynamics is applicable even to analysis of genome sequences.

In spite of all efforts, the authors could not avoid some, we hope small, differences in style, methods of presentation, notation, small repetitions, etc. Chapters 1, 2, 5 were written by RSI (first 5 sections of Chap. 5 are based on the introductory sections of the mentioned habilitation thesis of AK), chapters 6 and 7 by AK, and chapters 3, 4, 8 together with Appendices 1, 2 by MO.

Our book is written not only for mathematicians and mathematical physicists, but also for theoretical and experimental physicists, physical chemists, theoretical biologists, communication engineers, etc., i.e., all who are interested in entropy-information and open systems. We tried to keep the presentation as elementary as possible. However, in the quantum case, some advanced methods are needed. For the reader's convenience, the advanced methods of Hilbert space and operator algebras are explained and recapitulated in the two appendices. Because of the volume limitations we could not go into all proofs, but many essential proofs are given in full detail in the main text.

The authors wish to thank the following colleagues for their invaluable help in the course of preparation of this book: in Toruń Dr. M. Michalski, Dr. P. Staszewski and J. Jurkowski, in Tokyo Dr. N. Watanabe, Dr. H. Suyari, Dr. S. Akashi, Dr. T. Matsuoka and K. Inoue. We also thank European Commission for a support under the contract ERBCIPD-CT 940005.

Toruń–Tokyo, May 1996

R. S. Ingarden
A. Kossakowski
M. Ohya

Chapter 1

Introduction

"...we have no other method of cognition as construction of new and new theories. None of these theories, however, will contain our knowledge."
Andrzej Grzegorczyk, 'Philosophical aspects of mathematics' [114], p. 216.

"One should not mean that 'entropy' is a uniquely determined quantity. In fact, there exist various concepts that — quite often — refer to different logical levels (and have different merits) called 'entropy'."
Alfred Wehrl, 'The many facets of entropy' [338], p. 119.

"There are many faces to the mask of uncertainty." (p. 130) "... the theory of games ... has been of interest in enabling us to circumvent the rather awkward assumption that the statistics of the random effects are known. On the other hand, we certainly cannot take too seriously a theory that assumes that Nature is always directly opposed to our interests." (p. 190)
Richard Bellman, *Adaptive Control Processes* [34]

1.1. Philosophy

Here we use the word "philosophy" in the sense physicists frequently do when speaking about philosophy of some theory, meaning its fundamental idea, a clue. (Only marginally, and in some footnotes, we refer to philosophy in the usual sense of this word.) The main idea and aim of our book is the presentation of such modern methods of physics (and not only pure physics, also biophysics, astrophysics etc.) which are based on incomplete evidence, incomplete measurement, etc., of open complex systems. By an open system we mean a system which is non-isolated, i.e., which contacts with its environment, but is so relatively isolated that a probabilistic, adiabatic or thermodynamic description is possible. In physics, such an approach corresponds to the macroscopic or mesoscopic point of view, connected however somehow with the "microscopic world", i.e., not ignoring it completely as is done in a "phenomenological" treatment. In such cases, we can use concepts of probability, probabilistic entropy (information = uncertainty), fuzzy subsets, stochastic processes, statistical inference, temperature, phase, shape, pattern recognition, statistical decision functions, algorithmic information (algorithmic randomness and complexity), weighted entropy, cost functions, ergodic processes, deterministic chaos, etc. At first sight, it seems that all these concepts, and methods connected with them, can be called *statistical* or *stochastic*, but that would be too rough a simplification. Stochasticity (probabilistic randomness) exists in probability theory, probabilistic information, stochastic processes, fuzziness, but not in algorithmic information, deterministic chaos etc. The relations of all these concepts and methods are not yet completely investigated and at present it is difficult or impossible to present them all together. Nevertheless, in analogy to linguistics (cf., e.g., [271]) and logic (cf., e. g., [334]), we shall call all these methods *modal methods* since they have one property in common: they all are invented as a conscious estimation for the situation when a complete one-to-one physical description of a system is impossible, difficult, or undesirable. (The latter is particularly important for complex systems for too much details can make any orientation impossible.)

In general, a modal utterance has to express a relation of the speaking person to the subject of his own utterance. In particular, this relation can be also emotional and subjective, so it is in general a valuation. But this valuation can also have an objective aspect important for physics or biology, the other cases being interesting for psychology, ethics, etc. Actually, such methods are connected with the deepest feature of the human cognition which is its high selectivity, synthetic and operational character. We can never perceive all the real world. Full information is practically never available and almost never desirable (except for abstract cases in which we select for consideration an artificially limited world). Our brain is huge but finite, as computers are. Also our language, although very elastic, has only a limited number of words and concepts. In linguistics, the "generalizing", simplifying, estimating (or figurating, metaphoric, indirect etc.) methods of speaking are called *modalities*, from Latin *modus* 'measure, quantity, rhythm, limit, restriction, end, method, way' [344]. Modalities express various possible non-categorical attitudes to reality (i.e., except for the categorical indicative mood of saying only definitely "yes" or "no"). Here we use the concept of modality in a very wide sense, when grammarians and logicians usually restrict it to the limited problems they describe, as in logic — the present restriction to the concepts of possibility and necessity.[1] Especially rich in modal expressions is the Japanese language which has thousands ways of expressing modal moods and their combinations by agglutination of many (above 20 different) conjugating verbal suffixes to any verb, according to a special grammatical paradigm, as well as by analytic means of using "formal" (auxiliary) nouns and verbs (e.g., for the so-called aspects). In English, except for a few verbal moods and aspects and analytic expressions of modality, there are special "modals", i.e., "modal auxiliary verbs used to talk about events which are expected, possible, impossible, probable, improbable, necessary or unnecessary. They are: can, could, may, might, will, would, shall, should, must, ought, and need" [79] p.105. In the case of our book the problem arises how to formulate these, or other, modalities mathematically, scientifically, not only by the literary linguistic means. ("Scientifically" means here using, e.g., probability theory, mathematical statistics, information theory, fuzzy set theory, theory of algorithms, pattern recognition theory, artificial intelligence theory, game theory, decision theory, ergodic theory, control and system theory, deterministic chaos theory, etc.). In our book we cannot present all such methods in details, we will mainly concentrate on statistical methods and (statistical) information theory. The concept of entropy, however, is much more general and manysided (as Grad, Wehrl and Bellman said, it has many 'faces' or 'facets') and can be formulated in various theories mentioned above and on many logical levels, e.g., in probability theory, fuzzy set theory, algorithmic theory, topology. One can ask if there is any essential similarity between them, except for the joint name which can be explained historically? Yes, they have similar mathematical properties expressed by axioms, and by the property of modality. E.g., algorithmic information is connected, as probability, with the modality of possibility or impossibility (to shorten a message by algorithmic operations). So in some places of this book we shall briefly mention also the other, non-statistical modal methods.

Philosophers frequently make a sharp distinction between opinions and knowledge. E.g., Bertrand Russell said in his *Unpopular Essays* (this quotation is taken from [34], p.129): "Persecution is used in theology,

---

[1]In linguistics, many cases included by us into modality are usually called moods, categories, aspects, tenses, etc. Recently, perhaps under the influence of Chomsky's transformational grammer, also the term "conversion", "converted sentence", has been applied, cf. one of the most detailed Japanese grammars by Martin [226] p.33, where he uses the expression: "Step three: From nuclear or simplex sentence to converted sentence" and gives a graph of 22 types of basic modal expressions in Japanese. Cf. also an excellent comparative study of Japanese aspect by Majewicz [224]. Possibility (probability) and necessity (causality) are the most important modalities, but there are many other which cannot be ignored.

not in arithmetic, because in arithmetic there is knowledge, but in theology there is only opinion. So whenever you find yourself getting angry about a difference of opinion, be on the guard; you will probably find, on examination, that your belief is getting beyond what the evidence warrants." But modalities, in general, put a bridge between knowledge and opinion. In science, opinions and beliefs are on the same scale, necessary and unavoidable as in everyday life. The difference between scientific opinions and many everyday opinions (when the modality is omitted and the categorical style is used) is, however, that the estimation in science is explicitly or implicitly well-qualified by the degree of certainty, or other condition, contained in this estimation. The degree of certainty is just meant by the very concepts of probability, entropy, fuzziness, etc. All these scientific qualifications (or valuations) are intersubjective, i.e., they can be found and checked by definite procedures by anybody, and therefore they are independent of the subjective arbitrariness, under which so many non-scientific opinions suffer. Under these qualifications a modal opinion can be, in principle, as true as formulations of knowledge, or even as theorems of mathematics. But attention: we actually speak here about the truth in the metamathematical or metalogical sense, of the sentences of the type "on x"(*de dicto*), e.g., about the truth of probability, fuzziness, algorithmic randomness, etc., itself, not about the truth of the concrete case x (*de re*) to which the sentence refers. Thus we see that modality shifts the problem to a higher (meta-) logical level. The "world" in which we actually live is logically a multistore buiding as a skyscraper.[2]

[2]Philosophers of the phenomenological school speak in this connection about the "intentional" truth or about the "intentional world" which is not real but objective. We would rather consider all this world, or worlds, as a physical and biological reality having many layers or types of existence, various *modi existentiae* using a mediaeval term, but only under the mentioned "scientific" conditions. It seems that a modal, not categorical, formulation of a theory relaxes the ideological character of its philosophical interpretation connected with the controversy between idealism and realism. E.g., only in the categorical language such a strongly Platonistic formulation as that of K.Maurin, in [229], p. 784, is possible: "Mathematics is a language, is a Logos, and therefore — as today hermeneutics understands "language" — it is a creative process which only opens and co-creates the dimensions of reality. Without this language (without this mathematics) reality would be closed. Only word, language, Logos allows to see... Here we have to point out with stress that reality without Logos is *not* finished, actually it does not exist, i.e., without Logos, which is a binding material which creates a unity, it would desintegrate into a dust, disconnected atoms." We would like to say that, in principle, we accept this picture, but in a much weakened form of a modal formulation which changes decisively the statement from an idealistic one to a rather, but not exactly, realistic, empiristic conception of the world. (Mathematics in this understanding exists as a logical superstructure of the empiristic reality, its existence is only of different type, but it is definitely intersubjective.) Such philosophy is perhaps something in the middle, as a "middle path" of the Buddhist philosophy, between the two main extremes of philosophy. But we are not philosophers, and this is only a footnote. We can yet add that sometimes even myths or literary fantasies, science fiction etc. can be full of a real sense when considered as modal, not

The universal scientific method of avoiding bias or prejudice (subjectivity) in taking some opinion or judgement is the method of taking the maximum entropy under the condition of the existing evidence (experiment). The first, the most important example, and one of the main subjects of our study is (statistical) thermodynamics. We are interested both in its classical and quantum parts, as in its recent "higher order" generalization by one of the authors (R.S.I.). The latter can be also extended outside of physics, to biology, astrophysics, linguistics, etc., as well as to general information dynamics proposed by another co-author (M.O.). In physics, probability distributions depending on only one statistical parameter, the temperature, are usually considered. We shall show how this picture, the "Gibbs ensemble", can be generalized to many-parameter distributions with "temperatures" of many orders and types. Because of much wider application, even in physics, of such a general theory than only to the thermal phenomena (which is expressed by the name "thermo"-dynamics, from the Greek *thermos* 'hot,warm'), we also use in this book, and in its title, the name "information dynamics".

1.2. From philosophy to physics

Philosophy can be controversial, so it is usually inconclusive, or at least its conclusions are uncertain. But our philosophy, if at all, is rather only methodology or a trial of non-ideological analysis of the cognitive process in physics. Therefore, it is necessary to formulate as clear as possible the conclusions which follow from our analysis.

The first conclusion is that physical theories can be divided into such which are non-modal (in the sense to be deterministic, we abstract here from such "weaker modalities" as potentiality, causality, temporality, relativity, etc.), as classical mechanics, relativity theory, classical field theory (mainly electrodynamics), and such which are modal (non-deterministic in the general sense, based on estimation), as quantum mechanics, quantum field theory, statistical physics, biophysics, statistical astrophysics. The estimation can concern different logical levels of modality, as probability, amplitude of probability (including dynamical and Berry phases), fuzzy sets, entropy, etc. This estimation has to be based on an experimental evidence (measurements) and on a mathematical procedure of getting a unique result of estimation, to avoid subjectivity (as maximization of entropy). In quantum mechanics, the measurement can be microscopic, and if it is one of a complete commuting set of observables, then it

categorical, expressions. E.g., Leszek Kolakowski writes, [193], p. 39, "Therefore, factual history requires myth, philosophy generates myth, and therefore we have no right to consider ourselves as creators of the myth, but rather as its discoverers in each moment."

gives, up to an arbitrary phase, a unique pure state function (amplitude
of probability). But this case is rather exceptional and plays mostly a
rôle of a text-book example. In general, however, the system is much
too complicated to be measured completely, it is called then a *complex
system*. In this situation we can measure either only a small part of the
microscopic eigenvalues, or mean values (macroscopic or mesoscopic) of
some selected observables, or else, in the mixed case, both eigenvalues and
mean values. The first and the third cases are rather difficult for treatment,
so in our book we concentrate only on the second case. But we remark
that in the usual (classical) mathematical statistics (mainly of discrete
social phenomena) the first case is usually considered by means of the so-
called methods of "judging statistical hypotheses". This method is very
elegant and sophisticated, but in physics it is rarely applicable. The central
problem is how from a given incomplete experimental result (a *sample*) to
obtain a probability distribution. R.A. Fisher and many others developed
methods based on concepts of "statistic" and "estimator" (cf. [97], [206]).
Statistic is a random variable being a function of the observed collective
random variable $(X_1, X_2, ..., X_n)$, e.g., $(X_1 + ... + X_n)/n$. *Estimator* of an
unknown parameter a is a statistic $U(a)$ depending on this parameter or
the observed value $u(a)$ of $U(a)$ on a sample. We see that these definitions
are very general and depend on a guess (hypothesis). Kullback [206] used
information theory in the scheme of judging statistical hypotheses treating
information (entropy) as an estimator. From this it is only one step to the
principle of maximum information which was first formulated explicitly (for
the classical case) by E.T. Jaynes in 1957 [173]. (Kullback quoted this paper
in his book [206], but could not use it in his presentation, the paper was
too late.) In our book the principle of maximum information is one of the
principal methods and will be presented in detail below in Chapter 5 (we
start from the quantum case).

A probability distribution (statistical state) in classical statistics is an
element (a function) from the positive part of the real functional space
$\mathcal{L}^1$, which is infinite-dimensional. So the state is defined, in general, by
an infinite number of real coordinates (parameters). In classical mechanics
of finite systems a pure state, being a Dirac function, depends only on
a finite number of parameters (the positions and momenta of particles).
This rather paradoxical situation (that a statistical theory can be more
complicated than the corresponding deterministic one) may be explained by
the remark that in practical experiment we always have a finite number of
linearly independent measurements of mean values, much smaller than the
number of the degrees of freedom of the system. The "representative state"
which represents the maximum of entropy has exactly the same number
of Lagrange coefficients as the number of the mean value measurements

(or of the auxiliary conditions in a variational principle). In such a way the exactness of fixing a mixed state (statistical state) is, in general, much, much lower than that of a pure state. But the essential point is that we have a fixed and unique mathematical procedure which for any macroscopical or mesoscopic experiment gives a unique estimation of the statistical state (both in the classical and the quantum case). Of course, this does not mean that this estimation is good in all cases. We may say only that it is the best possible estimation with respect to the experimental evidence at our disposal. If further experiments appear to be in disagreeemt with the expectation based on our estimation (we can calculate, in principle, any mean value in the estimated state and compare with a measurement of this mean value), then we have to add the result of this measurement to the previous conditions and repeat the procedure of estimation again. Theoretically, this should be done so many times until we obtain a sufficient agreement with any further experiments, up to a given accuracy of measurements. Practically, however, in most cases of thermodynamics even the first measurement is sufficient if we measure a mean value of the total energy of the system, the system is near to be ideal (without strong interactions of particles), and the number of particles is very high (e.g. of the order of the Avogadro number). This fact is caused by the so-called central limit theorem of probability theory (the law of large numbers). This is the reason that usually only one parameter (temperature) is sufficient for the definition of a representative state (estimated state). But at present we have to go outside of the validity of the central limit theorem, especially for small systems in mesoscopic and laser physics, also in applications to biology (control by DNA, hormons, neurons, etc., in a hierarchical system of an organism), and probably in some cases of astrophysics (strong interactions of great masses). This explains the main subject of the present book, in parallel to the problems of quantum information channels which are also unconventional.

Sometimes a question is asked how the mean values can be measured? In the microscopic domain and in the classical mathematical statistics of population, economy, etc., this is not a real problem. We can easily calculate from the results of measurements the arithmetic mean value, although never the exact probabilistic mean value since the probability distribution is yet unknown. (The arithmetic mean value gives the probabilistic one only if the distribution is uniform.) Just because of this difficulty the method of estimators and judging of hypotheses was developed. But in the case of physical macroscopic or mesoscopic measurements the physical quantities are directly (phenomenologically) measured as mean values with respect to the (unknown) microscopic eigenvalues, with accuracy of the measurement. This is a physical principle which was perhaps first formulated by Paul Ehrenfest (1880–1933) in 1927 with respect to the relation between phenomenological and quantum physics. This enables the application of the mentioned method of maximum entropy by a given set of mean values.

Another question is which kind of entropy should be applied? As we will show in the next chapter, there are many types of entropy, even depending on one or many parameters (as Rényi's α-entropy), This depends on the case in consideration since each entropy has its special field of application. The parameter $\alpha \neq 1$ of Rényi's entropy can be interpreted by a type of generalized probability calculus used (mean values linear or nonlinear in probabilities), as we will explain later on, so it is usually out of application since we normally use only the linear case of probability theory.

Philosophically, the method of maximum information is the method of maximum uncertainty, so this method can be considered as a cognitive method of eliminating anything that has not been measured, any bias, any non-experimental favor. It gives reliable results if the set of mean values is reliable (sufficient for defining the macroscopic or mesoscopic situation under discussion).

What is macroscopic and what is mesoscopic? Macroscopic are physical systems of human dimensions having typically the number of molecules of the order of the Avogadro number ($\sim 10^{24}$). It is not so easy to answer the question what is mesoscopic? It depends on the choice of special systems in consideration which are much smaller than the macroscopic ones, but much larger than the usual microscopic elementary systems (small atomic and molecular systems), under condition that these special systems are accessible for statistical treatment and measurement. Large atomic and molecular systems, however, can be treated statistically as mesoscopic (cf. Thomas-Fermi theory of atoms and molecules). But especially interesting and important mesoscopic systems are *genomes* (half of DNA of a cell storing all genetic information of an organism in a haploid chromosome set descending from one parent) of living organisms, plants, fungi and animals, since they determine the growth and organization of these systems. Below we give a brief table of data concerning the most important genomes which are presently under investigation [94]

We see that the number of nucleotides in a genome varies between ca. 200 thousand and 4000 million, i.e., between ca. 10^5 and ca. 10^9. These numbers, although large with respect to 1, are so much smaller than the Avogadro number (ca. 10^{19} or 10^{24} depending on the respective volume, 1 cm^3 in normal conditions or 1 mol) that they represent a new quality which can be called a *mesoscopic domain*. Of course, this definition is relative and not very precise. The relativization is with respect to the assumed discrete elements ("particls") of the system considered as a set (not as a

Organism	Genome in mlns nukleotides	% of recognized sequentions	Year of finishing (expected)
Cow virus	0.19	100	1991
E. coli	4.7	60	(1996)
Yeast*	15.0	35	(1998)
Radish worm	100.0	1	
Nematoid	90.0	3	(1998)
Fly	170.0	2	
Mouse	3000.0	0.3	
Human	3600.0	0.6	(2005)

* Full sequences of chromosomes III – 1992; XI, II i VIII – 1994

TABLE 1.1.

whole with parts). They may be electrons, atoms, molecules, nucleotides, protein molecules, cells, humans, etc. E.g., a system can be also defined as a military army, or a population of a country. Then the elements are humans and the system is in the mesoscopic range. For cosmic dimensions, it is perhaps reasonable to introduce a *cosmoscopic domain*. If the respective cosmic systems relate to such microscopic particles as small atoms and molecules, then the order of magnitude of the number of elements may be, e.g., something in the range 10^{57}–10^{69} which is very large with respect to the Avogadro number. (The first number corresponds roughly to the number of molecules in the Sun, the second to that in a galactic.) In the cosmoscopic domain, the gravitational interaction dominates and spatially uniform Gibbs distribution is rather impossible, being unstable. (In Chap. 5 we shall come back to this problem). If we call a system with $10^5 - 10^9$ elements a *small system* (this concept will be theoretically founded in Chap. 5), it is possible to consider a hierarchy of small systems defining a chain of *big systems* with respect to the smallest physical elements (of the lowest system), e.g., small molecules or atoms. The selection of these systems depends on the accuracy of incomplete (modal) description, i.e., our means of observation and measurement. Each system gives a different picture of nature corresponding to one of nature's "metalogical" levels or layers, its many "faces".

Summing up and completing the picture, we may say that in today's physics the following basic modal concepts are used (or can be used):

1) relativity (with respect to an observer),
2) causality (as necessity),
3) potentiality (potential energy, gauges, etc., as possibility),

4) entropy (information, as uncertainty) of many types,

5) fuzzy sets (as yet not used in theoretical physics),

6) classical probability of finite systems,

7) quantum probability of finite systems of bosons and fermions,

8) general quantum and classical probability of infinite systems,

9) amplitude of quantum probability (with dynamical and Berry phases).

Actually, time (and space, together space-time) is also a kind of modality (in the general sense) since it is a relative element of relativity theory depending on an observer, so it may be considered as contained in 1). We should, however, point out that, for the purposes of the present book, we have in mind mainly the stationary states, i.e., statistical states independent of time, or the equilibrium states in thermodynamics (phenomenologically, they cannot be distinguished from the stationary states). By the methods presented in this book we may consider also (slow) evolution processes of statistical states which can be described by dynamical groups or semigroups (Markovian processes), but not more general stochastic processes which are defined on more general algebras (algebras of processes, not those of states). The latter case can be treated by stochastic equations, classical and quantum (cf., e.g., [124], [304], [309]), diffusion equations, etc., which form already a rather qualitatively different field of mathematics. Because of the limited size of the present book we only slightly touch this subject in Chap. 3, and do not go into details of this rather new and quickly developing field.

1.3. From physics to information theory

Historically, there have been initially two principal approaches to physical reality: either phenomenological, macroscopic, or microscopic, atomic. (In Antiquity these approaches have been represented by Aristotle and Democritus, respectively.) In older times, before the experimental discovery of such "microscopic" particles as molecules, atoms, electrons, quarks, etc., only the first approach was based on experiment, while the second, microscopical one was purely speculative, hypothetical, so not very serious. After the particles have been empirically discovered, however, the situation changed: the basic physical theory became mostly microscopic, while paradoxically, the phenomenological, macroscopic physics obtained a look of a hypothetical theory. Of course, strictly speaking, the latter impression is incorrect: the experiment, either microscopic or macroscopic, has the same basic importance for a theorist, and both types of theories are equally legitimate in their domain, if only they are in accordance with the corresponding experiment. Nevertheless, the situation of double theories became rather awkward and strange. This was rightly expressed by Sir

Arthur Eddington in the introduction to his popular lectures of 1928 "The Nature of the Physical World" [78]: "I have settled down to the task of writing these lectures and have drawn up my chairs to my two tables. Two tables! Yes; there are duplicates of every object about me — two tables, two chairs, two pens... One of them has been familiar to me from earliest years. It is a commomplace object which I call the world... Table No. 2 is my scientific table... My scientific table is mostly emptiness. Sparsely scattered in that emptiness are numerous electric charges..." [3]

At that time (around 1928) statistical theories of physics existed already since at least half a century, but they have not been always sufficiently understood and well-known. Statistical theories put a conceptual bridge between the two types of physics, the two "physical worlds", macroscopic and microscopic, actually making physics again into one. But this unified world has two logical levels, two floors, as we have said above:

(1) of elementary atomic events and phenomena, and
(2) of probability distributions, the thermodynamical states.

In the beginning of the 20th century there were, however, physicists

[3]It is perhaps interesting to remark that in the Western religious tradition, Egiptian, Jewish, Greek, Christian and other, the "second world" (the world after death) is imagined as similarly macroscopic as our present world. In contrast, the Eastern Buddhist tradition imagines this second world, anyhow in its final form, as emptiness, Nirvana, as contrasted to Samsara = the circle of existence. This position caused a tremendous philosophical discussion in Indian philosophy culminating in the person of Nagarjuna (2nd/3rd century) who formulated the so-called *Madhyamika* system (middle position or middle path: *madhyama pratipad* in Sanscrit). This is connected with the so-called "silence" of the Buddha who refused to answer the question if the soul is connected with the body or not, cf. [235], Chap. 2, and [269], Chap. 16. T.R.V. Murti, who is one of the leading specialists on this philosophy, characterizes it as: "This may be taken as the *Modal view of reality*. ... The terminology employed here is after the best Jaina epistemological treatises. Philosophical view, they say, are principally two — the *dravyarthika naya* (substance-view) and *paryayarthika naya* (modal view). ... The Vedanta is cited as the exponent of the extreme form of the Substance-view and Buddhism (*Tathagatamatam*) represents the exclusive Modal view." [235], p. 11. Nagarjuna considered Nirvana as, in a modal sense, identical with the phenomenal world of experience, Samsara, called also "dependent co-arising" *pratitya-samutpada* or else *Sunyata*, the relativity of things. A Japanese Buddhist philospher, Gadjin Nagao, writes [236], p. 4: "Emptiness is not, however, simply nothingness. It is also immediately and necessarily the being of dependent co-arising." How highly modal this identification is meant was expressed, with some exaggeration, by Kolakowski who wrote [193], p. 60: "In mad speculations Nagarjuna, commenting the words of the Buddha, announces that it is impossible to say about the world that it exists or does not exist, or that it exists and does not exist at the same time, or that it neither exists nor does not exist." We would like to comment that our position is not necessarily so extreme or identical with the exclusive Modal view: except of the modal theory of bodies, mind etc., we can also accept the substance theories of microscopic physics. But attention: modern quantum mechanics, as a probabilistic theory based on uncertainty relations (minimum entropy for products of conjugated observables), enables also, in principle, the radical modal view in microscopic physics.

thinking that physics should say only "yes" or "no" about sharply determined objective reality, about the "truth", not about probability ("similarity to truth") which seemed to them as "subjective" (incorrectly: just the sharp truth is not attainable in experiment). Even later some mathematicians still doubted the rigorous mathematical sense of probability since it was before 1933 when the basic article of Andrey Kolmogorov [185] laid down the modern foundations of probability theory in a rigorous mathematical sense. (The earlier similar papers of Steinhaus [310] and Lomnicki [222] were not so well-known and not so complete, cf. [297], Chap. 8.) Around that time even among statistical physicists themselves there were also some remains of doubts connected with the controversy between Zermelo and Boltzmann dating back to the beginning of the century about the statistical explanation of thermodynamics. It seemed that this discussion was not completely settled down by the famous article by P. and T. Ehrenfests of 1912 [82], cf. also [81]. Only the recent studies of deterministic chaos in nonlinear dynamical systems and the development of ergodic theory seem to fill the existing theoretical gap between classical mechanics and thermodynamics, at least in principle.

After Kolmogorov's booklet [185], probability theory was firmly based on mathematical measure theory, and Mark Kac said that "probability theory is a measure theory — with a soul." The soul is just the probabilistic interpretation of a (nonnegative) measure with norm 1. Since then the traditional notation of a *probability space* became $(\Omega, \mathcal{F}, P)$, where Ω is a set (of *elementary events*), $\mathcal{F}$ is a σ-field of subsets of Ω (a Boolean algebra or Boolean ring with unity of *events*) and P is a probability measure on $\mathcal{F}$, $P(\Omega) = 1$. We see that, anyhow for continuous probability measures, we have an essential relativization of the probability measure to an arbitrary σ-algebra $\mathcal{F}$. To diminish this arbitrariness, and to get the simplest properties, mathematicians usually assume in most typical cases that $\mathcal{F}$ is a Baire or Borel σ-algebra, and obtain, accordingly, a Baire or Borel probability measure. [4]

A *random variable* on a probability space $(\Omega, \mathcal{F}, \mathcal{P})$ is an $\mathcal{F}$-measurable

---

[4]We remind the definitions, cf. [350], p.18: "Let S be a locally compact space, e.g., an n-dimensional Euclidean space $\mathbb{R}^n$ or a closed subset of $\mathbb{R}^n$. The *Baire subsets* of S are the members of the smallest σ-ring of subsets of S which contains every compact G_δ-set, i.e., every compact set of S which is the intersection of a countable number of open sets of S. The *Borel subsets* of S are the members of the smallest σ-ring of subsets of S which contains every compact set of S. If S is a closed subset of a Euclidean space $\mathbb{R}^n$, the Baire and the Borel subsets of S coincide, because in $\mathbb{R}^n$ every compact (closed bounded) set is a G_δ-set. If, in particular, S is a real line $\mathbb{R}^1$ or a closed interval on $\mathbb{R}^1$, the Baire (=Borel) subsets of S may also be defined as the members of the smallest σ-ring of subsets of S which contains half open intervals $(a, b]$. (...) A non-negative *Baire (Borel) measure* on S is a σ-additive measure defined for every Baire (Borel) subset of S such that the measure of every compact set is finite. The Borel measure m is called *regular* if

function on Ω, and so on with other probabilistic concepts. (We cannot present here in detail the mathematical foundations of probability theory, cf. [91], [97], [291].) At the end of 1920's such a theory looked very fine and modern, but already in that time the new quantum mechanics with Born's probabilistic interpretation existed and it turned out that the quantum probability is somewhat different and alternative with respect to the brand-new classical measure-theoretic probability. (This fact is not yet noticed by many mathematicians and physicts up to now.) Actually, however, in this case the mathematicians were quick enough: already in 1932 the famous monograph [333] of Johann Von Neumann appeared in which a non-commuting quantum probability was defined as a *density operator*, i.e., a linear self-adjoint positive (nonnegative) operator of trace class with trace 1 in a complex separable Hilbert space of a physical system. (The physical idea was earlier expressed by Dirac and Landau, but a full mathematical theory in Hilbert space is due to Von Neumann.) Since then more than 60 years elapsed and quantum probability (and subsequently quantum groups, quantum geometry etc.) have been developed into a new, non-commuting, branch of mathematics and mathematical physics, cf. e.g. the recent review article of A. S. Holevo [134]. A new level of abstraction was achieved by the methods of operator algebras (C^*-algebras and Von Neumann algebras) which embrace both classical and quantum, finite and infinite systems, and have important applications in statistical mechanics of infinite systems (field theory), cf. [43] and our Chap. 3 and Chap. 4, and Appendix 2. (We call such theory a theory of general quantum systems.) Because of the dominating rôle of functional analysis of Hilbert space in the new quantum probability theory, Holevo [134] traversed Kac's saying as: "quantum theory of probability is a theory of operators in Hilbert space with a soul of statistical interpetation of quantum mechanics." But, although quantum probability is now rather functional-theoretic than measure-theoretic, the measure also appears on a higher level through the Gleason theorem [106].

Statistical thermodynamics is, however, much more than merely a probability theory. Already in phenomenological thermodynamics it was shown that heat cannot be interpreted as a substance, a fluid, and temperature as a level of this fluid. This basic discovery was done by Rudolf Clausius (1822–1888) by a careful analysis of the Carnot cycle (Sadi Carnot, 1796–1832, himself believed yet that heat is a substance), cf. [56]. In 1865

for each Borel set B we have

$$m(B) = \inf_{U \subseteq B} m(U), \tag{1.1}$$

where the infimum is taken over all open sets U containing B. We may also define the regularity for Baire measures in a similar way, but it turns out that a Baire measure is always regular. It is also proved that each Baire measure has a uniquely determined extension to a regular Borel measure." Cf. also [125], Sec. 51–52.

Clausius introduced the name *entropy* for the quantity

$$S = \int \frac{dQ}{T},$$ (1.2)

where Q is heat energy and T is (absolute) temperature. (The letter S introduced by Clausius for entropy is now mostly used by physicists for all types of entropy, much more frequently than the letter H introduced later by Boltzmann, which is now used sometimes by mathematicians. In physics, we have many other traditional symbols H, as for hamiltonian and Hilbert space, so we decided to comply with the physical tradition and use S as much as possible, if necessary with subscripts or indication of arguments. Only for mutual entropy we use the symbol I from "information", although all types of entropy are some kinds of information.) S should "denote [a quantity of] the momentary state of the system. The change of this quantity is a measure of the irreversibility of a thermodynamic process, i.e., of that part of the heat energy which will not be turned into mechanical work by the energy transformation" cf. [129], p. 92. S can only increase or be constant in a "spontaneous process" (the Second Principle of Thermodynamics). Therefore, Clausius called quantity S an "entropy" from the Greek *en* 'in' and *trope* 'way, direction', i.e., he gave it the interpretation of a "directional quantity". Because of some conceptual vagueness and ambiguity of the integral and its interpretation, formula (1.2) and also the name "entropy" became a torture for generations of school teachers and pupils, even for many specialists, physicists and mathematicians. ("Is this a normal integral?" they could ask. To obtain a more precise notation some wrote, instead of dQ, $d'Q$ or DQ as an "incomplete differential", or even Q alone, and one added a note that integration goes along a reversible process. Only in 1909 Constantin Carathéodory explained this question geometrically defining entropy on the level of differential equations [50]. But this explanation, although mathematically clear, was too advanced for many physicists. A simple physico-mathematical explanation is that heat work Q is not a function of state but of a process (a functional). In the phenomenological thermodynamical theory, the notation Q and formula (1.2) can at all be avoided, cf. e.g. the standard textbook of thermodynamics of Callen [48].)

In 1877, Ludwig Boltzmann (1844–1906) came to the same concept of entropy, already from the statistical point of view. He used probability density $f(\omega)$ of particle velocity ω in a gas (the concept introduced earlier by Maxwell) and defined a "quantity H" (treating this letter as the capital Greek *eta* = 'long e', in contradistinction to the capital Greek *epsilon* or

Latin $E =$ 'brief e' used for energy)

$$H = \int f(\omega) \ln f(\omega) d\omega, \tag{1.3}$$

cf. [41] p. 33. It turned out that H is actually the negative entropy

$$S = -kH, \quad k \geq 0, \tag{1.4}$$

where now one uses in the physical IS-scale $k = k_B = 1.38062 \pm 0.00006 \cdot 10^{-23}$ JK^{-1} as a unit factor called Boltzmann's constant. This is an expression introduced by Planck in 1902. In general, it is a positive constant connected with the arbitrary base of the logarithm. This constant defines a unit of entropy. We see that the physical macroscopic IS-unit of entropy, 1 J/K = (Joule)/(Kelvin), is 23 decimal orders of magnitudes greater than the "natural" or "microscopic" unit 1 nat connected with the natural logarithm $\ln = \log_e$ (1 J/K = 0.72431$\cdot 10^{23}$ nat). Boltzmann showed by means of his famous transport equation and "H–theorem" that S increases or is constant during particle collisions, which leads to an equilibrium when the maximum of entropy is reached. For the case of a finite probability distribution,

$$p_i \geq 0 \ (i = 1, ..., n), \qquad \sum_{i=1}^{n} p_i = 1, \tag{1.5}$$

one obtains by analogy

$$S = -k \sum_{i=1}^{n} p_i \ln p_i \geq 0, \tag{1.6}$$

which for the homogeneous "equilibrium distribution" of equal probabilities (the absolute maximum of entropy) $p_i = (1/n)$ gives

$$S = k \ln n, \quad (p_i = \frac{1}{n}, \ i = 1, ..., n). \tag{1.7}$$

Boltzmann treated n as the number of the microscopic states ("complexions"). Later Planck called n (not very luckily perhaps) "thermodynamic probability" and denoted by W. Boltzmann argued that "an actual number of objects is a more perspicuous concept than the mere probability" and that "one cannot speak of the permutation number of a fraction", cf. [212], p. 1. In Planck's form the formula (1.7) became famous, maybe also because at first sight it looks a little mysterious with its logarithm. The logarithm reminds the similarly "mysterious" Weber-Fechner law of psychophysiology, $S = a \ln I + b$, where S is the intensity of sensation, I is the strenght of a stimulus, and a, b are positive constants.

This parallel is perhaps not without a deeper reason in connection with information theory. The deeper sense of the logarithm will be explained in Chap. 2, Sec. 2.2. Friedrich Hasenoehrl (1874–1915), the successor of Boltzmann at his chair in Vienna, wrote that "The theorem that entropy is proportional to the logarithm of probability is one of the deepest, the most beautiful of theoretical physics, yes, of all natural sciences", cf. [129], p. 36. The formula (1.7) is engraved on the Boltzmann's tomb in Vienna. It seems that rather the more general, and earlier, formula (1.6) deserves this honor. Indeed, about 50 years later it was chosen as the foundation of information theory (now: entropy = *information*) by Claude Elwood Shannon [300], cf. also [182] and [90]. (Shannon used $k = 1/(\ln 2)$ which gives binary logarithm lb= $\log_2$ in (1.6) and the unit of entropy called 1 bit = 'binary unit', where 1 nat = 1.442695 bit, 1 bit = 0.6931472 nat, 1 J/K = $1.04496 \cdot 10^{23}$ bit.) But formula (1.7) is also very, very important: twenty years earlier than [300], Hartley [127] used the case (1.7) as the base of his more special theory of transmission of information (communication theory). (Hartley used decimal logarithm ld= $\log_{10}$ in (1.5), $k = (\ln 10)^{-1}$, which gives the unit 1 hartley = 1 det = 3.321928 bit, 1 bit = 0.3010300 det.) It is also natural that quantity (1.7) appeared again in the foundations of general algorithmic information theory formulated by Kolmogorov in 1965 [191] (and independetly by Solomonoff [305] and Chaitin [51]), in this case in connection with the idea of a recursive function. The same formula is also basic for the concept of topological information [13]. Kolmogorov said in [191] that there are three, logically independent, approaches in the quantitative definition of information: the combinatorial (Hartley's) one based on (1.7), the probabilistic (Boltznann-Shannon) one based on (1.6), and the algorithmic (Solomonoff-Kolmogorov-Chaitin) one based again on (1.7) and on recursive functions. The first and third interpretations are mainly connected with computer science, but the second one concerns mostly physics and general statistics. In all these cases entropy can be interpreted as a degree of that *uncertainty* before measurement (observation) which disappears in the act of measurement, being a measure of the *amount of information* obtained. This interpretation is expressed by the "formula"

$$\text{information gained} = \text{uncertainty removed,} \qquad (1.8)$$

cf. [40], p. 62. Now this interpretation makes the "soul" of information theory and all its applications. It seems that this interpretation, so simple and so striking, has been explicitly given only by Shannon, and not by his professor, Norbert Wiener, who in the same year called entropy — information cf. [342], Chap. III. Wiener refers to J. Von Neumann, who yet in his book [333], footnote 202, mentions the identification of entropy with the amount of information, but with the opposite sign, as an "negentropy" $N = -S = kH$, the name proposed later by Leon Brillouin [45], p. 116, and [46], p. 8, "according to a suggestion by Schrödinger" [298]. Von Neumann refers in turn to Leon Szilard's paper [317], a really pioneering work of genius about "Maxwell's demon". Information, as probability, according to our intuition, however, cannot be negative, anyhow for discrete and quantized systems, cf. (1.6). (For continuous systems information according to (1.3), but after adding the negative sign, can be negative when the respective probability density changes over one unit of the integration variable, so this information is actually relative. This case will be discussed in Chap. 2, Sec. 2.6, where it will be also shown that for the general physical case it is necessary to introduce in this formula a dimensional constant under the logarithm sign.)

To explain the difficulty with negentropy as information, we notice that what is here physically important is not $-S$, but the so-called *entropy defect*, i.e.,

$$D = k \log n - S, \tag{1.9}$$

which is always nonnegative (we have in mind the case (1.5)), and may represent "information at our disposal" or "free entropy", similarly as *free energy*

$$F = U - TS, \tag{1.10}$$

(where U is internal energy and T is absolute temperature) which is also nonnegative. When in a "spontaneous" process entropy is maximized, D and F are minimized. That such was also the intention of Schrödinger and Brillouin can be seen from the following quotation from [46], p. 116:

> An isolated system contains *negentropy* if it reveals a possibility for doing mechanical or electrical work: If the system is not in a uniform temperature T, but consists of different parts at different temperatures, it contains a certain amount of negentropy. This negentropy can be used to obtain some mechanical work done by the system, or it can be simply dissipated and lost by thermal conduction. A difference of pressure between different parts of the system is another case of negentropy. A difference of electrical potential represents another example. A tank of compressed gas in a room at atmospheric pressure, a vacuum tank in a similar room, a charged battery, any device that can produce high grade energy (mechanical work) or be degraded by some irreversible process (thermal conduction, electrical resistivity, friction, viscosity) is a source of negentropy.

The negentropy concept, although based on a good physical intuition, but mathematically not complete enough (lacking of the term S_{max}) contributed much to misunderstanding of information theory among physicists. They mostly try to distinguish between "physical entropy" and "informational entropy", as completely different entities, the first "objective" and the second "subjective". Actually, information (amount of information), as probability, is an abstract concept which can be applied both to subjective and objective (intersubjective) cases. (About the subjective information cf. Chap.5 of the book by G. Jumarie [178].) It depends on the definition of a system. Also in physics different systems can be considered, e.g., a system of molecules, and a system of galaxies, treating both a molecule as a particle, and a galaxy as a particle (in different theories, of course). But in all cases the essence of the concept of entropy is the same, the amounts of entropy can be compared between different cases (if taken in the same units of entropy), since entropy is always a real number. As we said, it is a modal concept, as that of probability, which depends on a valuation which can be approximately, but intersubjectively, measured in statistical experiments, if we use the principle of maximum entropy or other estimation criteria. New macroscopic experiments cause, in general, a diminishing of the constrained maximum entropy, so they cause a positive entropy defect with respect to the situation with no, or only one — energetic, experimental constrain. This defect D is just "our information", the actual one, about the system (in comparison to the non-observed or only slightly observed system), while the constrained entropy S can be considered as the entropy of the observed system itself, the potential one, contained yet in the system and possible to rectraction only by a microscopic, not macroscopic, experiment. In physics, however, we never can have a real experimental situation with zero entropy (the complete knowledge, the Third Principle of Thermodynamics). More generally, this is well expressed by the following saying of Einstein, cf. [178], motto:

> As far as the laws of mathematics refer to reality, they are not certain, and as far as they are certain, they do not refer to reality.

Probably only Shannon first understood that the change of sign of entropy is not necessary in the relation (1.8). An intelligent being (*intelligentes Wesen* of Szilard) experimentally can be replaced by a registering machine (machine with memory, an "artificial intelligence") and on a theoretical level can be objectivised by a Boolean algebra of events where a statistical description is performed. Therefore, in spite of a long "prehistory" of the concept of entropy, only 1948, the year of appearing of the fundamental paper by Shannon, seems to be the actual birth moment of the proper information theory. Still in 1925, R. A. Fisher formulated [96], cf. also [206], p. 26, the concept of "Fisher's information matrix" based on the Taylor expansion (with respect to the parameters of the considered probability densities) of *relative entropy* or *entropy gain* of two mutually absolutely continuous probability densities $f(\omega), g(\omega)$ (in mathematical statistics called sometimes *statistical hypotheses*)

$$I = \int f(\omega)[\ln f(\omega) - \ln g(\omega)]d\omega. \qquad (1.11)$$

Fisher's information matrix was used for statistical estimation theory and started the important differential-geometrical interpretations of statistics (Riemannian and Finslerian, cf. [206], p. 26, [22], [24], Sec. 4.5) which, however, will be avoided in the present book. Fisher's information matrix, however, was not yet the full information theory, although the relative entropy (entropy gain) became an important tool of information theory and of its physical and other applications.

It is nothing strange that the 20th century physics developed somehow into a part (or rather an application) of information theory. Experimental and theoretical study of a very complicated hierarchical reality of nature cannot avoid the question of information transmission between nature and a human being or a registering machine. For the first time this question appeared explicitly in relativity theory, where the concept of the observation frame became very essential for all the theory (just this aspect caused a heated discussion). As we mentioned, however, much earlier, through centuries, many other, simpler, modal concepts developed in physics, such as time and space, causality, potential energy, probability, entropy, temperature, gauge, phase, etc. All of them assume some "distance" position from nature itself as an object of investigation. Modalities give "perspectives" to reality and organize hierarchies of "modes of beings" into "ensembles" (sets of individual possibilities) and "metatheories" connected with various types of cognition and certainty, or various aims of our activity (cf. teleology in biology and technology). Linguistics, philosophy, logic, mathematics and metamathematics, even jurisprudence (deontology) frequently tried to analyze modality in many respects and situations. Because of the rather sharp mutual isolation of these various disciplines these investigations, however, did not meet the usual needs and interests of physicists. In the present book, we shall try to discuss the modal aspect of physical and, in general view, also of biological sciences, although the subject is too new and too vast to be treated exhaustively. It is also not the same to use something and to understand this thing theoretically. The future progress may concern especially the biological systems requiring an essential extension of the possibilities of physical methods, not only quantitative, but also qualitative, as we think, just in the modal sense. In particular, we hope that the theory of adaptive feedback control processes, cf. [34], connected with future extended "modal" thermodynamics may perhaps help to understand biological processes. Indeed, the latter cannot be completely explained merely by the reduction to the present "inanimate" physics and chemistry, as is the case in molecular biology. Molecular biology may help tremendously, but reductionism cannot explain the "active" and "global" behavior of the living systems, as was frequently stated in recent times, cf. e.g. [39], [284], [285], where extensive bibliographies can be found. Reductionism is not a good way towards understanding biology, anyhow. Thus we hope that physics will grow itself together with biology, and perhaps also with astrophysics, as the other extreme.

Chapter 2

Classical Entropy

2.1. Basic properties of information

For deeper understanding of entropy it is necessary to study its properties and to compare them with our intuitions about the "everyday" concept of the amount of information. In this connection it is usuful to quote the introductory words of Alfred Rényi to his Appendix about information theory in his excellent book [291]:

> Information theory deals with mathematical problems arising in connection with the storage, transformation, and transmission of information. In our everyday life we receive continuously various types of information (e.g. a telephone number); the informations received are stored (e.g. noted into a note-book), transmitted (told to somebody), etc. In order to use the informations it is often necessary to transform them in various fashions. Thus for instance in telegraphy the letters of the text are replaced by special signs; in television the continuous parts of the image are transformed into successive signals transmitted by electromagnetic waves. In order to treat such problems of communication mathematically, we need first of all a quantitative measure of information. [...] If we wish to introduce such a measure, we must abstract from form and content of the message. We have to work like the telegraph office, where only the number of words is counted in order to calculate the price of the telegram.

(Cf. also the popular book of Rényi published posthumously [292].) These words are perhaps sufficient to understand that from this point of view searching for a numerical measure of a "content" or "meaning" or "importance" of information has no definite way of solution, although the concepts of weighted entropy (cf. [35], [117], Chap.4, and Sec. 2.8 of the present book), and that of cost function can be reasonably introduced. The idea of "content" goes back to some concrete cases, while for the general concept of "information" we need an abstraction from any concrete form. Otherwise, we can express this difference in concepts using the word "message" (a quality) for the concrete case of a symbolic expression, and "information"= amount of information (a quantity) for entropy of this message (two messages can have the same information, but different content). The concept of content can be understood as a localization (address) in a *semantic space* (from Greek *semaino* 'to show by a sign, indicate, make known, point out', so also "semantics" = 'science about meaning' and "semiotics" or "semiology" = 'science about signes'. The concept of semantic space is important for logic and linguistics, and it is very probable that it will be important for physics. Recently, an attempt in this direction was done by G. Jumarie [177], [178], who used some analogy to special theory of relativity, but so far this bold attempt does not seem to be convincing enough. Cf. also the interesting and deep investigations on language and information of Yehoshua Bar-Hillel [32], especially, Chap. 15 of this book written with R. Carnap and concerning the so-called semantic information (but not very semantic in fact). These investigations go, however, in a different direction.

We said in Chap.1 that in the definitions of information (entropy) by Boltzmann and Shannon it is not clear at first sight why they contain the logarithm function. To explain this seeming puzzle, and to understand deeper the concept, it is useful to try to define information axiomatically. There are two levels of this procedure. First, we can try to formulate axioms for information considering the probability concept as well-known, then, in Sec. 2.2, without using this concept. The first method has been applied for the first time by Shannon himself in his pioneer paper [300], Sec. I.6 and Appendix 2, for the case of finite (discrete finite) information (1.6). Since, however, this was done by him only in a sketchy way, F. M. Reza in his book [294] p. 12 called this "a semiaxiomatic point of view". The first rigorous axiomatic formulation was probably given by Khinchin [182] (the original publication in Russian 1953), and then weakened by Faddeyev [87]. Since there are many other axiomatic systems, cf. [11], [12], [117], we will not present them all, but rather enumerate the most important properties of finite discrete information and then mention which combinations of them can be used as an axiomatic system. We use the excellent presentation by Rényi and Balatoni in [293]. They write that entropy (2.1), cf. below, "has all properties which one requires from a measure of uncertainty." As Shannon, these authors use bits as units of information, but we use an arbitrary logarithm base ≥ 2 applying the general notation "log". On the other hand, we use, according to the authors, the subscript 0 for S

indicating the discrete case,

$$S_0(\xi) = - \sum_{k=1}^{n} p_k \log p_k, \qquad (2.1)$$

where ξ denotes a classical (non-quantum) finite *discrete observable* (*discrete random variable*) defined by real values (eigenvalues) $x_k, k = 1, ..., n$, having probabilities p_k. In notation of Wehrl [338], this is the case CD (classical discrete), we may add letter F (finite, i.e., $n < \infty$), CDF. The basic properties of $S_0(\xi)$ are as follows:

1. From $P(\xi = x_i) = 1$ ($i = 1, ...n$, fixed), or using Kronecker's delta $p_k = \delta_{k,i}$, it follows that $S_0(\xi) = 0$. (We assume as usual that $0 \cdot \log 0 = 0$). If ξ is not a constant with probability 1, then $S_0(\xi) > 0$. If observables ξ and η have the same probability distribution, then $S_0(\xi) = S_0(\eta)$. Thus $S_0(\xi)$ depends only on the probability distribution $p = (p_1, p_2, ..., p_n)$ of ξ, and not on the eigenvalues x_i. Therefore, we can also write $S_0(\xi) = S_0(p)$. (Later on we will introduce a generalization of entropy, weighted entropy, for which this need not hold.) Therefore, if we introduce an invertible function $f(x)$, a bijection, on the set $X = \{x_1, ..., x_n\}$, then

$$S_0(f(\xi)) = S_0(\xi) \quad (f \text{— bijection}). \qquad (2.2)$$

2. If $f(x)$ is an arbitrary real function on X, then $S_0(f(\xi)) \leq S_0(\xi)$. If there exist two different values x_i and x_k with $p_i, p_k > 0$ and $f(x_i) = f(x_k)$, then $S_0(f(\xi)) < S_0(\xi)$. This theorem can be proved by means of the inequality

$$x \log x + y \log y \leq (x + y) \log(x + y) \quad (x, y > 0, \ x + y \leq 1). \qquad (2.3)$$

3. Let ξ and η be two CDF observables. Let $P(\xi = x_j) = p_j$ ($j = 1, ..., m$), $P(\eta = y_k) = q_k > 0$ ($k = 1, ..., n$), $P(\xi = x_j, \eta = y_k) = r_{jk}$ (*joint probabilities*), $P(\xi = x_j; \eta = y_k) = p_{j;k} = r_{jk}/q_k$ (*conditional probabilities* of $\xi = x_j$ under the condition $\eta = y_k$). Then we can define the *joint entropy* of ξ and η

$$S_0(\xi, \eta) = - \sum_{j=1}^{m} \sum_{k=1}^{n} r_{jk} \log r_{jk}, \qquad (2.4)$$

and the *conditional entropy* of ξ under the condition of a measurement (with any result) of η

$$S_0(\xi; \eta) = - \sum_{k=1}^{n} q_k \sum_{j=1}^{m} p_{j;k} \log p_{j;k}. \qquad (2.5)$$

Then we can easily prove that

$$S_0(\xi, \eta) = S_0(\eta) + S_0(\xi; \eta). \tag{2.6}$$

If ξ and η are *stochastically independent*, i.e., $r_{jk} = p_j q_k$ for all j and k (we denote this case by $\xi\eta$), then $S_0(\xi; \eta) = S_0(\xi)$ and we obtain

$$S_0(\xi, \eta) = S_0(\xi) + S_0(\eta) \qquad (\xi\eta). \tag{2.7}$$

Analogously, for any finite number of CDF observables $\xi_1, \xi_2, ..., \xi_n$ which are mutually stochastically independent, we have

$$S_0(\xi_1, \xi_2, ..., \xi_n) = \sum_{k=1}^{n} S_0(\xi_k) \qquad (\xi_j\xi_k, \; j \neq k). \tag{2.8}$$

4. Let $\xi_1, \xi_2, ..., \xi_n$ be CDF observables such that their values are mutually *alien* (German *fremd*), i.e., $P(\xi_i = x)P(\xi_j = x) = 0$ for $i \neq j$ and any real x. We construct a mixture η of these observables such that η is equal with probability q_k to ξ_k $(k = 1, ..., n)$ (such an observable may be called a *hierarchical observable*). Finally, let us denote by ζ the observable with values $k = 1, 2, ..., n$ with probabilities $q_1, q_2, ..., q_n$, respectively. Then we easily obtain using property 3

$$S_0(\eta) = \sum_{k=1}^{n} q_k S_0(\xi_k) + S_0(\zeta) = S_0(\zeta) + S_0(\eta; \zeta). \tag{2.9}$$

Below, in Sec. 2.2, we shall call this relation the *law of the broken choice*.

5. Entropy $S_0(\xi)$ depends *continuously* on probabilities in the following sense: for any positive integer n and any $\varepsilon > 0$ there exists a $\delta > 0$ such that

$$|S_0(\xi) - S_0(\eta)| < \varepsilon \tag{2.10}$$

if

$$|p_k - q_k| < \delta \quad (k = 1, 2, ..., n), \tag{2.11}$$

where $p_k = P(\xi = x_k)$, $q_k = P(\eta = y_k)$, $x_i \neq x_k$ for $i \neq k$ and $y_i \neq y_k$ for $i \neq k$.

6. When the number n of different values which a CDF observable ξ can assume is known, the maximal value of $S_0(\xi)$ is taken for the equal probability distribution $p_1 = p_2 = ... = p_n = 1/n$, i.e.,

$$S_0(\xi) \leq \log n = S_{0,\max}(\xi). \tag{2.12}$$

Khinchin [182] proved in 1953 that properties 1,3,5, and 6 are sufficient for fixing entropy (2.1) uniquely up to a positive multiplicative constant (representing an arbitrary choice of the unit of information). So the "enigma" of the logarithm has been explained by this axiom system. Many other weaker axiom systems have been found since then, cf. [87], [320], [181], [214]. Aczel enumerates in [11] more than 50 properties of information and discusses all known combinations of axioms. These problems are also discussed in books [12] and [117]. It seems that the properties discusssed above are sufficiently simple and intuitive to ascertain that the concept of "amount of information" has been well defined. We see that information is in a sense "logarithmic", namely, it is additive under multiplication (concatenation) of independent probabilities, as the logarithm function is.

2.2. Axioms without probability

All the axiom systems mentioned above are constructed under the silent assumption that information (entropy) is a function of the probability distribution or, in other words, that entropy is a property of the probability distribution. However, is such an assumption necessary? Now we know, after the introduction of algorithmic information and topological information in the later 1960's, that it is not necessary. But around 1960 it was not so obvious. One of the present authors (RSI) asked then this question and tried to construct an axiom system of information "without probability", i.e., without using the concept of probability. At that time he had an opportunity to discuss this problem in Budapest with one of the most outstanding experts of probability theory and information theory, Alfréd Rényi (who died untimely in 1970). Maybe, since Rényi was so intimately connected with these two branches of science in his activity, treating the latter as a part of the former, he could then not imagine that they could be disconnected. So Rényi was definitely against such a generalization: that in some systems information can exist while probability not. To some extent this position influenced the solution of this problem proposed by Urbanik and Ingarden in [136] and [137]. But anyhow such a solution was needed first. Namely, we have proved that it is possible to define information axiomatically, without using the concept of probability, in such a way that probability can be then defined by means of information so that the Boltzmann-Shannon expression for information can be proved as a theorem. In such a formulation the concept of information appears to be a primary concept of statistics, and that of probability a secondary one. Thus it is not necessary to consider information theory as a part of probability theory, also the reverse point of view is possible.

Actually, the problem had a history of 12 years (1961-1973), since the first publication [136] up to the final one [329], cf. also [117] Sec. 3.1 pp. 29–41 and 87–88. After the first publications, the first change consisted in an essential simplification of the axioms to make them more intuitive, [139], [142], and [146]. The second change concerned the so-called regularity of information, cf. [76], used in the proof of the theorem and previously too restrictively defined, with a proposition of correction. This correction was taken into account in the new complete proof given by Urbanik in [329]. Here we present this final stage of the art giving both formulations of the axioms, from [139] and [329], being equivalent and differing only inessentially. Since the proof of the equivalence with the Shannon definition is rather long and complicated, we omit it here.

We first quote few sentences from [329], p. 289, which explain our philosophical and mathematical position: "As it was pointed out in [136], information is seen intuitively as a much simpler and more elementary notion than that of probability. It gives a cruder and more global description of some physical situations than probability does. Therefore, information represents a more primary step of knowledge that that of cognition of probabilities. It should be noted that the axioms proposed in [136] and [137], in particular that which gives a connection between information of rings and their subrings, are very difficult to grasp in their intuitive meaning. On the other hand, the simplified axioms for information given in [139], [142] and [146] contain the weight factors which are nothing else than classical Laplacean probabilities for S-homogeneous rings. The aim of the present note is to propose a modification of the system of axioms of [137], which has rather simple intuitive meaning and the informational feature. Furthermore, we shall prove that the information determines the probability uniquely, except for some degenerate cases." Cf. also [328] (without proof).

We shall consider sets of finite non-trivial Boolean rings (of events) $A, B, C, ...$ containing the zero elements denoted by $0_A, 0_B, 0_C, ...,$ respectively. The rings are considered with the meet ($\cap$) and join ($\cup$) operations, so the *unit element* 1_A is the join of all elements of A. A Boolean ring A is *non-trivial* if $0_A \neq 1_A$. Non-trivial subrings will be briefly called *subrings*. Let $a \in A$ and $a \neq 0_A$. We shall denote by $a \cap A$ the subring of A consisiting of all elements of A contained in a. If $a_1, a_2, ..., a_m$ and $A_1, A_2, ..., A_n$ are elements and subrings of A, respectively, we shall denote by $[a_1, a_2, ..., a_m, A_1, A_2, ..., A_n]$ the least subring containing all elements $a_1, a_2, ..., a_m$ and all subrings $A_1, A_2, ..., A_n$.

A set $\mathcal{L}$ of finite non-trivial Boolean rings will be called a *Boolean ladder* if the following conditions are satisfied:

(L1) if $A \in \mathcal{L}$ and B is a subring of A, then $B \in \mathcal{L}$,

(L2) for any $A \in \mathcal{L}$ there exists such a $B \in \mathcal{L}$ that A is a proper subring of B.

To get a richer structure for our purpose we assume that $\mathcal{L}$ is equipped with the concept of the limit, i.e., it is a L^*-*space* in the sense of Fréchet, cf. [208] pp. 83–84. That means that $\mathcal{L}$ contains convergent sequences of elements $A_1, A_2, ..., A_n, ...$ having a unique limit element $A = \lim_{n \to \infty} A_n$, or $A_n \to A$, such that

(L3) if $k_1 < k_2 < ...$, then $\lim_{n \to \infty} A_{k_n} = A$,

(L4) if $A_n = A$ for every n, then $\lim_{n \to \infty} A_n = A$,

(L5) if a sequence $A_1, A_2, ...$ is not convergent to A, then there exists a subsequence $A_{k_1}, A_{k_2}, ...$ ($k_1 < k_2 < ...$) having no subsequence convergent to A.

Additionally we assume that $\mathcal{L}$ fulfilles the following axioms:

(L6) if $A_n \to A$, then for every subring B of A there exist subrings B_n of A_n, respectively, such that $B_n \to B$,

(L7) if $A_n \to A$, then each sequence of subrings of A_n contains a subsequence convergent to a subring of A.

We can make a comment that the above conditions, although they may look complicated, are in fact extremely simple and weak. It is not assumed that a Boolean ladder is a metric or pseudometric space, the L^* conditions (L3)–(L5) are weaker, while (L5) and (L6) express the obvious consequences of the fact that $\mathcal{L}$ with a ring A contains also all its subrings.

If we define a real function F on a Boolean ladder $\mathcal{L}$, $F : \mathcal{L} \to \mathbb{R}^1$, we get a decomposition of $\mathcal{L}$ into classes of equivalent rings. Two rings, A and B from $\mathcal{L}$, are said to be F-equivalent, symbolically $A \sim_F B$, if there exists an isomorphism h of A onto B such that $F(C) = F(h(C))$ for any subring C of A. Furthermore, a ring A from $\mathcal{L}$ is said to be F-homogeneous if for every automorphism g of A and every subring B of A we have $F(B) = F(g(B))$. The intuitive meaning of F-homogeneity is the maximal uniformity of the ring with respect to the function F, which means that all the isomorphic subrings of the ring have the same value of F.

Let $\mathcal{L}_{\mathcal{F}}$ denotes the set of all F-homogeneous rings from $\mathcal{L}$ with at least three atoms and all their subgroups. We say that a real-valued function F is *regular* on $\mathcal{L}$ if it is continuous and $\mathcal{L}_{\mathcal{F}}$ is dense in $\mathcal{L}$.

It can be easily proved that if a Boolean ladder $\mathcal{L}$ admits a real-valued function F, it is a pseudometric space with the pseudometric distance defined as proposed in [76]

$$\rho_F(A, B) = \begin{cases} \min_h \max_C |F(C) - F(h^{-1}(C))| & \text{if } n(A) > n(B), \\ \min_g \max_D |F(D) - F(g^{-1}(D))| & \text{if } n(A) \leq n(B) \end{cases}, \quad (2.13)$$

where $A, B \in \mathcal{L}$, $n(A)$ is the number of atoms of A, C is any Boolean subring of A, D is any subring of B, h is any homomorphism of A on B, and g is any homomorphism of B on A. By *pseudometric* we mean a non-negative symmetric function of A and B which disappears for $A = B$, but not necessarily only then.

Now we are able to formulate the definitions of information as proposed by Urbanik [328] [329] and Ingarden [139] [146]

DEFINITION 2.1 (Urbanik). A real-valued regular function $S = S_0$ defined on a Boolean ladder $\mathcal{L}$ is said to be *information* on $\mathcal{L}$ if it has the following properties:

(S1) *The law of the broken choice.* If $a_1, a_2, ..., a_n \in A, a_i \cap a_j = 0_A (i \neq j), a_1 \cup a_2 \cup ... \cup a_n = 1_A$ and $a_i \cap A \sim_S a_j \cap A \ (i, j = 1, 2, ..., n)$, then

$$S_0(A) = S_0([a_1, a_2, ..., a_n]) + S_0(a_1 \cap A). \qquad (2.14)$$

(S2) *The local character of information.* Let a, b and A be elements and a subring of a ring from $\mathcal{L}$, respectively. If a, b and 1_A are disjoint, then

$$S_0([a, b, A]) - S_0([a \cup b, A]) = S_0([a, b, 1_A]) - S_0([a \cup b, 1_A]). \qquad (2.15)$$

(S3) *Indistinguishability.* Isomorphic S_0-homogeneous rings from $\mathcal{L}_\mathcal{H}$ with at least three atoms are S_0-equivalent.

(S4) *The principle of increase of information (monotonicity)* . If A is S_0-homogeneous and B is an S_0-homogeneous proper subring of A, then

$$S_0(B) < S_0(A). \qquad (2.16)$$

This definition is equivalent with the following one (we try to use formulations as similar as possible):

DEFINITION 2.2 (Ingarden). A real-valued regular function S_0 defined on a Boolean ladder $\mathcal{L}$ is said to be *information* on $\mathcal{L}$ if it has the following properties:

(S1') *The law of the broken choice.* If A is an S_0–homogeneous ring of $\mathcal{L}$ with $N = N(A)$ atoms and $a_i \in A, N_i = N(a_i) \neq 0, a_i \cap a_j = 0_A (i, j = 1, 2, ..., n, i \neq j), a_1 \cup a_2 \cup ... \cup a_n = 1_a$, then

$$S_0(A) = S_0([a_1, a_2, ..., a_n]) + \sum_{i=1}^{n} \frac{N_i}{N} S_0(a_i \cap A). \qquad (2.17)$$

(S2') = (S3).
(S3') = (S4).

We see that Definition 2 is simpler and more symmetric than Definition 1. On the other hand, Definition 1 avoids using expressions $N_i/N = p_i$ which correspond to the Laplacian probabilities. In Definition 2, however, they are expressed by the direct properties of rings, so they are defined without explicit use of the concept of probability. By the way, our axiom (S1') has been already used (in a special case) by Shannon in his first axiomatic definition of information [300], Sec. I.6, Point 3.

We also see that the principal axiom of both definitions, the law of the broken choice, concerns the hierarchical systems (systems with hierarchical structure).

The final aim of the definitions is a proof, cf. [329], Sec. 4, of the

THEOREM 2.3. *Let S_0 be an information of a Boolean ladder $\mathcal{L}$. Then for every $A \in \mathcal{L}$ there exists a strictly positive probability measure p_A defined on A such that for every subring B of A the formula*

$$S_0(B) = -k \sum_{k=1}^{r} \frac{p_A(b_k)}{p_A(1_B)} \log \frac{p_A(b_k)}{p_A(1_B)} \qquad (2.18)$$

holds, where k is a positive constant and $b_1, b_2, ..., b_r$ are atoms of B. In particular,

$$S_0(A) = -k \sum_{k=1}^{n} p_A(a_k) \log p_A(a_k), \qquad (2.19)$$

where $a_1, a_2, ..., a_n$ are atoms of A. Moreover, the measure p_A is uniquely determined for all rings A with at least three atoms. Finally, p_A is uniquely determined on two-atomic rings up to a rearrangement of atoms.

The converse is also true, cf. [329], p. 294. We have to point out that in the present section information S_0 is treated as a function of a Boolean ring belonging to a Boolean ladder $\mathcal{L}$, while in the previous section it was considered as a function of an observable. Of course, it is no contradiction: the present treatment is, strictly speaking, a generalization of the previous concept of entropy. Traditionally, entropy is considered as a number, a functional of a probability distribution (of an observable). Now we consider it as a function on a Boolean ladder of Boolean rings and their subrings. Only because of this generalization we can say that entropy is a more primitive concept than that of probability. But, since the converse is also true, entropy and probability are as if equivalent, but different, as the two Janus faces of the same god, statistical modality.

It is obvious that by an appropriate generalization of axioms it is possible to generalize the concept of information itself to make it really free from that of probability. As an example we may take the Solomonoff-Kolmogorov-Chaitin algorithmic information $S(x)$ of an "object" x. For this information, instead of property (2.5) we have the more general one

$$S(x, y) = S(x) + S(x|y) + O(\log S(x, y)), \qquad (2.20)$$

where $O(v)$ is the well-known order symbol (i.e., $u = O(v)$ means that u is of the same order as v, $|u/v|$ being bounded), cf. [192], eq. (4'). As this example shows, the problem of generalization of the usual entropy concept is not a trivial one and requires special modal intuitions, cf. also [74], where some additional literature is quoted.

In recent decades, some applications of algorithmic information to physics and biophysics are developed, cf. [353], [354], [348], and the literature quoted there, but we cannot present them here because of a quite different mathematical background (recursive functions, theory of algorithms, cf. [352]).

2.3. Dimension of information

In this Chapter, we have used the notation S_0 for entropy in the discrete case, and the reader may ask what is the reason of the index 0. Indeed, it was not explained so far, but now we can say that it corresponds to the concept of dimension of information. The usage of this index beforehand should suggest that we consider only a special case of information, information of dimension zero which corresponds to a discrete probability distribution. The number of discrete elementary events can be finite (as was assumed up to now), but, in general, it can be also infinite. In the latter case information is not always convergent (finite). But it can diverge only in the proper sense, namely to the value $+\infty$ since all the terms in the sum are nonnegative. Therefore we can assume $+\infty$ as the value of entropy in the case of divergence. Unfortumately, it is impossible to go over from a discrete case to the continuous case by a limit procedure in a unique way. Indeed, the integral depends on the Lebesque measure chosen on $\mathbb{R}^1$ for integration. The concept of the "uniform" density on the infinite straight line cannot be uniquely defined, and the unit of measure is arbitrary. (For discussion of this problem cf. [117], pp. 16–28, and our Sec. 2.6, where we treat the question of units by a special constant.) Another proof of the discrepancy of two entropies is that the discrete entropy is always nonnegative, while the continuous entropy can be also negative. Indeed, the continuous entropy (2.21), cf. below, of the uniform probability distribution in the interval $(0, a)$, $a > 0$ (and 0 outside of this interval), is equal to $\log a$, so it is negative for $0 < a < 1$ and zero for $a = 1$. Therefore, it is better to treat the continuous entropy of Boltzmann as an essentially new type of entropy denoted by Rényi and Balatoni [293] by $(x \in \mathbb{R}^1)$

$$S_1(\xi) = -\int_{-\infty}^{+\infty} f(x) \log f(x) dx, \qquad (2.21)$$

i.e., as entropy of dimension one, where function $f(x)$ is a probability density of a real observable ξ. We assume here the existence of the probability density and the convergence of the integral (2.21). Here the integral can be essentially divergent, so entropy does not always exist.

The investigation of Rényi and Balatoni [293] shows that, except for S_0 and S_1, one can still introduce infinitely many entropies with positive, not necessarily integer, dimension d. We shall sketch this idea as follows.

Let ξ be a bounded, but besides arbitrary random variable (observable). We put

$$\xi^{(n)} = \frac{[n\xi]}{n}, \tag{2.22}$$

where $[x]$ denotes the integer part of x. Therefore,

$$\xi^{(n)} = \frac{k}{n} \qquad \text{for} \qquad \frac{k}{n} \leq \xi < \frac{k+1}{n}. \tag{2.23}$$

If $|\xi| < K$, then $\xi^{(n)}$ has no more than $2nK$ different values. Therefore, $S_0(\xi^{(n)}) \leq \log 2nK$ and

$$\limsup_{n\to\infty} \frac{S_0(\xi^{(n)})}{\log n} \leq 1. \tag{2.24}$$

If the limit value

$$d(\xi) = \lim_{n\to\infty} \frac{S_0(\xi^{(n)})}{\log n} \tag{2.25}$$

exists, we call it the *dimension of the probability distribution of ξ*. It is of course $0 \leq d(\xi) \leq 1$. Instead of $d(\xi)$ we briefly write d. The *d-dimensional entropy* of the distribution of ξ is defined by

$$\lim_{n\to\infty} [S_0(\xi^{(n)}) - d\log n] = S_d(\xi) \tag{2.26}$$

under the assumption that this limit exists. If the limits (2.25) and (2.26) do not exist (even in the improper sense, i.e., as $+\infty$ or $-\infty$), dimension and entropy of ξ cannot be defined (anyhow in this sense).

We recall the definition of the absolute continuity of a *distribution function* (called sometimes a *distribuant*, not to be mixed with probability distribution or density distribution, by which we always mean the probability density for continuous distributions!), namely, of $F(x) = P(\xi < x)$ for a random variable ξ, cf. [291], p. 175. (Not to be mixed with the absolute continuity of two probability measures mentioned in Chap. 1). A distribution function F is said to be *absolutely continuous* if for any $\varepsilon > 0$ there exists $\delta > 0$ such that for any system of disjoint intervals (a_k, b_k), $k = 1, 2, ..., n$, $a_k < b_k$, the inequality

$$\sum_{k=1}^{n} (b_k - a_k) < \delta \tag{2.27}$$

implies

$$\sum_{k=1}^{n} |F(b_k) - F(a_k)| < \varepsilon. \tag{2.28}$$

It is known that every absolutely continuous function is almost everywhere differentiable and is equal to the indefinite integral of its derivative. This is a necessary and sufficient condition of the absolute continuity of a function.

One can easily prove that for a bounded random variable ξ ($|\xi| < K$) with an absolutely continuous distribution function F and with a continuous probability density $f(x)$ the relations

$$d(\xi) = 1, \quad S_1(\xi) = -\int_{-K}^{+K} f(x) \log f(x) dx \tag{2.29}$$

hold. It is namely

$$S_0(\xi^n) = - \sum_{-nK < k < nK} f(z_k) \log f(z_k) \frac{1}{n} + \log n, \tag{2.30}$$

where

$$\frac{k}{n} \le z_k \le \frac{k+1}{n}. \tag{2.31}$$

If ξ has a discrete probability distribution and $P(\xi = x_k) = p_k, k = 1, 2, ..., x_j \neq x_k$ for $j \neq k$, then one obtains

$$d(\xi) = 0, \quad S_0(\xi) = - \sum_{k=1}^{\infty} p_k \log p_k. \tag{2.32}$$

Let us now combine two distributions: a discrete one $\{p_k\}$ and a continous one with a bounded and absolutely continuous distribution function F having a continuous density function $f(x)$, with probabilities $1 - d$ and d ($0 < d < 1$), respectively. Then one obtains the intermediate case, $d(\xi) = d$ and

$$S_d(\xi) = -(1-d) \sum_{k=1}^{\infty} p_k \log p_k - d \int_{-\infty}^{\infty} f(x) \log f(x) dx - (1-d) \log(1-d) - d \log d, \tag{2.33}$$

which corresponds to the hierarchical distribution mentioned above on the occasion of the law of the broken choice.

The presented theory can be somewhat generalized by a suitable axiomatic formulation, cf. [293], pp.126–132, but we avoid this matter here.

The theory can be also generalized for higher dimensions. We have the following

THEOREM 2.4. [293], p.133. *Let ξ and η be two arbitrary random variables, and let $d(\xi|\eta)$ be the dimension of the conditional distribution of ξ under the condition that η is known. One denotes by $S_{d(\xi|\eta)}(\xi|\eta)$ the entropy of the conditional distribution of ξ under the condition that η is known. Then one obtains, under the assumption that all the occurring quantities exist,*

$$d(\xi, \eta) = d(\eta) + d(\xi|\eta), \quad S_{d(\xi,\eta)} = S_{d(\eta)}(\eta) + S_{d(\xi|\eta)}(\xi|\eta). \tag{2.34}$$

If, in particular, ξ and η are statistically independent, then

$$d(\xi, \eta) = d(\xi) + d(\eta), \quad S_{d(\xi,\eta)}(\xi, \eta) = S_{d(\xi)}(\xi) + S_{d(\eta)}(\eta). \tag{2.35}$$

In particular, one obtains for n statistically independent variables $\xi = (\xi_1, \xi_2, ..., \xi_n)$ and their joint probability density $f(x)$, $x = (x_1, x_2, ..., x_n)$

$$S_n(\xi) = -\int_{-\infty}^{+\infty} ... \int_{-\infty}^{+\infty} f(x) \log f(x) d^n x. \tag{2.36}$$

2.4. Information gain and mutual information

To obtain a heuristic derivation of the information gain we use, following Rényi [291], Appendix, Sec.4, the same simple example which was the base of our axiom (S1') of Shannon's entropy, the law of the broken choice. Let E be a nonempty set of N elements and let us consider a partition of E into $n > 1$ disjoint nonempty subsets $E_1, E_2, ..., E_n$ having N_j, $j = 1, 2, ..., n$, elements, respectively, so $N = \sum_{j=1}^{n} N_j$. We assume, as in (S1'), that elements of E are equally probable (the corresponding ring is S_0-homogeneous), so

$$p_j = \frac{N_j}{N} > 0, \quad (j = 1, 2, ..., n), \tag{2.37}$$

are probabilities of finding an element of E in E_j. Labelling elements of E by an index k, e_k, $k = 1, 2, ..., N$, we may treat this index as a random variable ξ. The index $j = 1, 2, ..., n$ of the partition sets E_j will be treated as an observable η and the index of an element in E_η as an observable ζ. Then we can write the law of the broken choice as

$$S_0(\xi) = S_0(\eta) + S_0(\zeta|\eta) \tag{2.38}$$

where, of course,

$$S_0(\xi) = \log N, \quad S_0(\eta) = -\sum_{j=1}^{n} p_j \log_2 p_j, \tag{2.39}$$

and

$$S_0(\zeta|\eta) = \sum_{j=1}^{n} p_j \log N_j = \sum_{j}^{n} p_j \log p_j + \log N, \qquad (2.40)$$

which checks eq. (2.38).

Now let E' be a nonempty subset of E and $E'_j = E_j \cap E'$. If N'_j is the number of elements of E'_j and N' that of E', we have $N' = \sum_{j=1}^{n} N'_j$ and

$$q_j = \frac{N'_j}{N'}, \qquad (j = 1, 2, ..., n) \qquad (2.41)$$

are probabilities of finding an element of E' into E_j.

Now we ask the following question: suppose we know that an element of E chosen at random belongs to E', what amount of information will be furnished hereby about η? In other words, what is the *information gain* obtained by such an experiment? The initial (before the experiment, therefore called *a priori*) probability distribution of η was $p = (p_1, p_2, ..., p_n)$. After the experiment (measurement) told us that the chosen element belongs to E', η has the final (*a posteriori*) probability distribution $q = (q_1, q_2, ..., q_n)$. At first sight it seems that the information gain is equal to $S_0(p) - S_0(q)$. But this is impossible since the difference can be also negative, while, according to our intuition, the information gain should be always nonnegative (it will be zero only when $E = E'$ and then $p = q$).

Let the quantity we look for be denoted by $I_0(q|p)$. The statement $e_\zeta \in E'$ contains the information $\log(N/N')$. This information consists of two parts: 1) the *information gain (relative information)* $S_0(q|p)$ given by the statement about the value of η, 2) the information given by the statement about the value of ζ if η is already known. The second part is easy to calculate: if $\eta = j$, the information obtained is equal to $\log(N_j/N'_j)$, but since this information occurs with probability q_j, we have

$$\log \frac{N}{N'} = S_0(q|p) + \sum_{j=1}^{n} q_j \log \frac{N_j}{N'_j}. \qquad (2.42)$$

Since

$$\sum_{j=1}^{n} q_j = 1, \qquad \frac{N N'_j}{N' N_j} = \frac{q_j}{p_j}, \qquad (2.43)$$

we finally obtain

$$S_0(q|p) = \sum_{j=1}^{n} q_j \log \frac{q_j}{p_j}, \qquad p_j > 0 \quad (j = 1, 2, ...n). \qquad (2.44)$$

Nonnegativity is a very important property of information gain,

$$S_0(q|p) \geq 0. \tag{2.45}$$

PROOF: Let us consider the following function of $x > 0$ (here we assume for simplicity that in (2.45) $\log x = \ln x$, but for any other base of the logarithm the proof is similar)

$$f(x) = x - 1 - \log x, \quad f(1) = 0. \tag{2.46}$$

Since

$$f'(x) = \frac{x-1}{x} = 0 \quad \text{only for} \quad x = 1 \tag{2.47}$$

and

$$f''(x) = \frac{1}{x^2} \geq 0, \tag{2.48}$$

we obtain that function

$$f(x) \geq 0 \tag{2.49}$$

and vanishes only for $x = 1$. Now let us put

$$x = \frac{p_j}{q_j}. \tag{2.50}$$

One obtains from (2.49), multiplying by q_j and changing sign

$$q_j - p_j \leq q_j \log \frac{q_j}{p_j}. \tag{2.51}$$

Summing up from $j = 1$ to n one finally obtains (2.42). $\square$

Now, we introduce a very important quantity being a special case of information gain. Recalling our equation (2.6),

$$S_0(\xi, \eta) = S_0(\eta) + S_0(\xi; \eta), \tag{2.52}$$

one obtains (by a similar proof as above)

$$S_0(\xi; \eta) \leq S_0(\xi), \tag{2.53}$$

and therefore the *subadditivity* of S_0

$$S_0(\xi, \eta) \leq S_0(\xi) + S_0(\eta). \tag{2.54}$$

It is natural to consider the difference called a *mutual information* (*transinformation*)

$$I_0(\xi, \eta) = S_0(\xi) + S_0(\eta) - S_0(\xi, \eta) = S_0(\xi) - S_0(\xi; \eta). \tag{2.55}$$

Because of (2.4) and

$$p_j = \sum_{k=1}^{n} r_{jk}, \quad q_k = \sum_{j=1}^{m} r_{jk}, \tag{2.56}$$

one easily obtains

$$I_0(\xi, \eta) = \sum_{j}^{m} \sum_{k}^{n} r_{jk} \log \frac{r_{jk}}{p_j q_k} = S_0(r|p \otimes q), \tag{2.57}$$

where

$$r = (r_{11}, r_{12}, \ldots, r_{mn}), \quad p \otimes q = (p_1 q_1, p_1 q_2, \ldots, p_m q_n). \tag{2.58}$$

From (2.57) and (2.45) directly follows that always

$$I_0(\xi, \eta) \geq 0. \tag{2.59}$$

Furthermore, iff ξ and η are stochastically independent, i.e., $r = p \otimes q$,

$$S_0(\xi, \eta) = 0, \tag{2.60}$$

cf. (2.7). Therefore, if ξ and η are not independent, the value of η gives always information about ξ (and reciprocally). Kolmogorov, Gelfand and Yaglom, [186] [187] [103], called it *an amount of information contained in one object about another object*. They considered it as the basic concept of information theory, more important than the Shannon entropy itself. (Shannon actually mentioned $I_0(\xi, \eta)$ in Appendix 7 of [300], but the Russian authors originally did not know that since this Appendix was omitted in the first Russian translation of the paper.) Maybe this is a bit exaggerated, but in fact this quantity has marvellous mathematical properties and is of the greatest importance for the theory of information channels. We see from (2.57) that the mutual information is a symmetrical function of its arguments

$$I_0(\xi, \eta) = I_0(\eta, \xi). \tag{2.61}$$

Therefore, we have from (2.55)

$$I_0(\xi, \eta) = S_0(\xi) - S_0(\xi; \eta) = S_0(\eta) - S_0(\eta; \xi) = I_0(\eta, \xi), \tag{2.62}$$

so the information contained in ξ about η is equal to the information contained in η about ξ.

Rényi [291], p. 562 showed yet another connection of $I_0(\xi, \eta)$ with $S_0(\xi|\eta)$. Namely, denoting

$$p_{;k} = (p_{1;k}, p_{2;k}, \ldots, p_{m;k}) \tag{2.63}$$

and because of $r_{jk} = q_k p_{j;k}$, one obtains

$$I_0(\xi, \eta) = \sum_{k=1}^{n} q_k S_0(p_{;k}|p). \tag{2.64}$$

For physics, it is important to show the connection of information gain with the *entropy production* between states $p = (p_1, \ldots, p_n)$ and $q = (q_1, \ldots, q_n)$

$$D_0(q, p) = S_0(q) - S_0(p) = -D_0(p, q) \tag{2.65}$$

which was already mentioned above. Taking for q the state (probability distribution) with the absolute maximum of entropy

$$q = q_0 = \left(\frac{1}{n}, \frac{1}{n}, \ldots, \frac{1}{n} \right), \tag{2.66}$$

one obtains

$$S_0(p|q_0) = \sum_{k=1}^{n} p_k \log n p_k = \log n - S_0(p) = S_0(q_0) - S_0(p) = D_0(q_0, p). \tag{2.67}$$

In general, $S_0(p_{|k}|p) \neq D_0(p, p_{|k})$, but the averages of these quantities are equal. Indeed, we have

$$I_0(\xi, \eta) = S_0(\xi) - S_0(\xi; \eta) = \sum_{k=1}^{n} q_k D_0(p, p_{;k}). \tag{2.68}$$

Now using (2.64) one obtains

$$\sum_{k=1}^{n} q_k S_0(p_{;k}|p) = \sum_{k=1}^{n} q_k D_0(p, p_{;k}). \tag{2.69}$$

There is also a simple connection between $I_0(\xi, \eta)$ and $S_0(\xi)$. Namely, if $\xi = \eta$, $r_{jk} = p_j \delta_{jk}$, $p_j = q_j$, $m = n$ and one has

$$I_0(\xi, \xi) = -\sum_{j=1}^{n} p_j \log p_j = S_0(\xi). \tag{2.70}$$

Since all the (elementary) probabilities are functions of the (atomic) events $a_j, b_k, \ldots$, and the respective events form (finite) Boolean algebras $A, B, \ldots$, one can also write our (finite) mutual information, according to [186], [187], in the language of Boolean algebras with probabilities as a norm[1]

$$I_0(A, B) = \sum_{j=1}^{m} \sum_{k=1}^{n} P(a_j \cap b_k) \log \frac{P(a_j \cap b_k)}{P(a_j)P(b_k)}. \qquad (2.71)$$

In the case of arbitrary (not necessary finite) Boolean algebras A, B, one defines, according to [186], [187],

$$I(A, B) = \sup_{A_1 \subseteq A, B_1 \subseteq B} I_0(A_1, B_1), \qquad (2.72)$$

where the supremum is taken over all finite subalgebras $A_1 \subseteq A, B_1 \subseteq B$. The cited authors write symbolically the result in the form

$$I(A, B) = \int_A \int_B P(dAdB) \log \frac{P(dAdB)}{P(dA)P(dB)}. \qquad (2.73)$$

Since in the finite case I_0 is always nonnegative and finite, in the general case $I(A, B)$ is also nonnegative, but can be finite or equal to $+\infty$. We have the following simple properties of $M(A, B)$, cf. [187]:

1. $I(A, B) = I(B, A)$.
2. $I(A, B) = 0$ iff A and B are statistically independent.
3. If $[A_1 \cup B_1]$ and $[A_2 \cup B_2]$ are statistically independent ([A] denotes the smallest Boolean algebra containing set A), then

$$I([A_1 \cup A_2], [B_1 \cup B_2]) = I(A_1, B_1) + I(A_2, B_2). \qquad (2.74)$$

4. If $A_1 \subseteq A$, then $I(A_1, B) \leq I(A, B)$.
5. If algebra $A_1 \subseteq A$ is dense on the algebra A in the sense of metric $\rho(a, b) = P(ab' \cup a'b)$ (a' is the complement of a), then

$$I(A_1, B) = I(A, B). \qquad (2.75)$$

6. If $A_1 \subseteq A_2 \subseteq \ldots \subseteq A_n \subseteq \ldots$ and $A = \cup_n A_n$, then

$$I(A, B) = \lim_{n \to \infty} I(A_n, B). \qquad (2.76)$$

7. If $\lim_{n \to \infty} P_n(c) = P(c)$ for $c \in [A \cup B]$, then

$$I(A, B) \leq \liminf_{n \to \infty} I_n(A, B), \qquad (2.77)$$

[1]The mentioned Russian authors explicitly assume the existence of a fundamental (infinite) Boolean algebra A_{fund} containing all the other algebras and having the unit element e_{fund}, the zero element n_{fund}, and probability distribution P such that $P(e_{fund}) = 1$.

where $I(A, B)$ is the mutual information for probability distribution P, and $I_n(A, B)$ is the mutual information for probability distribution P_n.

The above consideration concerned an infinite generalization of the discrete finite case (denoted by index 0), which is more than only going to the limit $m \to \infty, n \to \infty$ (therefore we dropped the index). In particular, to get the case of continuous observables we have to assume that the fundamental Boolean algebra A_{fund} is a σ-algebra and the probability measure P is a σ-measure. Considering two observables ξ, η with value spaces X, Y, respectively, we can define probability σ-measures

$$P_\xi(a) = P(\xi \in a), a \subseteq X, \ P_\eta(b) = P(\eta \in b), b \subseteq Y,$$
$$P_{\xi,\eta}(c) = P((\xi, \eta) \in c), \tag{2.78}$$

where $c \subseteq X \times Y$. Then we obtain from (2.73)

$$I(\xi, \eta) = \int_X \int_Y P_{\xi,\eta}(dx\,dy) \log \frac{P_{\xi,\eta}(dx\,dy)}{P_\xi(dx)P_\eta(dy)}. \tag{2.79}$$

Using the Radon-Nikodym theorem we can write

$$P_{\xi,\eta}(c) = \int \int_c \alpha(x, y) P_\xi(dx) P_\eta(dy) + Q(c), \tag{2.80}$$

where $Q(c)$ is 0 or a singular measure with respect to $P_\xi \times P_\eta$. If $S(X \times Y) = 0$ (i.e., if $P_{\xi,\eta}$ is absolutely continuous with respect to $P_\xi \times P_\eta$), function $\alpha(x, y)$ is the Radon-Nikodym-Stieltjes derivative

$$\alpha(x, y) = \frac{dP_{\xi,\eta}(dx\,dy)}{d(P_\xi(dx)P_\eta(dy))}. \tag{2.81}$$

Thus we obtain the theorem: $I(\xi, \eta)$ can be finite only in the case when $Q(X \times Y) = 0$, and then

$$I(\xi, \eta) = \int_X \int_Y \alpha(x, y) \log \alpha(x, y) P_\xi(dx) P_\eta(dy) = \int_X \int_Y P_{\xi,\eta}(dx\,dy) \log \alpha(x, y). \tag{2.82}$$

If measures P_ξ, P_η, and $P_{\xi,\eta}$ can be expressed by means of probability densities

$$P_\xi(a) = \int_a \rho_\xi(x)dx, \ P_\eta(b) = \int_b \rho_\eta(x)dx, \ P_{\xi,\eta}(c) = \int \int_c \rho_{\xi,\eta}(x, y)dx\,dy, \tag{2.83}$$

where dx and dy denotes integration with respect to some given Lebesgue measures in X and Y, then one obtains the formula expressed by a Lebesgue, not Stieltjes, integral

$$I(\xi, \eta) = \int_X \int_Y \rho_{\xi,\eta}(x, y) \log \frac{\rho_{\xi,\eta}(x, y)}{\rho_\xi(x)\rho_\eta(y)} dx\,dy. \tag{2.84}$$

It is important to mention that if we define

$$I(\xi,\xi) = S(\xi),\tag{2.85}$$

in analogy to the finite case (2.70), we do not get, in general, our previous "differential entropy" $S_n(\xi)$ (n is here the number of dimensions of X and of integrations over x and y in (2.84), $dx = d^n x$, $dy = d^n y$,

$$S_n(\xi) = -\int_X \rho_\xi(x)\log\rho_\xi(x)d^n x,\tag{2.86}$$

corresponding to our previous (2.36). For the continuous case with density $\rho_\xi(x)$, $S(\xi)$ is always infinite, while $S_n(\xi)$ can be finite. Thus, in general, quantities $S_n(\xi)$ and $S(\xi)$ are not identical, although they may coincide in some special cases, as the discrete one ($n = 0$). Anyhow we obtain, cf. [187], on the one hand,

$$I(\xi,\eta) = S_n(\xi) - S_n(\xi;\eta),\tag{2.87}$$

and, on the other hand,

$$S(\xi,\eta) \geq S(\xi),\ S(\xi,\eta) \geq S(\eta),\ S(\xi,\eta) \leq S(\xi) + S(\eta),\tag{2.88}$$

while, when ξ and η are stochastically independent

$$S(\xi,\eta) = S(\xi) + S(\eta),\quad (\xi\eta).\tag{2.89}$$

(Note that the entropy S rather than the mutual entropy I appears on the left sides of expressions (2.88) and (2.89).)

Coming back to information gain $S_0(q|p)$, (2.44), we may also generalize it, in a similar way as mutual information I_0, (2.57), to infinite and continuous cases. Without going into mathematical details, we restrict ourselves to practically the most important case of two continuous probability densities $f(x), g(x), x \in X$ of any number of real dimensions ($X \subset \mathbb{R}^n$) such that the probability measures

$$\mu(E) = \int_E f(x)dx,\quad \nu(E) = \int_E g(x)dx,\quad (E \in \mathcal{F}),\tag{2.90}$$

are mutually absolutely continuous (that means that both measures can vanish exactly on the same set, cf. p. 68). We assume as Kullback that the measure of integration dx

$$\lambda(E) = \int_E dx,\quad (E \in \mathcal{F})\tag{2.91}$$

is the Lebesgue measure mutually absolutely continuous with μ and ν,

$$\lambda \equiv \mu,\quad \lambda \equiv \nu.\tag{2.92}$$

Then we may define the two following integrals as information gain

$$S(f|g) = S(\mu|\nu) = \int_X f(x) \log \frac{f(x)}{g(x)} \, dx \,, \qquad (2.93)$$

and

$$S(g|f) = \int_X g(x) \log \frac{g(x)}{f(x)} \, dx \,. \qquad (2.94)$$

Both integrals are nonnegatively finite or equal to $+\infty$, but, in general, they are not equal. In our book [24], Chap. 4, cf. also [159], we use this asymmetry to define an asymmetric (irreversible) thermodynamical time (processes in time). In equilibrium $f = g$ and the time is reversible. Therefore, for our physical application we do not need the concept of symmetrized information gain, called *divergence*, introduced by H. Jeffreys [174], [175], cf. also [206], p. 6,

$$J(f|g) = S(f|g) + S(g|f) = \int_X [g(x) - f(x)] \log \frac{g(x)}{f(x)} \, dx \,. \qquad (2.95)$$

The continuous information gain has properties very similar to those of the discrete one. It is interesting to note that there is also an analogue of the equation (2.67), where in the place of Shannon's entropy S_0 Boltzmann's entropy S_1 or S_n occurs. Let us take for the initial probability density $f_0(x)$ the uniform density in the the interval $[a, b], a > b, a, b \in \mathbb{R}^1$, and for the final density $f(x)$ an arbitrary density in this interval. Then one easily obtains, cf. [117], p. 28,

$$S(f|f_0) = \log(b - a) + \int_a^b f(x) \log f(x) dx = D_1(f_0, f). \qquad (2.96)$$

This is one of the arguments that the Boltzmann entropy S_1 or S_n has a good mathematical sense in the continuous case, although it is, maybe, not so general and elegant as that of the discrete Schannon case.

We refer the readers interested in further details on the information gain and its statistical applications to the excellent book of Kullback [206] which starts with the formula (2.93). We remark that because of such a nice elaboration of this concept by Solomon Kullback, the information gain is sometimes called *Kullback information* or *Kullback entropy*. Some authors call it also *Kullback-Leibler entropy* because of the earlier publication [205]. On the other hand, Guiaşu [117] calls the information gain the *variation of information* because of the equation (2.96).

2.5. Coarse graining and sufficiency

We would like to sketch briefly the idea of sufficiency which is very important in mathematical statistics and was introduced by R. A. Fisher in

[95]. If we introduce in the sample space $X \subseteq \Omega$ a finite or infinite partition (which physicists usually call a *coarse graining*)

$$E_i \in \mathcal{F} \quad (i = 1, 2, \ldots), \quad E_i \cap E_j = 0 \quad (i \neq j), \quad X = \bigcup_i E_i, \quad (2.97)$$

the following theorem holds, cf. [206], p. 16,

$$S(g|f) = S(\nu|\mu) \geq \sum_i \mu(E_i) \log_2 \frac{\mu(E_i)}{\nu(E_i)}, \quad (2.98)$$

with the equality sign iff

$$\frac{f(x)}{g(x)} = \frac{\mu(E_i)}{\nu(E_i)}[\lambda], \quad x \in E_i \quad (i = 1, 2, \ldots), \quad (2.99)$$

where $[\lambda]$ (read "modulo λ") means that the sentence is true except for a set E such that $E \in \mathcal{F}$ and $\lambda(E) = 0$. The theorem, as many previous ones, is a consequence of the convexity of the function $x \log x$ and the well-known Jensen inequality [176].

In mathematical statistics such a partitioning (coarse graining) of sample space X into sets of equivalent x's is called simply a *statistic* (E. L. Lehmann [215]). *Sufficient partitioning*, or *sufficient statistic*, is such a partition that the condition (2.99) of the equality sign is satisfied, i.e., that the information is preserved (in general it is lost in a coarse graining). Fisher [95], p. 316 formulated it as: "the statistic chosen should summarise the whole of the relevant information supplied by the sample", cf. also [206], p. 18.

In general, a partitioning or a statistic causes a transformation of probability spaces. So as we can speak about a measure-preserving transformation, we can also speak about an information-preserving transformation. In physics and technology, we frequently use a dissipative coarse graining (with a lose of information). If we like, however, to simplify mathematically the problem but to preserve the information which we already have, a sufficient statistic is needed. Then information (in our case the information gain) is invariant with respect to the transformation (for more details about this invariance cf. section 2.4 in [206], p. 18).

Other type of transformation of the probability space is simply a coordinate transformation in X. In general, information is not invariant with respect to such transformations, and this is natural (since information is defined with respect to a given reference, given densities etc.). But there are special classes of transformations (as in physics — symplectic or canonical, or unitary transformations) with respect to which information can be preserved. E.g., Boltzmann's information S_n is preserved with respect to the mentioned transformations in closed systems. That makes this information physically important, in spite of the mentioned criticism of Kolmogorov (which is right in other respects, of strictly mathematical character). On the other hand, the mentioned classes of transformation may be called *sufficient transformations* with respect to the considered class of information functions.

2.6. Examples

Let us now discuss some simple examples in order to prepare the physical discussion and applications. We shall consider continuous Gaussian probability distributions and their simple generalizations as mathematically and physically the simplest cases.

1) Let us start with the simplest case of the Boltzmann entropy S_1 for the Gaussian (normal) one-dimensional probability density,

$$f(x) = \frac{1}{\sigma\sqrt{2\pi}} \exp\left(-\frac{x^2}{2\sigma^2}\right), \quad x \in (-\infty, +\infty), \quad \sigma > 0, \qquad (2.100)$$

where σ is a parameter called the *standard deviation* (and its square the *variance*). We obtain

$$
\begin{aligned}
S_1(f) &= \frac{1}{\sigma\sqrt{2\pi}} \int_{-\infty}^{+\infty} \exp\left(-\frac{x^2}{2\sigma^2}\right) \log\left(\frac{1}{\sigma\sqrt{2\pi}} \exp\left(-\frac{x^2}{2\sigma^2}\right)\right) dx \\
&= \log(\sigma\sqrt{2\pi}) + \frac{1}{2} = \log(\sigma\sqrt{2\pi e}).
\end{aligned}
\qquad (2.101)
$$

We see that $S_1(f)$ is equal to zero for

$$\sigma = \frac{1}{\sqrt{2\pi e}} \approx 0.24198, \qquad (2.102)$$

while, for smaller positive σ, S_1 is negative, and for greater σ, S_1 is positive. Of course, such a result is nonsense from the point of view of physics or technology since σ should have dimension of length, $[x]$, and not be dimesionless as in (2.102). The reason is clear: the formula (2.21) for the Boltzmann entropy S_1 (in contradistinction to (2.84) and (2.94)) is dimensionally incorrect since it contains density with dimension $[x^{-1}]$, and not a pure number, under the logarithm sign. Above we have seen that such a continuous entropy has also a drawback from the point of view of information theory: it does not express the "absolute" information of the continuous case which is always infinite, and has only some weaker properties. (It is rather a generalized form of the Kullback relative information, cf. below and [194]). No improvement is introduced if we write this formula in the form proposed by W. Ochs [245], Eq. (1.1)

$$S(\mu|m) = -c \int_\Omega dm(\frac{d\mu}{dm} \log \frac{d\mu}{dm}), \qquad (2.103)$$

where c is a positive constant, μ is an m-absolutely continuous probability measure, m — a Lebesgue measure previously denoted symbolically by the Leibniz infinitesimal (dx), and $\frac{d\mu}{dm}$ is the Radon-Nikodym derivative equal to our density f. Ochs's formula (2.103) nicely presents the dependence on the measure m, so the Boltzmann entropy appears as a relative entropy, in a sense (actually, its generalization with the opposite sign). But, since $d\mu/dm = f$ has dimension $[x^{-1}]$ as before (in spite of its notation which is only an analogy to the usual derivative), to make $S(\mu|m)$ dimensionally correct, we have to introduce an additional dimensional constant a under the logarithm sign (it may be, in particular, e.g., 1 cm):

$$S(\mu|m) = -c \int_\Omega dm \left(\frac{d\mu}{dm} \log a \frac{d\mu}{dm}\right). \qquad (2.104)$$

Thus in our case we obtain

$$S(f) = -\frac{1}{\sigma\sqrt{2\pi}} \int_{-\infty}^{+\infty} \exp\left(-\frac{x^2}{2\sigma^2}\right) \log\left[\frac{a}{\sigma\sqrt{2\pi}} \exp(-\frac{x^2}{2\sigma^2})\right] dx. \qquad (2.105)$$

Now we can write instead of (2.102)

$$\sigma = \frac{a}{\sqrt{2\pi e}} \approx 0.24198a \qquad (2.106)$$

which is physically correct as being homogeneous with respect to units.

Constant a should have an experimental meaning, e.g., as a constant of nature or a parameter of experiment. In the case when x is *action* (a physical quantity equal to energy×time), a can be taken to be the Planck constant $h \approx 6.626196 \times 10^{-34}$ J.s. When x is geometrically even-dimensional ($2n$-dimensional, $n = 1, 2, \ldots$), and is given in symplectic coordinates, while X is a symplectic space representing *phase space* in physics (position-momenta space $\{(q, p)\}$ with the volume element $d^n q d^n p$), we may take as our constant the Planck constant in the n-th power, h^n, since the physical dimension [position× momentum]= [action]. Therefore, such an expression for entropy has a physical meaning in any canonical (symplecitc) coordinate system since Jacobian of a canonical (symplectic) transformation is equal to 1. (The calculation of such entropy for important physical cases will be given in Chap. 5.)

Such an assumption can be interpreted as introducing cells in the phase space with volume h^n (a specific coarse graining, in view of quantum theory defining a sufficient statistic in the sense of Sect. 2.5), i.e., a quasi-classical approximation of quantum mechanics. Then probability density $f(x)$ cannot change appreciably in the limits of one cell, and the Boltzmann entropy with this improvement cannot be negative and appears to be approximatively equal to the Shannon entropy. Thus we obtained a reasonable approximation of the quantum (Von Neumann) entropy (in the eigen-representation of a density operator) which is always discrete and nonnegative (cf. Chap. 3). Our assumption corresponds exactly to the notation common among physicists that one divides the integral representing the Boltzmann entropy in phase space by h^n. Indeed, introducing dimensionless probability density $f'(q,p) = h^n f(q,p)$ and dimensionless positions q' and momenta p', one has $d^n q' d^n p' = h^{-n} d^n q d^{-n} p$. Thus the problem is actually well-known, although maybe usually not in this form. The modification, howewer, could not be introduced by Boltzmann himself in the XIX-th century before the Planck constant was discovered. Now the Boltzmann entropy is actually replaced by Von Neumann entropy being a Hilbert-space generalization of the Shannon entropy. It is, however, limited to the case of action and phase space. But a similar situation can occur in engineering. E.g., in communication engineering one applies the so-called sampling theorem ([340], [301], cf. also [346], Sec. 2.4 and [206], p. 9) which consists in discretization of time of transmitted signals. Signal is not an arbitrary function of time which has infinite number of degrees of freedom, but is transmitted by a finite frequency bandwidth W and has finite duration T. According to the sampling theorem the number of degrees of freedom of such a signal is $2TW$, and therefore a "cell" in time is $(2W)^{-1}$ with dimension s (s=second). Thus we may put in this case $a = (2W)^{-1}$. On the other hand, in the essentially continuous case, if such occurs in a concrete problem, omitting the Boltzmann "absolute" entropy in information theory, as proposed by Kolmogorov-Gelfand-Yaglom and Kullback, is legitimate. Our assumption on a is not a "regularization" of the infinite continuous entropy, but, actually, its discretization.

As mentioned, we may also look at the introduction of a dimensional constant under the logarithm sign in the Boltzmann (Gibbs) information as a generalization of the Kullback entropy (information gain) for a non-normalizable reference measure with constant density, instead of the normalized density in the denominator under the logarithm (cf. Kossakowski [194], but in his paper the problem of the dimensional constant is not mentioned).

2) Using the definite integral [165], p. 230

$$\int_0^{+\infty} x^p \exp(-x^q)dx = \frac{1}{q}\Gamma\left(\frac{p+1}{q}\right), \quad \mathrm{Re}\,(q) > 0 \tag{2.107}$$

we can calculate the following one-dimensional *generalized Gauss function* and its entropy ($\sigma > 0, x \in (-\infty, +\infty)$) (which will appear in Chap. 5):

$$f(x) = \frac{n}{\sigma\sqrt{2}\Gamma(\frac{1}{2n})} \exp\left(-\frac{x^{2n}}{2^n\sigma^{2n}}\right), \quad n = 1, 2, \ldots, \tag{2.108}$$

and (using now the physical constants h and k)

$$S(f) = k\left[\log\left(\frac{\sigma\sqrt{2}\Gamma(\frac{1}{2n})}{nh}\right) + \frac{1}{2n}\right] = k\log\left(\frac{\sigma\sqrt{2e^{1/n}}\Gamma(\frac{1}{2n})}{nh}\right) \tag{2.109}$$

Now entropy is zero for

$$\sigma = \frac{n\exp(-\frac{1}{2n})}{\sqrt{2}\Gamma(\frac{1}{2n})}h. \tag{2.110}$$

E.g., for $n = 2$ one has

$$\sigma \approx 0.30374h. \tag{2.111}$$

3) The Gauss density can be also modified in another direction, namely for the interval $x \in [0, +\infty)$ (these functions can be called *generalized semi-Gaussian densities* as defined the "semi" interval). . Using again (2.107) one obtains

$$f(x) = \frac{n}{\sigma\Gamma(\frac{1}{n})} \exp\left(-\frac{x^n}{\sigma^n}\right), \quad n = 1, 2, \ldots. \tag{2.112}$$

Then the entropy is (in physical units)

$$S(f) = k\left[\log\left(\frac{\sigma\Gamma(\frac{1}{n})}{nh}\right) + \frac{1}{n}\right] = k\log\left(\frac{\sigma e^{\frac{1}{n}}\Gamma(\frac{1}{n})}{nh}\right). \tag{2.113}$$

Entropy vanishes for

$$\sigma = \frac{nh}{\Gamma(\frac{1}{n})}\exp\left(-\frac{1}{n}\right), \tag{2.114}$$

i, e., for $n = 1$

$$\sigma \approx 0.36788h. \tag{2.115}$$

4) We now calculate information gain for the following densities ($\sigma, \tau > 0, x \in (-\infty, +\infty)$):

$$f(x) = \frac{1}{\sigma\sqrt{2\pi}}\exp\left(-\frac{x^2}{2\sigma^2}\right), \quad g(x) = \frac{1}{\tau\sqrt{2\pi}}\exp\left(-\frac{x^2}{2\tau^2}\right). \tag{2.116}$$

We obtain

$$S(g|f) = \int_{-\infty}^{+\infty} g(x) \log \frac{g(x)}{f(x)} dx = \log\left(\frac{\sigma}{\tau}\right) + \frac{1}{2}\left[\left(\frac{\tau}{\sigma}\right)^2 - 1\right]. \quad (2.117)$$

We see that now there is no problem with dimension under the logarithm and the result depends only on the relation of σ and τ. For $\sigma = \tau$ we have $S(g|f) = 0$, while for

$$\frac{\sigma}{\tau} = 2: \quad S(g|f) \approx 0.45895 \text{ bit} > 0, \quad (2.118)$$

$$\frac{\sigma}{\tau} = \frac{1}{2}: \quad S(g|f) \approx 1.16419 \text{ bit} > 0. \quad (2.119)$$

5) Let us now calculate information gain for the two following generalized semi-Gaussian densities ($\sigma, \tau > 0, n, m = 1, 2, \ldots$)

$$f(x) = \frac{n}{\sigma\Gamma(\frac{1}{n})} \exp\left(-\frac{x^n}{\sigma^n}\right), \quad g(x) = \frac{m}{\tau\Gamma(\frac{1}{m})} \exp\left(-\frac{x^m}{\tau^m}\right). \quad (2.120)$$

One obtains

$$S(g|f) = \log\left(\frac{m\sigma\Gamma(\frac{1}{n})}{n\tau\Gamma(\frac{1}{m})}\right) - \frac{1}{m} + \frac{\tau^n\Gamma(\frac{n+1}{m})}{\sigma^n\Gamma(\frac{1}{m})}. \quad (2.121)$$

If $m = n$,

$$S(g|f) = \log\left(\frac{\sigma}{\tau}\right) - \frac{1}{n} + \frac{\tau^n}{n\sigma^n}. \quad (2.122)$$

When $\sigma = \tau$, $S(g|f) = 0$.

6) We repeat the calculation of 5) with the "full" generalized Gaussian densities, i.e., in the full interval $(-\infty, +\infty)$ ($\sigma, \tau > 0, n, m = 1, 2, \ldots$):

$$f(x) = \frac{n}{\sigma\sqrt{2}\Gamma(\frac{1}{2n})} \exp\left(-\frac{x^{2n}}{2^n\sigma^{2n}}\right), \quad g(x) = \frac{m}{\tau\sqrt{2}\Gamma(\frac{1}{2m})} \exp\left(-\frac{x^{2m}}{2^m\tau^{2m}}\right). \quad (2.123)$$

One obtains

$$S(g|f) = \log\left(\frac{m\sigma\Gamma(\frac{1}{2n})}{n\tau\Gamma(\frac{1}{2m})}\right) - \frac{1}{2m} + 2^{m-n}\frac{\tau^{2m}\Gamma(2n+12m)}{\sigma^{2n}\Gamma(\frac{1}{2m})}. \quad (2.124)$$

For $m = n$ we have

$$S(g|f) = \log\left(\frac{\sigma}{\tau}\right) - \frac{1}{2n} + \frac{\tau^{2n}}{2n\sigma^{2n}}, \quad (2.125)$$

similarly as (2.122). When $\sigma = \tau$, $S(g|f) = 0$.

7) To consider a non-trivial generalization to higher dimensions we first take the case of the mutual entropy of the bivariate normal density, cf. [206] p. 8, $\sigma, \tau > 0$, $|\rho| \le 1$, $(x, y) \in \mathbb{R}^2$,

$$f(x,y) = \frac{1}{2\pi\sigma\tau\sqrt{1-\rho^2}} \exp\left\{-\frac{1}{2(1-\rho^2)}\left(\frac{x^2}{\sigma^2} - 2\rho\frac{xy}{\sigma\tau} + \frac{y^2}{\tau^2}\right)\right\}. \quad (2.126)$$

with its two marginal normal densities

$$g(x) = \frac{1}{\sigma\sqrt{2\pi}} \exp\left(-\frac{x^2}{2\sigma^2}\right), \quad h(y) = \frac{1}{\tau\sqrt{2\pi}} \exp\left(-\frac{y^2}{2\tau^2}\right). \quad (2.127)$$

We calculate

$$S(gh|f) = I(\xi, \eta) = \int\limits_{-\infty}^{+\infty}\int\limits_{-\infty}^{+\infty} f(x,y) \log \frac{f(x,y)}{g(x)h(y)} dx dy = -\frac{1}{2}\log(1-\rho^2).$$
$$(2.128)$$

We see that I is independent of σ and τ and depends only on the correlation coefficient ρ. When $|\rho|$ changes from 0 to 1, I changes from 0 to $+\infty$. We note, besides, that in (2.128), as in (2.117). a pure number stays under the logarithm sign. Thus no additional constant is necessary, the quantity is well defined from the point of view of physical dimension theory (dimensional homogeneity).

8) Example 7) has an important application in Shannon's communication theory, cf. [206], pp. 8–9. Let us interpret x (random quantity ξ) as an input signal, and y (random quantity η) as an output signal in a *communication channel*. The output signal is considered as the sum of the input signal and a statistically independent noise n, $y = x + n$, and, therefore,

$$f(x,y) = g(x)h(y|x) = g(x)h(y-x). \quad (2.129)$$

$I(\xi, \eta)$ is a measure of the correlation between the output and input signals, i.e., it is the information about x contained in y, and therefore, I is an important characteristic of the communication channel. Assuming the normal character of the probability distributions we remark that the bivariate normal density (2.126) from example 7) can be written in the form

$$f(x,y) = \frac{1}{\sigma\sqrt{2\pi}} \exp\left(-\frac{x^2}{2\sigma^2}\right) \frac{1}{\tau\sqrt{2\pi(1-\rho^2)}} \exp\left[-\frac{1}{2\tau^2(1-\rho^2)}\left(y - \frac{\rho\tau}{\sigma}x\right)^2\right].$$
$$(2.130)$$

From comparison of (2.129) and (2.130) we see that $h(y|x) = h(y - x)$ if

$$\rho \frac{\tau}{\sigma} = 1, \quad \rho^2 = \frac{\sigma^2}{\tau^2} = \frac{S}{S + \mathcal{N}}, \tag{2.131}$$

where $S = Ex^2$ is the mean transmitted *signal power* and $\mathcal{N} = En^2$ is the mean *noise power*, cf. [213] (E denotes a mean value). We obtain

$$I(\xi, \eta) = -\frac{1}{2} \log \left(1 - \frac{S}{S + \mathcal{N}}\right) = \frac{1}{2} \log \left(1 + \frac{S}{\mathcal{N}}\right). \tag{2.132}$$

This corresponds to a single observation. Since according to the sampling theorem in a signal of time duration T and frequency bandwith W there is $2WT$ independent sample values, and mutual information is additive for independent observations, we thus obtain the channel (mutual) information for the whole signal

$$I_{W,T} = WT \log \left(1 + \frac{S}{\mathcal{N}}\right), \tag{2.133}$$

which is the well-known communication engineering formula obtained first by Shannon [300], Sec. 25. If we divide (2.133) by T we obtain the *channel capacity*

$$C = W \log \left(1 + \frac{S}{\mathcal{N}}\right), \tag{2.134}$$

an important property of information channels also introduced by Shannon [300].

9) Let us now consider the multidimensional normal distribution called usually *multivariate distribution*, cf. [231]. Shannon, [300], Sec. 20, calculated the Boltzmann "absolute" entropy of a multivariate Gaussian density

$$f(x_1, x_2, \ldots, x_n) = \frac{|a_{ij}|^{\frac{1}{2}}}{(2\pi)^{\frac{n}{2}}} \exp \left(-\frac{1}{2} \sum_{i,j=1}^{n} a_{ij} x_i x_j\right), \tag{2.135}$$

where $|a_{ij}|$ is the determinant of matrix a_{ij}, in the form

$$S_n(f) = \log \left[(2\pi e)^{\frac{n}{2}} a^{-n} |a_{ij}|^{-\frac{1}{2}}\right], \tag{2.136}$$

where we introduced a constant a with dimension of length to make this formula dimensionally correct. Thus the "absoluteness" is, actually, relative to this constant, in accordance with our discussion showing that the Boltzmann entropy is a type of relative entropy, a generalization of the Kullback entropy for nonnormalizable densities. Formula (2.136) is a generalization of (2.101).

10) The multidimensional case of the mutual entropy for Gaussian multivariate densities was evaluated by Gelfand and Yaglom [103]. They wrote the multivariate densities in the form usual in multivariate analysis (cf. [206], Chap. 9, [231]), $x = (x_1, \ldots, x_k), y = (y_1, \ldots, y_l), z = (x, y) = (x_1, \ldots, x_k, y_1, \ldots, y_l)$,

$$f(z) = \frac{1}{(2\pi)^{\frac{k+l}{2}}|C|^{\frac{1}{2}}} \exp\left(-\frac{1}{2}(C^{-1}z, z)\right), \qquad (2.137)$$

and

$$g(x) = \frac{1}{(2\pi)^{k/2}|A|^{1/2}} \exp\left(-\frac{1}{2}(A^{-1}x, x)\right),$$

$$h(x) = \frac{1}{(2\pi)^{l/2}|B|^{1/2}} \exp\left(-\frac{1}{2}(B^{-1}y, y)\right). \qquad (2.138)$$

A, B, C are non-degenerated matrices (with nonvanishing determinants),

$$C = \begin{pmatrix} A & D \\ D^T & B \end{pmatrix}, \qquad (2.139)$$

where T denotes the transposition of a matrix, and

$$A = (E\xi_i\xi_j) = (a_{ij}), \;\; B = (E\eta_i\eta_j) = (b_{ij}), \;\; D = (E\xi_i\eta_j) = (d_{ij}). \quad (2.140)$$

Then one has

$$I(\xi, \eta) = \frac{1}{2}\log\frac{|A||B|}{|C|}. \qquad (2.141)$$

Under weaker assumptions (when the distributions are not necessarily nondegenerated) there exists a linear coordinate transformation $(x, y) \to (x', y')$ such that for some $m \le \min(k, l)$, $j = 1, 2, \ldots, m$, (by nondegeneration $m = \min(k, l)$) one obtains the correlation coefficients

$$r_j = r(\xi'_j, \eta'_j) = \frac{E\xi'_j\eta'_j}{\sqrt{E\xi'^2_j E\eta'^2_j}} \qquad (2.142)$$

and

$$I(\xi, \eta) = -\frac{1}{2}\sum_{j=1}^{m}\log(1 - r_j^2). \qquad (2.143)$$

Since

$$|r_j| = |\cos\alpha_j| \le 1, \quad (j = 1, \ldots, m) \qquad (2.144)$$

one can write

$$I(\xi, \eta) = - \log | \sin \alpha_1 \sin \alpha_2 \ldots \sin \alpha_m |, \qquad (2.145)$$

where α_j's have the meaning of the stationary angles between the linear subspaces spanned by the vectors ξ, η.

Further examples will be given in Chap. 5 on the occasion of discussion of physical problems.

2.7. Incomplete observables and Rényi's entropies

In literature, many other types of entropy-like functions of probabilities are known, in addition to those presented above, cf. e.g. [291], [11], [12]. (Especially, Indian mathematicians contributed much to the definition of entropies with many indices, cf. [118], [119].) All of them have weaker properties (e.g., are non-additive, etc.) than the entropies presented above, and have smaller practical value. We shall not try to present them all, we only briefly discuss the most important from them — the *Rényi entropies*, [290], [291], which found a direct interpretation in coding theory with an arbitrary cost function, given by L. Lorne Campbell in 1965, [49], and are frequently used in abstract considerations and proofs in mathematical and physical investigations.

Let us begin with the problem of an incomplete observable which led Rényi to his generalization, cf. [291], p. 569. In a Kolmogorov probability space $(\Omega, \mathcal{F}, \mathcal{P})$ a random variable (observable) is defined as an $\mathcal{F}$-measurable function on Ω. Let us now define an *incomplete (partial) random variable* as an $\mathcal{F}$-measurable function $\xi = \xi(\omega)$ defined on a subset Ω_1 of Ω such that $\Omega_1 \in \mathcal{F}$ and $P(\Omega_1) > 0$. Thus ordinary random variables are special cases of incomplete random variables. If ξ is an incomplete (real) random variable assuming finite discrete values $x_i, i = 1, 2, \ldots, n$, with probabilities $p_i \geq 0$, we have

$$\sum_{i=1}^{n} p_i = s \leq 1, \qquad (2.146)$$

s is not necessarily equal to 1.

Two incomplete random variables, ξ and η, are said to be *independent* if for any two sets, $A, B \in \mathcal{F}$, the events $\xi \in A$ and $\eta \in B$ are independent (i.e., the probability of the joint event is equal to the product of their probabilities). The probability distribution of an incomplete random variable is called *incomplete probability distribution*. The *direct product* of two discrete incomplete probability distributions, $p = \{p_i\}, i = 1, 2, \ldots, m$, and $q = \{q_j\}, j = 1, 2, \ldots, n$, is defined as the incomplete distribution

$p \otimes q = \{p_i q_j\}$. To every incomplete distribution $p = (p_1, \ldots, p_n)$ there can be assigned an ordinary distribution $p' = (p'_1, \ldots, p_n)$ by putting

$$p'_i = \frac{p_i}{s}, \quad (i = 1, 2, \ldots, n), \tag{2.147}$$

Distribution p' is said to be a *complete conditional distribution*.

For $0 < s < 1$ an incomplete random variable can be "objectively" interpreted either as such (complete) random variable whose some possible values lie outside of the range of our measuring instrument (e.g., energy which is unbounded from above, but bounded from below, while the range of any energy measuring instrument is bounded from both sides), or as a quantity at all undefined outside of some domain of independent parameters, e.g., of temperature and pressure (properties of condensation phases or chemical compounds). A simpler case occurs when we have at our disposal only incomplete statistical measurement (for some limited domain of independent variables ω) and have no time or money, etc., to measure elsewhere. This is a more "subjective", but very real and frequent situation. Anyhow we have here a high step of incompleteness. (From the point of view of deterministic physics also a complete statistical measurement is incomplete, as we discussed this in Chap. 1.) Now the essential point is the measurement of probabilities, not eigenvalues (which in statistics are considered as well-known). Physically, probabilities are measured from frequencies or from intensities (e.g., of spectral lines), also indirectly from mean values, as it will be shown in Chap. 5.

Rényi tried first to define information gain $S(q|p)$ for finite incomplete distributions p and q, assuming that they have the same number of terms and p has only positive probabilities ($p_i > 0$ for all i), while q can also have probabilities equal to 0. He then assumed the following postulates, cf. [291] p. 570:

1. If $p = p^{(1)}p^{(2)}$ and $q = q^{(1)}q^{(2)}$ (multiplication means here stochastic independence), then

$$S(q|p) = S(q^{(1)}|p^{(1)}) + S(q^{(2)}|p^{(2)}). \tag{2.148}$$

2. If $p_j \leq q_j$ ($j = 1, 2, \ldots, n$), then $S(q|p) \geq 0$. If $p_j \geq q_j$ ($j = 1, 2, \ldots, n$), then $S(q|p) \leq 0$.

Remark: It follows that $S(p|p) = 0$. For complete distributions p and q, Point 2 says nothing than this since then from $p_j \leq q_j, (j = 1, 2, \ldots, n)$ it follows that $p_j = q_j, (j = 1, 2, \ldots, n)$.

3. Denoting by e_p a distribution with a single term $\{p\}$ ($0 < p \leq 0$), let us fix the unit of information gain (a bit) by

$$S(e_1|e_{1/2}) = 1. \tag{2.149}$$

Remark: It follows from 2 and 3 that $(0 < p \leq 1, 0 \leq q \leq 1)$

$$S(e_q | e_p) = \log_2 \frac{q}{p}, \quad S(e_1 | e_p) = -\log_2 p. \tag{2.150}$$

In the following, however, we shall use, as before, the notation "log" instead of "$\log_2$" to enable to interpret it as a logarithm with any base for an arbitrary unit of entropy.

4. $S(q|p)$ depends only on the distribution function (distribuant)

$$F(q, p, x) = \sum_{\log(\frac{q_j}{p_j}) < x} q'_j \quad (-\infty < x < +\infty). \tag{2.151}$$

5. If F and G are two different distribution functions and

$$G(x) \geq F(x) \quad (-\infty < x < +\infty), \tag{2.152}$$

then

$$S[G(x)] < S[F(x)]. \tag{2.153}$$

Remark: This contains 2. But 2 is necessary since for the formulation of 5 we used the formula (2.150) proved under the assumption of 2.

6. If F_1, F_2, F_3 are distribution functions and $S[F_2] = S[F_3]$, then for every $t \in [0, 1]$ we have the "quasi-linearity condition"

$$S[tF_1 + (1-t)F_2] = S[tF_1 + (1-t)F_3] \tag{2.154}$$

and $S[tF_1 + (1-t)F_2]$ is a continuous function of t (the assumption of continuity is actually not necessary, but simplifies the proof of the following theorem).

Rényi proved, cf. [291], pp. 574–578, the theorem that if $S(q|p)$ satisfies 1 to 6, then

— either there exists a real number $\alpha \neq 1$ such that

$$S(q|p) = S^{(\alpha)}(q|p) = \frac{1}{\alpha - 1} \log \left(\frac{1}{\sum_{j=1}^n q_j} \sum_{j=1}^n \frac{q_j^\alpha}{p_j^{\alpha - 1}} \right), \tag{2.155}$$

— or

$$S(q|p) = S^{(1)}(q|p) = \frac{1}{\sum_{j=1}^n q_j} \sum_{j=1}^n q_j \log \frac{q_j}{p_j}, \tag{2.156}$$

while

$$\lim_{\alpha \to 1} S^{(\alpha)}(q|p) = S^{(1)}(q|p). \tag{2.157}$$

Thus we obtain in (2.156) our previous information gain slightly generalized for incomplete distribution, while (2.159) is an essential generalization called the *Rényi's information gain of order* α. As a closer investigation shows, cf. [291] p. 581, only the positive orders, $\alpha > 0$, have sense in information theory, so we restrict the range of α only to $(0, +\infty)$.

In particular, one has for the uniform distribution e_n of n terms

$$S^{(\alpha)}(q|e_n) = \log n + \frac{1}{\alpha - 1} \log \left(\frac{\sum_{j=1}^n q_j^\alpha}{\sum_{j=1}^n q_j} \right) \quad (\alpha \neq 1), \tag{2.158}$$

and

$$S^{(1)}(q|e_n) = \log n - \frac{\sum_{j=1}^n q_j \log q_j}{\sum_{j=1}^n q_j}. \tag{2.159}$$

Putting

$$S^{(\alpha)}(p) = -\frac{1}{\alpha - 1} \log \left(\frac{\sum_{j=1}^n p_j^\alpha}{\sum_{j=1}^n p_j} \right) \quad (\alpha \neq 1),$$

$$S^{(1)}(p) = -\frac{\sum_{j=1}^n p_j \log p_j}{\sum_{j=1}^n p_j} \tag{2.160}$$

we obtain

$$S^{(\alpha)}(q|e_n) = S^{(\alpha)}(e_n) - S^{(1)}(q) \tag{2.161}$$

and in (2.159) and (2.160) the corresponding generalizations of the Shannon entropy.

For a complete distribution p our definition of $S^{(\alpha)}$ gives

$$S^{(\alpha)}(p) = -\frac{1}{\alpha - 1} \log \sum_{j=1}^n p_j^\alpha \quad (\alpha \neq 1), \tag{2.162}$$

while

$$\lim_{\alpha \to 1} S^{(\alpha)}(p) = S^{(1)}(p) = S_0(p). \tag{2.163}$$

The quantity (2.162) is called the *Rényi entropy (information) of order* α, $\alpha > 0$, $\alpha \neq 1$.

Rényi proved the following theorems, cf. [291], pp. 580–583:

1) Let p be an incomplete distribution, $\sum_{j=1}^n p_j = s \leq 1$. Then $S^{(\alpha)}(p)$ is a positive, monotonly decreasing function of α and one has $S^{(0)}(p) = \log \frac{n}{s}$. Thus for a complete distribution

$$0 \leq S^{(\alpha)}(p) \leq \log n \quad (\alpha \geq 0). \tag{2.164}$$

2) If p and q are any incomplete distributions ($\sum_{j=1}^n p_j = s \leq 1$, $\sum_{j=1}^n q_j = t \leq 1$), then $S^{(\alpha)}(q|p)$ is an increasing function of

α. Since $S^{(0)}(q|p) = \log \frac{t}{s}$, for complete distributions p and q one has the inequality

$$S^{(\alpha)}(q|p) \geq 0 \quad (\alpha \geq 0). \tag{2.165}$$

Thus only for incomplete distributions $S^{(\alpha)}(q|p)$ $(\alpha \geq 0)$ can be negative.

3) If ξ and η are two real random variables with discrete finite distributions p and q, respectively, and if r is a two-dimensional distribution of the pair (ξ, η), then the subadditivity inequality

$$S^{(\alpha)}(r) \leq S^{(\alpha)}(p) + S^{(\alpha)}(q) \tag{2.166}$$

holds for every p and q with the mentioned properties iff $\alpha = 1$.

4) If p, q and r are discrete finite incomplete distributions such that

$$r_j = \sqrt{p_j q_j} \quad (j = 1, 2, \ldots, n), \tag{2.167}$$

or equivalently

$$\log \frac{r_j}{p_j} + \log \frac{r_j}{q_j} = 0 \quad (j = 1, 2, \ldots, n), \tag{2.168}$$

then the equation

$$S^{(\alpha)}(r|p) + S^{(\alpha)}(r|q) = 0 \tag{2.169}$$

holds for every distribution fulfilling (2.167) iff $\alpha = 1$.

We see from 3) and 4) that among all Rényi's entropies only that with $\alpha = 1$, i.e., the Shannon entropy, has exceptionally simple properties which are important in variational procedures. This causes that Shannon's entropy and connected with them Kullback's (information gain), Kolmogorov's (mutual entropy) and Boltzmann's (Gibbs') entropies have the exceptional importance in physics and engineering. But this does not mean that Rényi's entropies for $\alpha > 1$ have no importance. We already mentioned that Campbell [49] showed their close connection with the generalized coding theorem of Shannon and, therefore, with the algorithmic information of "higher order". This means that algorithmic information is defined not only by the shortest length of a message but also by a prescribed cost function (up to now it was silently assumed that the cost of the message is proportional to its length, in general it can grow with length faster than linearly). It seems that in future the limitation of information by the growing costs of it can be essential in communication engineering and in economy. This problem may also appear in theoretical biology. Recently Rényi's entropies occurred in investigations of new abstract theory of evolution of classical chaotic (nonlinear) dynamical systems together with the Kolmogorov-Sinai and topological entropies, cf., e.g., [230]. Rényi's entropy can help the estimation of Lyapunov exponents and dimension spectra and can be easier connected with experimental parameters than the Kolmogorov-Sinai entropy.

Rényi extended his entropies also to the discrete infinite and continuous cases, cf. [291], Chap. 9, Sec. 4. It can be also extended to the quantum case, cf. Chap. 3,

2.8. Weighted entropy

The last kind of entropy we would like to consider briefly in this Chapter is the weighted entropy introduced by M. Belis and S. Guiaşu in 1968, [35], cf. also [116] and [117], Chap. 4. Their definition concerns the discrete finite case and goes as follows:

Let $p = (p_1, p_2, \ldots, p_n)$ be a probability distribution, and $w_i \geq 0$ $(i = 1, 2, \ldots n)$ be a *weight* of the i-th elementary event proportional to its importance or significance (e.g., in biology or engineering). Then we define as a *weighted entropy*

$$W(w, p) = W_0(w, p) = -\sum_{i=1}^{n} w_i p_i \log p_i. \qquad (2.170)$$

The weighted entropy has the following properties:

1) $W(w, p) \geq 0$.

2) If $w_1 = \ldots = w_n = w$, then $W(w, p) = wS(p) = wS_0(p)$ which gives the Shannon entropy expressed in other units than that of $S(p)$ (if $w \neq 1$).

3) If $p_i = \delta_{ij}$ $(i, j = 1, 2, \ldots, n)$ (Kronecker's delta), then $W_0(w, p) = 0$ independently of w.

4) $W(\lambda w, p) = \lambda W(w, p)$ $(\lambda \geq 0)$.

5) $W(w_1, \ldots, w_{n-1}, w', w'', p_1, \ldots, p_{n-1}, p', p'') =$
$W(w_1, \ldots, w_n, p_1, \ldots, p_n) + p_n W(w', w'', \frac{p'}{p_n}, \frac{p''}{p_n})$, where

$$w_n = \frac{p'w' + p''w''}{p' + p''}, \qquad p_n = p' + p''. \qquad (2.171)$$

The quoted theorems are either obvious or very easy to prove.

Guiaşu gives two examples of weight functions. The first example is

$$w_i = -\frac{p_i}{\log p_i} = \frac{p_i}{S(e_1|e_{p_i})} \qquad (i = 1, 2, \ldots, n) \qquad (2.172)$$

which gives

$$W(w, p) = \sum_{i=1}^{n} p_i^2 \qquad (2.173)$$

called by O. Onicescu [268] an *information energy*. The second example is

$$w_i = \frac{q_i}{p_i} \quad (i = 1, 2, \ldots, n), \tag{2.174}$$

where q and p are two probability distributions (two statistical hypotheses), p being called "objective" and q — "subjective". Then

$$W(w, p) = -\sum_{i=1}^{n} q_i \log p_i. \tag{2.175}$$

called an *objective-subjective entropy* and introduced independently by P. Weiss [339] and M. Bongard [42].

From *Jensen's inequality*

$$\sum_{i=1}^{n} q_i \log p_i \leq \log \left(\sum_{i=1}^{n} q_i p_i \right) \tag{2.176}$$

it follows

$$W\left(\frac{q}{p}, p\right) \geq S(p) \tag{2.177}$$

with the equality sign iff $q = p$.

In general, weights w can be independent of probabilities p.

It is interesting to calculate the maximum value of the weighted entropy for arbitrary probability distribution and given weights. It may be shown by direct calculation, cf. [117] p. 69–71, that if p is an arbitrary probability distribution and w is a given weight distribution with positive weights, the weighted entropy is maximum iff

$$p_i = \exp\left(-\frac{\alpha}{w_i} - 1\right) \quad (i = 1, 2, \ldots, n) \tag{2.178}$$

where α is the solution of the equation

$$\sum_{i=1}^{n} \exp\left(-\frac{\alpha}{w_i} - 1\right) = 1. \tag{2.179}$$

The maximum weighted entropy is equal to

$$W(w, p)_{\max} = \alpha + \sum_{i=1}^{n} w_i \exp\left(-\frac{\alpha}{w_i} - 1\right). \tag{2.180}$$

The last remark is about the *weighted entropy of a random variable* ξ with positive eigenvalues x_i $(i = 1, 2, \ldots, n)$ and probabilities $p_i > 0$. Guiaşu defines it as

$$W(\xi) = -\sum_{i=1}^{n} x_i p_i \log_2 p_i. \tag{2.181}$$

We see that this generalized entropy of a random variable depends both on eigenvalues and on probabilities, while the Shannon entropy $S(\xi)$ depends only on probabilities.

It seems that, similarly as the Rényi entropies, the weighted entropy will find applications in theoretical biology and econometrics.

2.9. Recapitulation and extensions

To facilitate the comparison of the classical case with the quantum case discussed in the following Chap. 3, we give below lists of the most important properties of the most important entropies, namely, the Shannon absolute entropy $S(p)$ and the Kullback relative entropy $S(p|q)$. Most of these properties have been discussed above, but some of them are mentioned for the first time, as extensions of our previous material. These extensions are mostly easy to prove, but for more difficult cases and further extensions we refer to the monographic literature about information theory, as e.g. [182], [90], [206], [294], [92], [345], [102], [291], [117], [278].

To simplify the formulation we define yet the set of all possible (complete) finite probability distributions (states)

$$\Delta_n = \left\{ p = \{p_1, \ldots, p_n\}: \sum_{i=1}^{n} p_i = 1, \ p_i \geq 0 \ (i = 1, \ldots, n) \right\}. \quad (2.182)$$

Denoting as before a random variable by $\xi = \{x_1, \ldots, x_n\}$ with probabilities $p \in \Delta_n$, we write for a complete discrete event system $(\xi, p) = (\xi, p)_n$. If we have two complete discrete event systems, $(\xi, p)_n$ and $(\eta, q)_m$, we denote a compound (joint) system of pairs (ξ, η) by $(\xi\eta, r)_{mn}$, where $r_{ij} = P(\xi = x_i, \eta = y_j)$, $i = 1, \ldots, n$, $j = 1, \ldots, m$.

THEOREM 2.5. *For any $p \in \Delta_n$ we have the following properties of the Shannon entropy $S(p)$:*

(1) *Positivity: $S(p) \geq 0$, $= 0$ iff $p_i = \delta_{ij}$ $(j = 1, \ldots, n)$.*

(2) *Concavity: $S(\lambda p + (1 - \lambda)q) \geq \lambda S(p) + (1 - \lambda)S(q)$ for any $q \in \Delta_n$ and any $\lambda \in [0, 1]$.*

(3) *Symmetry: For any permutation τ of indices of p,*

$$S(p_1, p_2, \ldots, p_n) = S(p_{\tau(1)}, p_{\tau(2)}, \ldots, p_{\tau(n)}).$$

(4) *Additivity: For any $q \in \Delta_m$, let $r = p \otimes q = \{p_i q_j\} \in \Delta_{nm}$. Then*

$$S(p \otimes q) = S(p) + S(q).$$

(5) *Subadditivity: For any $r = \{r_{ij}\} \in \Delta_{nm}$ such that $\sum_{j=1}^{m} r_{ij} = p_i$ and $\sum_{i=1}^{n} r_{ij} = q_j$,*

$$S(r) \leq S(p) + S(q).$$

(6) Continuity: $S(p_1, p_2, \ldots, p_n)$ *is a continuous function in each* p_k.

(7) Monotonicity: $S(1/n, 1/n, \ldots, 1/n)$ *is a monotonically increasing function of* $n \in \mathbb{N}$ *(log n) and* $\max S(p), S(q) \leq S(r)$ *holds for the distributions* p, q *and* r *given in (5).*

(8) Expansibility: $S(p_1, p_2, \ldots, p_n, 0) = S(p_1, p_2, \ldots, p_n)$.

(9) Mixing: Let $A = (a_{ij})$ $(i, j = 1, 2, \ldots, n)$ *be a doubly stochastic matrix (i.e.,* $a_{ij} \geq 0$, $\sum_{i=1}^{n} a_{ij} = 1$, $\sum_{j=1}^{n} a_{ij} = 1$) *and* $q = Ap$. *Then*

$$S(q) \geq S(p).$$

(10) Joining: For two complete event systems (ξ, p) *and* (η, q),

$$S(\xi\eta) = S(\eta) + S(\xi; \eta) = S(\xi) + S(\eta; \xi) \leq S(\xi) + S(\eta).$$

THEOREM 2.6. *For any* $p, q \in \Delta_n$ *we have the following properties of the Kullback relative entropy* $S(p|q)$:

(1) Positivity: $S(p|q) \geq 0$, $= 0$ *iff* $p = q$.

(2) Joint Convexity: $S(\lambda p + (1-\lambda)q | \lambda r + (1-\lambda)t) \leq \lambda S(p|q) + (1-\lambda)S(q|t)$ *for any* $t, r \in \Delta_n$ *and* $\lambda \in [0, 1]$.

(3) Symmetry: for any permutation π *of indices,*

$$S(p_1, p_2, \ldots, p_n | q_1, q_2, \ldots, q_n) = S(p_{\pi(1)}, \ldots, p_{\pi(n)} | q_{\pi(1)}, \ldots, q_{\pi(n)}).$$

(4) Additivity: $S(p \otimes r | q \otimes t) = S(p|q) + S(r|t)$ *for any* $t, r \in \Delta_m$.

(5) Continuity: $S(p_1, p_2, \ldots, p_n | q_1, q_2, \ldots, q_n)$ *is continuous for each variable* p_i *and* q_j.

(6) Expansibility: $S(p_1, p_2, \ldots, p_n, 0 | q_1, q_2, \ldots, q_n) = S(p|q)$.

THEOREM 2.7. *For any two probability measures* μ, ν *mutually absolutely continuous on the same sample space* Ω *with* σ-*field* $\mathcal{F}$ *we have the following properties of the Kullback relative entropy* $S(\mu|\nu)$:

(1) Positivity: $S(\mu|\nu) \geq 0$, $= 0$ *iff* $\mu = \nu$.

(2) Joint Convexity: $S(\lambda\mu + (1 - \lambda)\nu | \lambda\rho + (1 - \lambda)\sigma) \leq \lambda S(\mu|\rho) + (1 - \lambda)S(\nu|\sigma)$ *for any probability measures* ρ, σ *on* Ω, *mutually absolutely continuous with* μ *and* ν, *respectively, and* $\lambda \in [0, 1]$.

(3) Symmetry: For an ivertible mapping j *from* $\mathcal{F}$ *to* $\mathcal{F}$ *such that* $\mu \circ j$ *is a probability measure and* $j(\Omega) = \Omega$,

$$S(\mu \circ j | \nu \circ j) = S(\mu|\nu).$$

(4) Additivity: For any two probability measures ρ *and* σ *on* Ω, *mutually absolutely continuous,*

$$S(\mu \otimes \rho | \nu \otimes \sigma) = S(\mu|\nu) + S(\rho|\sigma).$$

(5) Lower Semicontinuity: When $\mu_n \to \mu$, $\nu_n \to \nu$ *in norm,* μ_n *being mutually absolutely continuous,*

$$S(\mu|\nu) \leq \lim_{n\to n} S(\mu_n|\nu_n).$$

(6) Monotonicity: For σ-*subfields* $\mathcal{G}$ *and* $\mathcal{H}$ *of* σ-*field* $\mathcal{F}$ *with* $\mathcal{G} \subset \mathcal{H}$,

$$S_{\mathcal{G}}(\mu|\nu) \leq S_{\mathcal{H}}(\mu|\nu).$$

Remark: For simplicity, to enable the formulation of the inverse relative entropy, we assumed, following Kullback [206], the mutual absolute continuity of the two measures. (This relation is denoted by Kullback by $\equiv$). For a single relative entropy a weaker asumption is, of course, possible: μ is *absolutely continous with respect to* ν, $\mu \ll \nu$, if $\mu(E) = 0$ for all $E \in \mathcal{F}$ for which $\nu(E) = 0$.

Mutual entropy, $I(p, q)$ or $I(\mu, \nu)$, is a special case of Kullback's entropy, so it has the same properties. The Boltzmann-Gibbs "differential" entropy $S(f)$, cf. also [105], can be regarded, as we mentioned, as a generalization of Kullback's entropy for nonnormalizable measures ν, cf. [194], so it has weaker properties, e.g., can be negative. But, as we have shown, for quasi-classical approxiomation of the quantum case it can be approximated by Shannon's entropy which is then directly generalized by Von Neumann's quantum entropy.

In this Chapter we did not discuss two classical entropies which will be needed below, in Chap. 4 and Chap. 8, for the respective quantum generalizations: (1) the Kolmogorov-Sinai (KS)-entropy of a measure-preserving transformation of the sample space (Chap. 4), (2) the Kolmogorov ε-entropy (Chap. 8). It is better to define both these concepts in contexts of the mentioned Chapters concerning information dynamics and fractals, respectively.

Chapter 3

Quantum Entropy and Quantum Channel

3.1. Entropy of density operators

In this section, we discuss the entropy in usual quantum dynamical systems (QDS for short) described by Hilbert space terminology. A formulation of quantum mechanical entropy, started by Von Neumann around 1930, 20 years ahead of Shannon, now becomes a fundamental tool to analyze physical phenomena. *Von Neumann entropy* is defined for a density operator (state) $\rho \in \mathfrak{S}(\mathcal{H})$, the set of all density operators on a Hilbert space $\mathcal{H}$, by

$$S(\rho) = -\operatorname{tr}\rho \log \rho. \tag{3.1}$$

That is, for any CONS (complete orthonormal system) $\{x_k\}$ in $\mathcal{H}$,

$$S(\rho) = -\sum_k \langle x_k, \rho \log \rho x_k \rangle \tag{3.2}$$

The value of "trace" does not depend on the choice of CONS $\{x_k\}$.

Since the spectral set of ρ is discrete, we write the spectral decomposition of ρ as

$$\rho = \sum_n \lambda_n P_n,$$

where λ_n is an eigenvalue of ρ and P_n is the projection from $\mathcal{H}$ onto the eigenspace associated with λ_n. Therefore, if every eigenvalue λ_n is not degenerate, then the dimension of the range of P_n is one (we denote this by $\dim P_n = 1$). If a certain eigenvalue, say λ_n, is degenerate, then P_n can be further decomposed into one-dimensional orthogonal projections:

$$P_n = \sum_{j=1}^{\dim P_n} E_j^{(n)}, \tag{3.3}$$

where $E_j^{(n)}$ is one dimensional projection expressed by $E_j^{(n)} = | x_j^{(n)} \rangle \langle x_j^{(n)} |$ with the eigenvector $x_j^{(n)}$ ($j = 1, 2, \ldots, \dim P_n$) for λ_n. By relabeling the indices j, n of $\{E_j^{(n)}\}$, we write

$$\rho = \sum_n \lambda_n E_n \qquad (3.4)$$

with

$$\lambda_1 \geq \lambda_2 \geq \cdots \geq \lambda_n \geq \cdots \qquad (3.5)$$

$$E_n \perp E_m \quad (n \neq m). \qquad (3.6)$$

We call this decomposition a *Schatten decomposition* of ρ. Now, in (3.5), the eigenvalue of multiplicity n is repeated precisely n times. For example, if the multiplicity of λ_1 is 2, then $\lambda_1 = \lambda_2$. Moreover, this decomposition is unique if and only if no eigenvalue is degenerate. In the sequel, when we write $\rho = \sum_n \lambda_n E_n$, it is a Schatten decomposition of ρ, if not otherwise stated.

Before discussing some of the fundamental properties of $S(\rho)$, we fix a few notations. For two Hilbert spaces $\mathcal{H}_1$ and $\mathcal{H}_2$, let $\mathcal{H} = \mathcal{H}_1 \otimes \mathcal{H}_2$ be the tensor product Hilbert space of $\mathcal{H}_1$ and $\mathcal{H}_2$ and denote the tensor product of two operators A and B acting on $\mathcal{H}_1$ and $\mathcal{H}_2$, respectively, by $A \otimes B$. The reduced states ρ_1 in $\mathcal{H}_1$ and ρ_2 in $\mathcal{H}_2$ for a state ρ in $\mathcal{H}$ are given by the partial traces, which are denoted by $\rho_k = \mathrm{tr}_j \rho$ ($j \neq k; j, k = 1, 2$). That is, for instance, $\rho_1 = \mathrm{tr}_2 \rho \equiv \sum_n \langle y_n, \rho y_n \rangle$ for any CONS $\{y_n\}$ of $\mathcal{H}_2$.

THEOREM 3.1. *For any density operator $\rho \in \mathfrak{S}(\mathcal{H})$, we have the following properties of Von Neumann entropy:*

(1) Positivity: $S(\rho) \geq 0$.
(2) Symmetry: Let $\rho' = U\rho U^*$ for an unitary operator U. Then

$$S(\rho') = S(\rho).$$

(3) Concavity: $S(\lambda \rho_1 + (1-\lambda)\rho_2) \geq \lambda S(\rho_1) + (1-\lambda)S(\rho_2)$ for any $\rho_1, \rho_2 \in \mathfrak{S}(\mathcal{H})$ and any $\lambda \in [0, 1]$.
(4) Additivity: $S(\rho_1 \otimes \rho_2) = S(\rho_1) + S(\rho_2)$ for any $\rho_k \in \mathfrak{S}(\mathcal{H}_k)$.
(5) Subadditivity: For the reduced states ρ_1, ρ_2 of $\rho \in \mathfrak{S}(\mathcal{H}_1 \otimes \mathcal{H}_2)$,

$$S(\rho) \leq S(\rho_1) + S(\rho_2).$$

(6) Lower Semicontinuity: If $\|\rho_n - \rho\|_1 \to 0$, (i.e. $\mathrm{tr}|\rho_n - \rho| \to 0$) as $(n \to \infty)$, then

$$S(\rho) \leq \liminf_{n \to \infty} S(\rho_n).$$

In order to prove this theorem, we prepare a lemma.

LEMMA 3.1. *Let f be a convex C^1 function on a certain domain and $\rho, \sigma \in \mathfrak{S}(\mathcal{H})$. Then*

(1) Klein's inequality: $\mathrm{tr}\{f(\rho) - f(\sigma) - (\rho - \sigma)f'(\sigma)\} \geq 0.$

(2) Peierls' inequality: $\sum_k f(\langle x_k, \rho x_k \rangle) \leq \mathrm{tr} f(\rho)$ *for any CONS $\{x_k\}$ in $\mathcal{H}$.*

(Remark: $\rho = \sum_n \lambda_n E_n \Rightarrow f(\rho) = \sum_n f(\lambda_n) E_n.$)

Proof. (1): Let $\{x_n\}$ and $\{y_m\}$ be two CONS containing all eigenvalues x_n and y_m of ρ and σ, respectively such that $\rho x_n = \lambda_n x_n, \sigma y_m = \mu_m y_m$. It is easy to see

$$\mathrm{tr}\{f(\rho) - f(\sigma) - (\rho - \sigma)f'(\sigma)\}$$

$$= \sum_n \left\{ f(\lambda_n) - \sum_m |\langle x_n, y_m \rangle|^2 f(\mu_m) \right.$$

$$\left. - \sum_m |\langle x_n, y_m \rangle|^2 (\lambda_n - \mu_m) f'(\mu_m) \right\}$$

$$= \sum_{n,m} |\langle x_n, y_m \rangle|^2 \{f(\lambda_n) - f(\mu_m) - (\lambda_n - \mu_m) f'(\mu_m)\},$$

which is non-negative because f is convex.

(2): Let us take σ in (1) the operator satisfying

$$\langle x_k, \sigma x_j \rangle = \langle x_k, \rho x_j \rangle \delta_{kj}.$$

Then

$$\mathrm{tr} f(\sigma) = \sum_k f(\langle x_k, \rho x_k \rangle),$$

$$\mathrm{tr}(\rho - \sigma) f'(\sigma) = 0,$$

which imply

$$0 \leq \mathrm{tr}\{f(\rho) - f(\sigma) - (\rho - \sigma)f'(\sigma)\}$$
$$= \mathrm{tr} f(\rho) - \sum_k f(\langle x_k, \rho x_k \rangle).$$

$\square$

Proof of Theorem 3.1: (1) For the Schatten decomposition of $\rho : \rho = \sum_n \lambda_n E_n$, it is easy to get

$$S(\rho) = -\sum_n \lambda_n \log \lambda_n,$$

which is non-negative.

(2)

$$
\begin{aligned}
S(\rho') &= S(U\rho U^*) = -\mathrm{tr}\, U\rho U^*(\log U\rho U^*) \\
&= -\mathrm{tr}\, U\rho U^* U(\log\rho)U^* = -\mathrm{tr}\, U\rho(\log\rho)U^* \\
&= -\mathrm{tr}\, U^* U\rho(\log\rho) = S(\rho).
\end{aligned}
$$

(3) Put $\eta(t) = -t\log t$. Since $\eta(t)$ is concave, for the CONS containing all eigenvectors of $\lambda\rho_1 + (1-\lambda)\rho_2$, we obtain

$$
\begin{aligned}
S(\lambda\rho_1 + (1-\lambda)\rho_2) &= \sum_n \eta(\lambda\langle x_n, \rho_1 x_n\rangle + (1-\lambda)\langle x_n, \rho_2 x_n\rangle) \\
&\geq \lambda \sum_n \eta(\langle x_n, \rho_1 x_n\rangle) + (1-\lambda)\sum_n \eta(\langle x_n, \rho_2 x_n\rangle) \\
&\geq \lambda \sum_n \langle x_n, \eta(\rho_1) x_n\rangle + (1-\lambda)\sum_n \langle x_n, \eta(\rho_2) x_n\rangle \\
&= \lambda S(\rho_1) + (1-\lambda)S(\rho_2),
\end{aligned}
$$

where the last inequality comes from Peierls' inequality.

(4) Let $\rho_1 = \sum_n \lambda_n E_n$ and $\rho_2 = \sum_n \mu_n F_n$ be Schatten decompositions. Then $\rho_1 \otimes \rho_2 = \sum_{n,m} \lambda_n \mu_m E_n \otimes F_m$ is a Schatten decomposition of $\rho_1 \otimes \rho_2$. Hence

$$
\begin{aligned}
S(\rho_1 \otimes \rho_2) &= -\sum_{n,m} \lambda_n \mu_m \log \lambda_n \mu_m \\
&= S(\rho_1) + S(\rho_2).
\end{aligned}
$$

(5) Applying Klein's inequality to the function $f(t) = -\eta(t) = t\log t$, we have

$$
\mathrm{tr}\,\rho\log\rho - \mathrm{tr}\,\rho\log\rho_1 \otimes \rho_2 \geq \mathrm{tr}\,\rho - \mathrm{tr}\,\rho_1 \otimes \rho_2 = 1 - 1 = 0,
$$

whose left hand side is equal to $S(\rho_1) + S(\rho_2) - S(\rho)$, which concludes the inequality (5).

(6) Define

$$
S_\alpha(\rho) = \frac{1}{1-\alpha}\log(\mathrm{tr}\,\rho^\alpha)
$$

for any $\alpha \in \mathbb{R}^+ \equiv \{\alpha \in \mathbb{R};\ \alpha \geq 0\}$ with $\alpha \neq 1$ ($S_\alpha(\rho)$ is called the α-th Rényi entropy in QDS). Substituting $\rho = \sum_n \lambda_n E_n$,

$$
S_\alpha(\rho) = \frac{1}{1-\alpha}\log\left(\sum_n \lambda_n^\alpha\right) \to -\sum_n \lambda_n \log \lambda_n = S(\rho) \quad (\alpha \to 1).
$$

Since it is easily shown that $S_{\alpha'}(\rho) \leq S_{\alpha}(\rho)$ $(1 < \alpha \leq \alpha')$,

$$S(\rho) = \lim_{\alpha \to 1} S_{\alpha}(\rho) = \sup\{S_{\alpha}(\rho); \alpha > 1\}.$$

We may assume that every eigenvalue of $\rho_n - \rho$ is less than 1 for sufficiently large n because of $\|\rho_n - \rho\|_1 \to 0$ as $n \to \infty$. Therefore,

$$| \operatorname{tr}\rho_n{}^{\alpha} - \operatorname{tr}\rho^{\alpha} | \leq \operatorname{tr} | \rho_n - \rho |^{\alpha} \leq \operatorname{tr} | \rho_n - \rho | \to 0 \quad (n \to \infty),$$

which means that $S_{\alpha}(\rho)$ is a continuous function w.r.t. the trace norm $\| \cdot \|_1$ for any $\alpha \geq 1$. $S(\rho)$ is the supremum of a continuous function, hence it is lower semicontinuous. $\qquad\square$

For the entropy in CDS (classical dynamical systems), the monotonicity $\max\{S(p), S(q)\} \leq S(r)$ is satisfied. This monotonicity is not always satisfied in QDS. Indeed, take

$$x = \frac{(x_1 \otimes x_2 + y_1 \otimes y_2)}{\sqrt{2}},$$

where x_k is in $\mathcal{H}_k$ and $\langle x_k, y_k \rangle = 0$ $(k = 1, 2)$, and put

$$\rho = | x \rangle \langle x | \quad \text{and} \quad \rho_k = \operatorname{tr}_j \rho \quad (j \neq k; j, k = 1, 2).$$

Then $S(\rho) = 0$ because ρ is a pure state (i.e., $\rho^2 = \rho$), and $S(\rho_k) = \log 2$ because of $\rho_k = (| x_k \rangle \langle x_k | + | y_k \rangle \langle y_k |)/2$.

In addition to the above properties, we have the following additive property, whose proof is rather tedious, so we omit it here (see [337], [217], [256]): Let $\mathcal{H} = \mathcal{H}_1 \otimes \mathcal{H}_2 \otimes \mathcal{H}_3$ and denote the reduced states $\operatorname{tr}_k \rho$ and $\operatorname{tr}_{ij} \rho$ by ρ_{ij} and ρ_k, respectively. Then

[Strong Subadditivity]   (1) $S(\rho) + S(\rho_2) \leq S(\rho_{12}) + S(\rho_{23})$,

(2) $S(\rho_1) + S(\rho_2) \leq S(\rho_{13}) + S(\rho_{23})$.

3.2. Entropy in C^*-systems

In this section, we briefly discuss the entropy of a state in *general quantum dynamical systems* (GQDS), i.e., C^*-dynamical systems (see Appendix 2) introduced in [249]. Let $(\mathcal{A}, G, \alpha) = (\mathcal{A}, \mathfrak{S}, \alpha(G))$ be a C^*-dynamical system and $\mathcal{S}$ be a weak* compact and convex subset of the set $\mathfrak{S}$ of all states on $\mathcal{A}$, $\operatorname{ex}\mathcal{S}$ be the set of all extreme points of $\mathcal{S}$. Here G is a group and α is

-automorphic action on $\mathcal{A}$. From the Krein-Milman theorem, $\mathcal{S}$ is equal to the weak closure of the convex hull of ex$\mathcal{S}$. We are interested in the following three cases for the set $\mathcal{S}$: (1) $\mathcal{S} = \mathfrak{S}$, (2) $\mathcal{S} = I(\alpha)$, the set of all α-invariant states, (3) $\mathcal{S} = K(\alpha)$, the set of all KMS (Kubo-Martin-Schwinger) states, (cf. Appendix 1) [43], [44] :

Every state $\varphi \in \mathcal{S}$ has a maximal measure μ pseudosupported on ex$\mathcal{S}$ such that

$$\varphi = \int_{\mathcal{S}} \omega d\mu, \tag{3.7}$$

where ω can be regarded as extreme point in $\mathcal{S}$. The measure μ giving the above decomposition is not unique, so we denote the set of all such measures by $M_\varphi(\mathcal{S})$. The uniqueness occurs only when $\mathcal{S}$ is a *Choquet simplex*, i.e., if for every $\varphi \in \mathcal{S}$ $M_\varphi(\mathcal{S})$ is a base of a convex cone $K = \{a\mu; a \geq 0, \mu \in M_\varphi(\mathcal{S})\}$ such that K is a lattice for the order "$\geq$" defined by: $\mu_1 \geq \mu_2 \Leftrightarrow \mu_1 - \mu_2 \in M_\varphi(\mathcal{S})$ (we say then that $M_\varphi(\mathcal{S})$ is a *singleton*), cf. [55]. Take

$$D_\varphi(\mathcal{S}) \equiv$$

$$\left\{ \mu \in M_\varphi(\mathcal{S}); \exists\{\mu_k\} \subset \mathbb{R}^+ \text{and} \{\varphi_k\} \subset \text{ex}\mathcal{S} \text{ s.t. } \sum_k \mu_k = 1, \mu = \sum_k \mu_k \delta(\varphi_k) \right\},$$

where $\delta(\varphi)$ is the delta measure concentrated on $\{\varphi\}$, and put

$$H(\mu) = -\sum_k \mu_k \log \mu_k$$

for a measure $\mu \in D_\varphi(\mathcal{S})$. Then the *entropy of a general state* $\varphi \in \mathcal{S}$ w.r.t. $\mathcal{S}$ is defined by

$$S^{\mathcal{S}}(\varphi) = \begin{cases} \inf\{H(\mu); \mu \in D_\varphi(\mathcal{S})\} & \text{if } D_\varphi(\mathcal{S}) \neq \emptyset \\ +\infty & \text{if } D_\varphi(\mathcal{S}) = \emptyset, \end{cases} \tag{3.8}$$

This entropy is an extension of Von Neumann's entropy as it will be shown below. It depends on the choice of the set $\mathcal{S}$ chosen, which comes from the fact that the entropy is a quantity describing how much pure states (c.f. elementary events in CDS) are mixing in the reference system (here $\mathcal{S}$). Hence it represents the uncertainty of the state measured from the reference system $\mathcal{S}$. Three interesting entropies $S^{\mathfrak{S}}(\varphi)$ ($= S(\varphi)$ for short), $S^{I(\alpha)}(\varphi)$ ($= S^I(\varphi)$ for short) and $S^{K(\alpha)}(\varphi)$ ($= S^K(\varphi)$ for short) are generally different even for $\varphi \in K(\alpha)$.

THEOREM 3.2. *When $\mathcal{A} = B(\mathcal{H})$ and $\alpha_t = Ad(U_t)$ (i.e, $\alpha_t(A) = U_t^* A U_t$ for any $A \in \mathcal{A}$) with a unitary operator U_t, for any state φ given by $\varphi(\cdot) = \text{tr}\rho\cdot$ with a density operator ρ, we have the following statements:*

(1) $S(\varphi) = -\text{tr}\rho\log\rho$;

(2) If φ is an α-invariant faithful state and every eigenvalue of ρ is nondegenerate, then $S^I(\varphi) = S(\varphi)$;

(3) If $\varphi \in K(\alpha)$, then $S^K(\varphi) = 0$.

Sketch of the proof: (1): Let $\rho = \sum_k \lambda_k \rho_k$ be a decomposition of ρ into extremal states ρ_k (i.e., $\rho_k^2 = \rho_k$). It is easily seen that $-\sum_k \lambda_k \log \lambda_k$ attains the minimum value when the above extremal decomposition is the Schatten decomposition of ρ. Hence $S(\varphi) = -\text{tr}\rho \log \rho$.

(2): Since φ is α-invariant, the equality $[U_t, \rho] = 0$ holds for all $t \in \mathbb{R}$. From the assumptions, we have $[U_t, E_k] = 0$ for each E_k of the Schatten decomposition $\rho = \sum_k \lambda_k E_k$. Thus E_k is α-invariant for every k, by which we obtain $S(\varphi) \geq S^I(\varphi)$. The converse inequality is shown by the ergodic extremal α invariant decomposition of φ.

(3): The KMS state is unique for $\mathcal{A} = B(\mathcal{H})$, so $S^K(\varphi) = 0$.   □

Some relations among $S(\varphi)$, $S^I(\varphi)$ and $S^K(\varphi)$ exist. For instance, we have

THEOREM 3.3. For any $\varphi \in K(\alpha)$, we have the following inequalities:

(1) $S^K(\varphi) \leq S^I(\varphi)$;

(2) $S^K(\varphi) \leq S(\varphi)$;

(3) If our dynamical system $(\mathcal{A}, \alpha(R))$ is G-abelian on φ, then

$$S^K(\varphi) \leq S^I(\varphi) \leq S(\varphi).$$

Remark: $(\mathcal{A}, \alpha(R))$ is called G-abelian if $E_\varphi \pi_\varphi(\mathcal{A})'' E_\varphi$ is an abelian Von Neumann algebra, where E_φ is the projection from $\mathcal{H}$ onto the set of all $U_\varphi(t)$-invariant vectors.

(4) If our dynamical system $(\mathcal{A}, \alpha(R))$ is η-abelian, then

$$S^K(\varphi) \leq S^I(\varphi).$$

Remark: $(\mathcal{A}, \alpha(R))$ is called η-abelian if the equality

$$\lim_{T \to \infty} \frac{1}{T} \int_0^T \varphi(C^*[\alpha_t(A), B]C)dt = 0$$

holds for any $A, B, C \in \mathcal{A}$.

By the way, most of the properties of Von Neumann entropy $S(\rho)$ also hold for our entropy $S(\varphi)$ under some additonal conditions (see [249] for details). Moreover, when a physical system has a phase transition, for instance, the entropy $S^K(\varphi)$ may change w.r.t. some parameters such as the temperature, so that our entropy will be used to study some phase transitions in physical systems. We call the above entropy $S^S(\varphi)$ the S-entropy (or *mixing entropy*).

Connes, Narnhofer and Thirring introduced the entropy of a subalgebra $\mathfrak{M}$ of Von Neumann algebra $\mathcal{A}$ [59], [60]. The *CNT-entropy* $H_\varphi(\mathfrak{M})$ is defined as follows: For a state φ and a subalgebra $\mathfrak{M}$

$$H_\varphi(\mathfrak{M}) \equiv \tag{3.9}$$

$$\sup\left\{\sum_j \mu_j S(\varphi_j|_\mathfrak{M}, \varphi|_\mathfrak{M}); \varphi = \sum_j \mu_j \varphi_j (\text{finite decomposition of } \varphi),\right\}$$

where $S(\cdot, \cdot)$ is the relative entropy for C^*-algebra according to the definition of Araki or Uhlmann (cf. the next Section) and $\varphi|_\mathfrak{M}$ is the restricton of φ to $\mathfrak{M}$. The relation between S-entropy and CNT-entropy has been studied in [233], whose main results are summarized in the following theorem.

THEOREM 3.4. *(1) For a normal state φ on a Von Neumann algebra $\mathcal{A}$,*

$$S^\mathfrak{S}(\varphi) = H_\varphi(\mathcal{A}). \tag{3.10}$$

(2) Let $(\mathcal{A}, G, \alpha)$ be a G-finite Von Neumann algebraic dynamical system and φ be a G-invariant state. Then

$$S^{I(\alpha)}(\varphi) = H_\varphi(\mathcal{A}^\alpha), \tag{3.11}$$

where $\mathcal{A}^\alpha$ is the fixed point subalgebra of $\mathcal{A}$.
(3) Let $(\mathcal{A}, G, \alpha)$ be a C^*-algebraic dynamical system and φ be a G-invariant state. Then

$$S^{I(\alpha)}(\varphi) \geq H_\varphi(\mathcal{A}^\alpha). \tag{3.12}$$

In general, $S^{I(\alpha)}(\varphi)$ is strictly greater than $H_\varphi(\mathcal{A}^\alpha)$ as shown in the following example. Let $A^\mathbf{Z}$ be the compact Hausdorff space of all doubly infinite sequences in a finite set $A = \{a, b, c, d\}$, where the topology of $A^\mathbf{Z}$ is the product topology of discrete topology of A. Let $\mathcal{A} = C(A^\mathbf{Z})$ be the abelian C^*-algebra of continuous functions on $A^\mathbf{Z}$, α be the action of integers $G = \mathbf{Z}$ on $\mathcal{A}$ defined by $(\alpha_n(f))(x) \equiv f(T^{-n}x)$ $(n \in \mathbf{Z})$, where T denotes the shift of a sequence; $T : \{x_n\} \to \{x_{n+1}\}$. Then $(\mathcal{A}, \mathfrak{S}, \alpha)$ becomes a C^*-dynamical system. Since we can easily construct a point y in the spectrum $A^\mathbf{Z} = \mathrm{Sp}(\mathcal{A})$ of $\mathcal{A}$ such that the orbit $\alpha_G(y)$ is dense in $A^\mathbf{Z}$, the fixed point algebra is trivial: $\mathcal{A}^\alpha = \mathbf{C}$. Let φ_1 (resp. φ_2) be the G-invariant state on $\mathcal{A}$ identified with the product measure $\mu_1 = \otimes_{-\infty}^{+\infty} p$

(resp. $\mu_2 = \otimes_{-\infty}^{+\infty} q$), where p (resp. q) is the probability distribution on A given by

$$\begin{pmatrix} a & b & c & d \\ 1/2 & 1/2 & 0 & 0 \end{pmatrix}, \quad \text{resp.} \quad \begin{pmatrix} a & b & c & d \\ 0 & 0 & 1/2 & 1/2 \end{pmatrix}$$

Put $\varphi = (\varphi_1 + \varphi_2)/2$. Then φ_1 and φ_2 are ergodic measures, that is, $\varphi_1, \varphi_2 \in \text{ex} I(\alpha)$, and hence we have

$$S^{I(\alpha)}(\varphi) = \log 2 > 0 = H_\varphi(A^\alpha).$$

This result tells us that the S-entropy distinguishes states more sharply than the CNT-entropy.

In spite of the above example, if we treat only compact groups, then we have the equality between two entropies.

THEOREM 3.5. *Let* (A, G, α) *be a* C^*-*dynamical system with a compact group* G. *Then*

$$S^{I(\alpha)}(\varphi) = H_\varphi(A^\alpha). \tag{3.13}$$

Along the same line as for the S-entropy $S^S(\varphi)$, we define Rényi's type S-entropy as

$$S_\alpha^S(\varphi) = \inf \left\{ \frac{\log \sum_{k=1}^\infty \mu_k^\alpha}{1 - \alpha} ; \{\mu_k\} \in D_\varphi(S) \right\}$$

where α is a positive number with $\alpha \neq 1$. Let us consider the dimension

$$d^S(\varphi) = \inf \left\{ \alpha > 0; S_\alpha^S(\varphi) < \infty \right\}$$

for a density operator $\rho \in S$ on a Hilbert space $\mathcal{H}$. Akashi [15] found a theorem characterizing states in quantum systems by this dimension. When we consider the case $S = \mathfrak{S}(\mathcal{H})$, we abbreviate $d^{\mathfrak{S}(\mathcal{H})}(\varphi)$ to $d(\varphi)$ for simplicity.

For any positive number $p \leq 1$, let $\ell^p(\mathcal{H})$ be the set of all sequences consisting of the elements of $\mathcal{H}$ defined by

$$\ell^p(\mathcal{H}) = \left\{ \{x_k \in \mathcal{H}; k \in \mathbb{N}\} ; \inf\{r > 0 ; \sum_{k=1}^\infty \|x_k\|^{2r} < \infty\} = p, \right.$$

$$\left. \langle x_i, x_j \rangle = \|x_i\|^2 \delta_{i,j}, i, j \in \mathbb{N} \right\},$$

where $\delta_{i,j}$ is Kroneker's delta for all $i, j \in \mathbb{N}$. Then, for any positive number $p \leq 1$, we define the quasi σ-strong operator topology (denoted by $\tau_{\sigma s}^p$) with the degree of continuity p by the locally convex topology over $B(\mathcal{H})$ determined by the family of the following semi-norms: $A \in B(\mathcal{H}) \mapsto \{\sum_k \|Ax_k\|^2\}^{1/2}$, where $\{x_k\}$ runs over all elements of $\ell^p(\mathcal{H})$. He proved the following theorem.

THEOREM 3.6. *(1) For any non-negative numbers p and q satisfying $0 \leq p < q \leq 1$, $\tau_{\sigma s}^q$ is strictly finer than $\tau_{\sigma s}^p$.*
(2) Let φ be a normal state on $B(\mathcal{H})$ satisfying $0 < d(\varphi) < 1$. Then, for any positive number ε satisfying $\varepsilon < d(\varphi)$, the semi-norm on $B(\mathcal{H})$ $A \mapsto \sqrt{\varphi(A^*A)}$ is not continuous on $(B(\mathcal{H}), \tau_{\sigma s}^{d(\varphi)-\varepsilon})$, but continuous on $(B(\mathcal{H}), \tau_{\sigma s}^{d(\varphi)+\varepsilon})$.

For any positive number $p < 1$, $\tau_{\sigma s}^p$ is coaser than the σ-strong operator topology and finer than the strong operator topology.

This theorem might be used to discuss symmetry breaking in physical systems.

3.3. Relative entropy

The *relative entropy* of two states (density operators) ρ and σ has been defined by Umegaki, [325], as

$$S(\rho, \sigma) = \mathrm{tr}\rho(\log \rho - \log \sigma). \tag{3.14}$$

Umegaki and Lindblad, [325], [218], [219], studied the fundamental properties of this relative entropy corresponding to those of Shannon's type relative entropy (Kulback-Leibler information) in CDS, cf. Sec. 2.4. There are several trials to extend the relative entropy in general quantum systems and to apply it to some other fields, [130], [131], [75]. Here we review Araki's [26], [27] and Uhlmann's [322] definitions of the relative entropy and state some fundamental properties of the relative entropy.

DEFINITION 3.7 (Araki). Let $\mathfrak{N}$ be σ-finite (this condition is easily taken off) Von Neumann algebra acting on a Hilbert space $\mathcal{H}$ and let φ, ψ be normal states on $\mathfrak{N}$ given by $\varphi(\cdot) = \langle x, \cdot x \rangle$ and $\psi(\cdot) = \langle y, \cdot y \rangle$ with $x, y \in \mathcal{K}$ (a positive natural cone). The operator $S_{x,y}$ is defined by

$$S_{x,y}(Ay + z) = s^{\mathfrak{N}'}(y)A^*x, \quad A \in \mathfrak{N}, \quad s^{\mathfrak{N}}(y)z = 0, \tag{3.15}$$

on the domain $\mathfrak{N}y + (I - s^{\mathfrak{N}'}(y))\mathcal{H}$, where $s^{\mathfrak{N}}(y)$ is the projection from $\mathcal{H}$ to $\{\mathfrak{N}'y\}^-$, the $\mathfrak{N}$-support of y and $\mathfrak{N}$ is the commutant of $\mathfrak{N}$. Using this $S_{x,y}$, the relative modular operator $\Delta_{x,y}$ is defined as $\Delta_{x,y} = (S_{x,y})^* S_{x,y}^-$, whose spectral decomposition is denoted by $\int_0^\infty \lambda de_{x,y}(\lambda)$ ($S_{x,y}^-$ is the closure of

$S_{x,y}$). Then the *Araki relative entropy* is given by

$$S(\psi,\varphi) = \begin{cases} \int_{+0}^{\infty} \log \lambda d\langle y, e_{x,y}(\lambda)y \rangle & \text{if } \psi \ll \varphi \\ +\infty & \text{otherwise,} \end{cases} \qquad (3.16)$$

where $\psi \ll \varphi$ means that $\varphi(A^*A) = 0$ implies $\psi(A^*A) = 0$ for $A \in \mathfrak{N}$.

DEFINITION 3.8 (Uhlmann). Let $\mathcal{L}$ be a complex linear space and p, q be two seminorms on $\mathcal{L}$. Moreover, let $H(\mathcal{L})$ be the set of all positive hermitian forms α on $\mathcal{L}$ satisfying $|\alpha(x,y)| \leq p(x)q(y)$ for all $x, y \in \mathcal{L}$. Then the quadratical mean $QM(p,q)$ of p and q is defined by

$$QM(p,q)(x) = \sup\left\{\alpha(x,x)^{1/2}; \alpha \in H(\mathcal{L})\right\}, \quad x \in \mathcal{L}, \qquad (3.17)$$

and we know that there exists a function $p_t(x)$ of $t \in [0,1]$ for each $x \in \mathcal{L}$ satisfying the following conditions:

(1) For any $x \in \mathcal{L}$, $p_t(x)$ is continuous in t;
(2) $p_{1/2} = QM(p,q)$;
(3) $p_{t/2} = QM(p,p_t)$;
(4) $p_{(t+1)/2} = QM(p_t,q)$.

This seminorm p_t is denoted by $QI_t(p,q)$ and is called the quadratical interpolation from p to q. It is shown in [322] that for any positive hermitian forms α, β there exists a unique function $QF_t(\alpha,\beta)$ of $t \in [0,1]$ with values in the set $H(\mathcal{L})$ such that $QF_t(\alpha,\beta)(x,x)^{1/2}$ is the quadratical interpolation from $\alpha(x,x)^{1/2}$ to $\beta(x,x)^{1/2}$. The *relative entropy functional* $S(\alpha,\beta)(x)$ of α and β is defined as

$$S(\alpha,\beta)(x) = -\liminf_{t\to 0} \frac{1}{t}\{QF_t(\alpha,\beta)(x,x) - \alpha(x,x)\} \qquad (3.18)$$

for $x \in \mathcal{L}$. Let $\mathcal{L}$ be a *-algebra $\mathcal{A}$ and φ, ψ be positive linear functionals on $\mathcal{A}$ defining two hermitiam forms φ^L, ψ^R such as $\varphi^L(A,B) = \varphi(A^*B)$ and $\psi^R(A,B) = \psi(BA^*)$. Then the *Uhlmann relative entropy* of φ and ψ is defined by

$$S(\psi,\varphi) = S(\psi^R,\varphi^L)(I). \qquad (3.19)$$

If $\mathcal{L}$ is a Von Neumann algebra $\mathfrak{N}$ and φ, ψ are positive normal linear functionals on $\mathfrak{N}$, then the Uhlmann relative entropy is shown, [131], to be equal to the Araki relative entropy. For a C^*-algebra $\mathcal{A}$ and two positive linear functionals φ, ψ on $\mathcal{A}$, Uhlmann's definition can be directly applied. Further, by considering the GNS representation $\pi(= \pi_{\varphi+\psi})$ of the functional $\varphi+\psi$ and the canonical extensions $\tilde{\varphi}$, $\tilde{\psi}$ of φ, ψ to the Von Neumann algebra $\pi(\mathcal{A})''$, we have [131]:

THEOREM 3.9. *Under the above notations, the relative entropy $S(\psi, \varphi)$ of Uhlmann in a C^*-system is equal to $S(\tilde{\psi}, \tilde{\varphi})$ for its canonically extended Von Neumann system.*

Therefore, we can use both definitions for states in C^*-systems. Let us show that the expression of the relative entropy for two density operators ρ and σ is obtained from the Uhlmann expression. For normal states φ, ψ on a Von Neumann algebra $B(\mathcal{H})$ such that $\varphi(\cdot) = \mathrm{tr}\rho\cdot$ and $\psi(\cdot) = \mathrm{tr}\sigma\cdot$ with density operators ρ and σ, we get

$$QF_t(\psi^R, \varphi^L)(I, I) = \mathrm{tr}\rho^{1-t}\sigma^t, \tag{3.20}$$

hence

$$
\begin{aligned}
S(\psi, \varphi) &= S(\psi^R, \varphi^L)(I) \\
&= -\liminf_{t \to 0} \frac{1}{t} \left\{ QF_t(\psi^R, \varphi^L)(I, I) - \psi^R(I, I) \right\} \\
&= -\liminf_{t \to 0} \frac{1}{t} \mathrm{tr} \left(\rho^{1-t}\sigma^t - \rho \right) \\
&= \mathrm{tr}\rho(\log \rho - \log \sigma). \tag{3.21}
\end{aligned}
$$

Here we summarize the fundamental properties of the relative entropy. For the notational simplicity, we formulate the next theorem using the Von Neumann algebraic terminology. Namely, let φ, ψ be normal states and $\{\varphi_n\}$, $\{\psi_n\}$ be the sequences of normal states on a Von Neumann algebra $\mathfrak{N}$.

THEOREM 3.10. *(1) Positivity:* $S(\psi, \varphi) \geq 0$.
(2) Joint convexity: $S(\lambda\psi_1 + (1 - \lambda)\psi_2, \lambda\varphi_1 + (1 - \lambda)\varphi_2) \leq \lambda S(\psi_1, \varphi_1) + (1 - \lambda)S(\psi_2, \varphi_2)$ *for any* $\lambda \in [0, 1]$.
(3) Additivity: $S(\psi_1 \otimes \psi_2, \varphi_1 \otimes \varphi_2) = S(\psi_1, \varphi_1) + S(\psi_2, \varphi_2)$.
(4) Lower semicontinuity: If $\lim_{n \to \infty} \|\psi_n - \psi\| = 0$ *and* $\lim_{n \to \infty} \|\varphi_n - \varphi\| = 0$, *then* $S(\psi, \varphi) \leq \liminf_{n \to \infty} S(\psi_n, \varphi_n)$. *Moreover, if there exists a positive number* λ *satisfying* $\psi_n \leq \lambda\varphi_n$, *then* $\lim_{n \to 0} S(\psi_n, \varphi_n) = S(\psi, \varphi)$.
(5) Monotonicity: For a linear mapping Λ^* *from a state space* $\mathfrak{S}$ *on* $\mathcal{A}$ *to another state space* $\overline{\mathfrak{S}}$ *on* $\overline{\mathcal{A}}$,

$$S(\Lambda^*\psi, \Lambda^*\varphi) \leq S(\psi, \varphi).$$

(6) Lower bound: $\|\psi - \varphi\|^2/4 \leq S(\psi, \varphi)$.

The proofs of (1)–(4) can be seen in [26], [27], and the proof of (5) is essentially given in [322], while that of (6) is given in [131]. The relative entropy is related to the concept of sufficiency in statistics, and it can be used to classify some equilibrium states and stationary states [130], [131], [274], [323], [115]. See also [256].

3.4. Channel and lifting

In the course of information transmission or more generally in some physical processes, an initial (input) state changes to another state called a final (output) state under the external or internal effect of the system. This state change is described by a channeling transformation [246] (call it "channel" for simplicity). In this section, we discuss some fundamental facts about channels.

Let $(\mathcal{A}, \mathfrak{S})$ be an input system and $(\overline{\mathcal{A}}, \overline{\mathfrak{S}})$ be an output system with the mathematical structure of those considered in §3.2. That is, $\mathcal{A}$ (resp. $\overline{\mathcal{A}}$) is a C^*-algebra or the set of all bounded operators on a Hilbert space $\mathcal{H}$ (resp. $\overline{\mathcal{H}}$), and $\mathfrak{S}$ (resp. $\overline{\mathfrak{S}}$) is the set of all potitive linear functionals on $\mathcal{A}$ (resp. $\overline{\mathcal{A}}$) or the set of all density operators in $\mathcal{H}$ (resp. $\overline{\mathcal{H}}$). We denote the set of all density operators on $\mathcal{H}$ by $\mathfrak{S}(\mathcal{H})$. A map $\Lambda^* : \mathfrak{S} \to \overline{\mathfrak{S}}$ is called a *channel*. Moreover,

(1) Λ^* is a *linear channel* if Λ^* is affine.
(2) Λ^* is a *completely positive (CP) channel* if its dual map $\Lambda : \overline{\mathcal{A}} \to \mathcal{A}$ satisfies

$$\sum_{i,j=1}^{n} B_i^* \Lambda\left(A_i^* A_j\right) B_j \geq 0$$

for any $\{B_j\} \subset \mathcal{A}, \{A_i\} \subset \overline{\mathcal{A}}$ and any $n \in \mathbb{N}$.

When some external effects have to be taken into account in a physical process, such as noise or reservoir effect, it is convenient to extend the system $\mathcal{A}$ to $\mathcal{A} \otimes \mathcal{B}$, where $\mathcal{B}$ describes the external system. In such cases, the concept of lifting, a special channel, is useful. A *lifting* from $\mathcal{A}$ to $\mathcal{A} \otimes \mathcal{B}$ is a continuous map

$$\mathcal{E}^* : \mathfrak{S}(\mathcal{A}) \to \mathfrak{S}(\mathcal{A} \otimes \mathcal{B})$$

where we denote the set of states on $\mathcal{A}$ by $\mathfrak{S}(\mathcal{A})$ and that on $\mathcal{A} \otimes \mathcal{B}$ by $\mathfrak{S}(\mathcal{A} \otimes \mathcal{B})$.

(3) A lifting $\mathcal{E}^*$ is *linear* if it is affine.
(4) A lifting $\mathcal{E}^*$ is *nondemolition* for a state $\varphi \in \mathfrak{S}(\mathcal{A})$ if $\mathcal{E}^*\varphi(A \otimes I) = \varphi(A)$ for any $A \in \mathcal{A}$.

Given a channel $\Lambda^* : \mathfrak{S} \to \overline{\mathfrak{S}}$ and $\varphi \in \mathcal{S} \subseteq \mathfrak{S}$, take a decomposition such that

$$\varphi = \int_S \omega \, d\mu$$

Then the *compound state* [247]

$$\mathcal{E}_\mu^* \varphi = \int_S \omega \otimes \Lambda^* \omega \ d\mu$$

is a nondemolition lifting. This compound state exhibits the correlation between φ and $\Lambda^* \varphi$ like a joint probability measure in CDS, which is explained in the next section in more detail.

We take $\mathcal{A} = \overline{\mathcal{A}} = B(\mathcal{H})$, hence $\mathfrak{S} = \overline{\mathfrak{S}} = \mathfrak{S}(\mathcal{H})$ and let $\mathcal{B}$ be an algebra describing the external effect on another Hilbert space $\mathcal{K}$.

Then the state space of the total system is denoted by $\mathfrak{S}(\mathcal{H} \otimes \mathcal{K})$. We have several examples of channels and liftings encountered in physics and engineering.

⟨1⟩ Unitary evolution:
$\rho \to \Lambda_t^* \rho = A d U_t(\rho) \equiv U_t \rho U_t^*, t \in \mathbb{R}$, where U_t is a unitary operator on $\mathcal{H}$ generated by the Hamiltonian H of the system, i.e., $U_t = \exp(-itH)$.

⟨2⟩ Semigroup evolution:
$\rho \to \Lambda_t^* \rho = V_t \rho V_t^*, t \in \mathbb{R}^+$, where $\{V_t; t \in \mathbb{R}^+\}$ is a one-parameter semigroup on $\mathcal{H}$.

⟨3⟩ Measurement:
When we measure an observable $A = \sum_n a_n P_n$ (spectral decomposition) in a state ρ, the state ρ changes to a state $\Lambda^* \rho$ by this measurement according to the rule $\rho \to \Lambda^* \rho = \sum_n P_n \rho P_n / \text{tr} \left(\sum_n P_n \rho P_n \right)$.

⟨4⟩ Open System:
When two systems $\mathcal{A} = B(\mathcal{H})$ and $\mathcal{B} = B(\mathcal{K})$ are interacting through some time evolution U_t (e.g. $U_t = \exp(-itH)$, total Hamiltonian H on $\mathcal{H} \otimes \mathcal{K}$), an initial state $\rho \otimes \sigma, \rho \in \mathfrak{S}(\mathcal{H}), \sigma \in \mathfrak{S}(\mathcal{K})$, evolves by U_t such as

$$\mathcal{E}_t^* \rho \equiv U_t(\rho \otimes \sigma) U_t^*$$

which is a lifting of ρ. Moreover, a channel is constructed by taking the partial trace of $\mathcal{E}_t^* \rho$ with respect to $\mathcal{K}$, namely

$$\Lambda_t^* \rho \equiv \text{tr}_\mathcal{K} \mathcal{E}_t^* \rho.$$

Using this lifting, we construct a Markov chain in Section 3.6.

⟨5⟩ Isometric Lifting and Attenuation Channel:
Let V be an isometry from $\mathcal{H}$ to $\mathcal{H} \otimes \mathcal{K}$. Then $\mathcal{E}^* \rho \equiv V \rho V^*, \rho \in \mathfrak{S}(\mathcal{H})$ is a lifting from $\mathfrak{S}(\mathcal{H})$ to $\mathfrak{S}(\mathcal{H} \otimes \mathcal{K})$. An important example of this isometric

lifting comes from the attenuation process of optical communication, that is, the above isometry V is defined as

$$V_{\alpha,\beta}|\theta\rangle = |\alpha\theta\rangle \otimes |\beta\theta\rangle$$

for a coherent state $|\theta\rangle$ with $|\alpha|^2 + |\beta|^2 = 1$. Therefore, the attenuation channel is given by

$$\Lambda^*_{\alpha,\beta}\rho = \text{tr}_{\mathcal{K}}\mathcal{E}^*_{\alpha,\beta}\rho \equiv \text{tr}_{\mathcal{K}}V_{\alpha,\beta}\rho V^*_{\alpha,\beta}$$

Note that $V^*_{\alpha,\beta}V_{\alpha,\beta} = I$, and

$$
\begin{aligned}
\left\langle \theta \left| V^*_{\alpha,\beta}V_{\alpha',\beta'} \right| \theta \right\rangle \cdot &= \langle \alpha\theta, \alpha'\theta \rangle \langle \beta\theta, \beta'\theta \rangle \\
&= \exp\left\{ -|\alpha - \alpha'|^2 |\theta|^2 \right\} \exp\left\{ -|\beta - \beta'|^2 |\theta|^2 \right\} \\
&= \exp\left\{ -\left(|\alpha - \alpha'|^2 + |\beta - \beta'|^2\right)|\theta|^2 \right\}
\end{aligned}
$$

This lifting is called a beam splitting, which is applied to construct a new Markov chain [93].

$\langle 6 \rangle$ General Quantum Communication Process:

Quantum communication process is described by the following scheme [247].

$$\nu \in \mathfrak{S}(\mathcal{K})$$

$$\downarrow$$

$$\mathfrak{S}(\mathcal{H}) \ni \rho \xrightarrow{\hspace{3cm}} \bar{\rho} = \Lambda^*\rho \in \mathfrak{S}(\mathcal{H})$$

$$\downarrow$$

$$\text{Loss}$$

$$
\begin{array}{ccc}
\mathfrak{S}(\mathcal{H}) & \xrightarrow{\Lambda^*} & \mathfrak{S}(\mathcal{H}) \\
{\scriptstyle \gamma^*}\downarrow & & \uparrow{\scriptstyle a^*} \\
\mathfrak{S}(\mathcal{H} \otimes \mathcal{K}) & \xrightarrow[\pi^*]{} & \mathfrak{S}(\mathcal{H} \otimes \mathcal{K})
\end{array}
$$

The above maps γ^*, a^* are given as

$$
\begin{aligned}
\gamma^*(\rho) &= \rho \otimes \nu, & \rho \in \mathfrak{S}(\mathcal{H}), \\
a^*(\theta) &= \text{tr}_{\mathcal{K}}\theta, & \theta \in \mathfrak{S}(\mathcal{H} \otimes \mathcal{K}),
\end{aligned}
$$

where ν is a noise coming from the outside of the system. The map π^* is a certain channel determined by physical properties of the combined system. Hence the lifting and the channel for the above process are given as

$$\mathcal{E}^* \rho \equiv \pi^* \left(\rho \otimes \nu \right) = \left(\pi^* \circ \gamma^* \right) (\rho),$$
$$\Lambda^* \rho \equiv \mathrm{tr}_K \pi^* \left(\rho \otimes \nu \right) = \left(a^* \circ \pi^* \circ \gamma^* \right) (\rho).$$

The following channels are useful [247] to study in quantum processes. Let two C^*–triples $(\mathcal{A}, \mathfrak{S}, \alpha(\mathbb{R}))$ and $\left(\overline{\mathcal{A}}, \overline{\mathfrak{S}}, \overline{\alpha}(\mathbb{R}) \right)$ be an input and an output system, respectively, and Λ^* be a channel from $\mathfrak{S}$ to $\overline{\mathfrak{S}}$.

(1) Λ^* is said to be *stationary* if $\Lambda \circ \overline{\alpha}_t = \alpha_t \circ \Lambda$ for any $t \in \mathbb{R}$.
(2) Λ^* is said to be *ergodic* if Λ is stationary and $\Lambda^*(exI(\alpha)) \subset exI(\overline{\alpha})$, where $exI(\alpha)$ is the set of extremal invariant states.
(3) Λ^* is said to be *orthogonal* if for any two orthogonal states ρ_1, $\rho_2 \in \mathfrak{S}$ denoted by $(\rho_1 \perp \rho_2)$ we also have $\Lambda^* \rho_1 \perp \Lambda^* \rho_2$.
(4) Λ^* is said to be *deterministic* if Λ^* is orthogonal and bijective.
(5) Λ^* is said to be *chaotic for a subset* S of $\mathfrak{S}$ if $\Lambda^* \varphi = \Lambda^* \psi$ for any pair $\varphi, \psi \in exS$, and it is said to be *chaotic* if $S = \mathfrak{S}$.

3.5. Mutual entropy

When a state φ changes to another state $\overline{\varphi}$ under a physical transformation, we ask how much information of φ is correctly transmitted to $\overline{\varphi}$, and the amount of this information is expressed by the mutual entropy (information) invented by Shannon in CDS. We like to formulate this mutual entropy in GQDS for two states φ and $\overline{\varphi} = \Lambda^* \varphi$ with a channel Λ^*. We first set the compound state of the initial state φ and the final state $\overline{\varphi}$ expressing the correlation existing between these two states as an extension of the compound measure (joint distribution) in CDS.

The compound state Φ on the tensor product C^*-algebra $\mathcal{A} \otimes \overline{\mathcal{A}}$ of two states φ on $\mathcal{A}$ and $\overline{\varphi}$ on $\overline{\mathcal{A}}$ should satisfy the following properties:

(c.1) $\Phi(A \otimes I) = \varphi(A)$ for any $A \in \mathcal{A}$;
(c.2) $\Phi(I \otimes B) = \overline{\varphi}(B)$ for any $B \in \overline{\mathcal{A}}$;
(c.3) the expression of Φ contains the classical expression as a special case;
(c.4) Φ indicates the correspondence between each elementary state of φ and that of $\overline{\varphi}$.

There are several states satisfying the above two conditions (c.1) and (c.2). For instance, the direct product state Φ_0 of φ and $\overline{\varphi}$ given by

$$\Phi_0 = \varphi \otimes \overline{\varphi} \tag{3.22}$$

is such a state, which corresponds to the direct product measure in CDS. We call a state satisfying the conditions (c.1) and (c.2) a *quasicompound* *state.* Let us formulate the "true" compound state having all of the above conditions. Such a compound state is given through the decomposition (3.7) of the state φ. For a state φ in a weak $*$ compact convex subset S of $\mathfrak{S}$ and a channel Λ^*, let μ be an extremal decomposition measure of φ. Then a *compound state* Φ_μ of φ and $\Lambda^*\varphi$ with respect to S and μ was introduced by Ohya in [248], [250] as

$$\Phi_\mu^S = \int_S \omega \otimes \Lambda^* \omega d\mu, \qquad (3.23)$$

where $\omega \in S$. This state obviously satisfies (c.1), (c.2) and (c.4) because the measure μ is pseudosupported by ex$S$. The condition (c.3) is indeed satisfied: When A and $\overline{A}$ are abelian algebras with measurable spaces $(\Omega, \mathcal{F})$ and $(\overline{\Omega}, \overline{\mathcal{F}})$, respectively, and φ is a probability measure on Ω, the extremal decomposition of φ is unique and given by

$$\varphi = \int_\Omega \delta_x d\varphi, \qquad (3.24)$$

where δ_x is the Dirac measure concentrated at a point $x \in \Omega$. Put $\lambda(x, Q) = \Lambda^* \delta_x(Q)$ for any $x \in \Omega$ and $Q \in \overline{\mathcal{F}}$. Then λ is the Markov kernel defining the classical channel, and we have

$$
\begin{aligned}
\Phi_\varphi(P, Q) &= \int_\Omega \delta_x(P) \Lambda^* \delta_x(Q) d\varphi \\
&= \int_\Omega 1_P(x) \lambda(x, Q) d\varphi = \int_P \lambda(x, Q) d\varphi
\end{aligned}
$$

for any $P \in \mathcal{F}$, $Q \in \overline{\mathcal{F}}$. Thus our compound state defined by (3.23) is a desired one. This compound state might play a similar role as the joint probability in CDS although the joint probability does not exist in QDS [326]. Moreover, this compound state will be related to the filtering theory of Belavkin [33].

Now let us formulate the mutual entropy representing the information transmitted from an initial state $\varphi \in \mathfrak{S}$ to the final state $\Lambda^*\varphi \in \mathfrak{S}$. The mutual entropy with respect to an initial state $\varphi \in S$, the decomposition measure μ and a channel Λ^* is defined by

$$I_\mu^S(\varphi; \Lambda^*) = S(\Phi_\mu^S, \Phi_0), \qquad (3.25)$$

where $S(\cdot, \cdot)$ is the relative entropy for two states in C^*-algebra. The *mutual entropy* with respect to an initial state $\varphi \in S$ and a channel Λ^* is now defined by Ohya

$$I^S(\varphi; \Lambda^*) = \limsup \left\{ I_\mu^S(\varphi; \Lambda^*); \mu \in F_\varphi(S; \varepsilon) \right\}, \qquad (3.26)$$

where $F_\varphi(S;\varepsilon)$ is the subset of the set $M_\varphi(S)$ such that $F_\varphi(S;\varepsilon) = \{\mu \in D_\varphi(S); S^S(\varphi) \le H(\mu) < S^S(\varphi) + \varepsilon < +\infty\}$ $(F_\varphi(S;0) = \{\mu \in D_\varphi(S); S^S(\varphi) = H(\mu)\})$ or $F_\varphi(S;\varepsilon) = M_\varphi(S)$ when $S^S(\varphi) = \infty$. The above sets $M_\varphi(S)$, $D_\varphi(S)$ and the functional $H(\mu)$ are those introduced in §3.2. Note that the mutual entropy (3.25) should be used when the decomposition measure is fixed. In the sequel we use the simple notations Φ_μ, $I_\mu(\varphi;\Lambda^*)$ and $I(\varphi;\Lambda^*)$ when $S = \mathfrak{S}$. Before discussing the fundamental properties of the mutual entropy, we introduce another mutual type entropy for an initial state φ and a final state ψ. We call it the quasimutual entropy and denote it by $I^0(\varphi,\psi)$. Let $\mathfrak{S}_{qc}$ be the set of all quasicompound states in $\mathfrak{S}$ for φ and ψ, and let Ψ_0 be $\varphi \otimes \psi$. Here it is not necessary that ψ is connected to φ through a channel. We define the *quasimutual entropy* for φ and ψ by

$$I^0(\varphi,\psi) = \sup\{S(\Psi,\Psi_0); \Psi \in \mathfrak{S}_{qc}, \Psi \le \Psi_0\}. \qquad (3.27)$$

Other weak versions of the mutual entropy are discussed in §4.3 of chapter 4 as the complexities in quantum systems.

In the rest of this section, we assume that $\mathcal{A}$ and $\overline{\mathcal{A}}$ are von Neumann algebras acting on Hilbert spaces $\mathcal{H}$ and $\overline{\mathcal{H}}$, respectively, and the states denoted by φ, φ_k and ψ are normal states on $\mathcal{A}$. Furthermore, let $\mathcal{K}$ and $\overline{\mathcal{K}}$ be positive natural cones for $\mathcal{A}$ and $\overline{\mathcal{A}}$, respectively.

Two states φ_1 and φ_2 are said to be *orthogonal* each other (denoted by $\varphi_1 \perp \varphi_2$) if their supports $s(\varphi_1)$ and $s(\varphi_2)$ are orthogonal, where the *support* $s(\varphi)$ of φ means the smallest projection E satisfying $\varphi(I - E) = 0$. The measure $\mu \in M_\varphi(S)$ is said to be orthogonal if $(\int_Q \omega d\mu) \perp (\int_{S/Q} \omega d\mu)$ is satisfied for every Borel set Q in S. A channel is called *normal* if it sends a normal state to a normal state. We have the following convenient expression of the mutual entropy [252]:

THEOREM 3.11. *For a normal channel Λ^* and a normal state φ, if a measure μ is in the set $F_\varphi(\mathfrak{S};\varepsilon) \cap D_\varphi(\mathfrak{S})$ and is orthogonal, then for any $\varepsilon > 0$*

$$I_\mu(\varphi;\Lambda^*) = \int_\mathfrak{S} S(\Lambda^*\omega, \Lambda^*\varphi)d\mu < S(\varphi) + \varepsilon.$$

When $\mathcal{A}$ is the full algebra $B(\mathcal{H})$, any normal state φ is described by a density operator ρ such as $\varphi(A) = \mathrm{tr}\rho A$ for any $A \in \mathcal{A}$. Then our entropy $S(\varphi)$ defined by (3.8) is equal to that of Von Neumann: $S(\varphi)(= S(\rho)) = -\mathrm{tr}\rho\log\rho$. Every Schatten decomposition $\rho = \sum_n \lambda_n E_n, E_n = |x_n\rangle\langle x_n|$(i.e.,
λ_n is the eigenvalue of ρ and x_n is its associated eigenvector) provides every orthogonal measure in $D_\varphi(\mathfrak{S})$ defining the entropy $S(\varphi)$. Since the Schatten decomposition of ρ is not unique unless every eigenvalue λ_n is nondegenerate, the compound state Φ given by (3.23) is expressed as

$$\Phi_E(Q) = \mathrm{tr}\sigma_E Q, \quad Q \in \mathcal{A} \otimes \overline{\mathcal{A}},$$

with

$$\sigma_E = \sum_n \lambda_n E_n \otimes \Lambda^* E_n,$$

where E represents a Schatten decomposition $\{E_n\}$. Then the mutual entropy for φ and the channel Λ^* is given by

$$I(\varphi; \Lambda^*) = \sup \{I_E(\varphi; \Lambda^*); \ E = \{E_n\} \text{ of } \rho\}, \quad (3.28)$$

with

$$I_E(\varphi; \Lambda^*) = S(\sigma_E, \sigma_0) = \mathrm{tr}\sigma_E(\log \sigma_E - \log \sigma_0), \quad (3.29)$$

where $\sigma_0 = \rho \otimes \Lambda^* \rho$. This form of the mutual entropy has been first defined in [247] to study optical communication processes. The expression (3.29) can be written in the following form

$$I_E(\varphi; \Lambda^*) = \sum_n \lambda_n S(\Lambda^* E_n, \Lambda^* \rho).$$

Since every Schatten decomposition is discrete and orthogonal, for a state φ given by a density operator ρ such as $\varphi(A) = \mathrm{tr}\rho A$, we have the following fundamental inequality [252], and it has been used to study several quantum communication processes [258, 335].

THEOREM 3.12. $0 \le I(\varphi; \Lambda^*) \le \min\{S(\varphi), S(\Lambda^*\varphi)\}$.

Proof. Since φ is defined by $\varphi(\cdot) = \mathrm{tr}\rho \cdot$, the compound state σ_E is given by

$$\sigma_E = \sum_k \lambda_k E_k \otimes \Lambda^* E_k$$

for a shatten decomposition $\rho = \sum_k \lambda_k E_k$. The positivity of $S(\sigma_E, \sigma_0)$ implies that of $I(\rho; \Lambda^*)$. From the definition of $S(\sigma_E, \sigma_0)$, we obtain

$$S(\sigma_E, \sigma_0) = S(\Lambda^* \rho) - \sum_n \lambda_n S(\Lambda^* E_n) \le S(\Lambda^* \rho),$$

from which the inequality $I(\rho; \Lambda^*) \le S(\Lambda^*\rho)$ follows. Let us show another inequality $I(\rho; \Lambda^*) \le S(\rho)$. Since Λ is a completely positive and identity preserving map, it is known by the monotonicity of the relative entropy that $S(\Lambda^*\rho_1, \Lambda^*\rho_2) \le S(\rho_1, \rho_2)$ for any ρ_1 and ρ_2 in $\mathfrak{S}(\mathcal{H})$. Hence, we have

$$
\begin{aligned}
S(\sigma_E, \sigma_0) &= \sum_n \lambda_n S(\Lambda^* E_n, \Lambda^* \rho) \\
&\le \sum_n \lambda_n S(E_n, \rho) \\
&= \sum_n \lambda_n (\mathrm{tr} E_n \log E_n - \mathrm{tr} E_n \log \rho) \\
&= -\mathrm{tr} \sum_n \lambda_n E_n \log \rho = -\mathrm{tr}\rho \log \rho = S(\rho)
\end{aligned}
$$

Taking the supremum over E, we get $I(\rho; \Lambda^*) \leq S(\rho)$.   □

This theorem means that the information correctly transmitted Λ from the input system to the output system is less than that carried by the initial state. When we send an information through a channel, we have to consider the efficiency of the communication. This efficiency is measured by the mutual entropy; namely, we ask what channel takes the mutual entropy larger. This efficiency is often expressed as the capacity of channel Λ^*, that is,

$$C(\Lambda^*) = \sup\{I(\rho; \Lambda^*); \rho \in \mathfrak{S}_0\}$$

where $\mathfrak{S}_0$ is a subset of $\mathfrak{S}(\mathcal{H})$ subject to certain conditions.

THEOREM 3.13. *For a state φ given by $\varphi(\cdot) = \mathrm{tr}\rho \cdot$ and a channel Λ^*, we have*

(1) if Λ^* is deterministic, then $I(\varphi; \Lambda^*) = S(\varphi)$;
(2) if Λ^* is chaotic, then $I(\varphi; \Lambda^*) = 0$;
(3) if Λ^* is ergodic and φ is stationary for a time evolution $\alpha_t \equiv AdU_t$, and if every eigenvalue of ρ is nonzero and nondegenerate, then $I(\varphi; \Lambda^*) = S(\Lambda^*\varphi)$.

Proof. (1) Since Λ^* is deterministic, for each elementay event E_k of ρ, $\Lambda^* E_k$ is a pure state and $\Lambda^*\rho = \sum_k \lambda_k \Lambda^* E_k$ is an extremal decomposition of $\Lambda^*\rho$ into pure states. Hence, $S(\Lambda^*\rho) = S(\rho) = -\sum_k \lambda_k \log \lambda_k$. We have $S(\sigma_E, \sigma_0) = S(\Lambda^*\rho)$ because $\Lambda^* E_k$ is pure (so $S(\Lambda^* E_k) = 0$). Therefore we get $I(\rho; \Lambda^*) = S(\rho)$.

(2) Since Λ^* is chaotic, the compound state σ_E is equivalent to σ_0. Indeed

$$\sigma_E = \sum_k \lambda_k E_k \otimes \Lambda^* E_k = \sum_k \lambda_k E_k \otimes \Lambda^* \rho = \rho \otimes \Lambda^* \rho = \sigma_0.$$

Thus we get

$$I(\rho; \Lambda^*) = \sup_E S(\sigma_E, \sigma_0) = S(\sigma_0, \sigma_0) = 0.$$

(3) When ρ is α-invariant, every pure state E_k appearing in the decomposition is also α-invariant. Let us prove this fact. Since ρ is α-invariant, $[u_t, \rho] = 0$ holds for all $t \in \mathbb{R}$. From the assumptions for the eigenvalues of the state ρ, the equality $[u_t, E_k] = 0$ holds for all t and k. Thus E_k is a α-invariant state, hence ergodic. Since it is assumed that Λ^* is ergodic, $\Lambda^* E_k$ is an ergodic state for τ_t. We now show that this ergodic state $\Lambda^* E_k$ is also pure. Suppose that $\Lambda^* E_k$ is not pure; then we have an extremal decomposition of $\Lambda^* E_k$ into pure states such as $\Lambda^* E_k = \sum_n \mu_n^k \theta_n^k$. The τ-invariant of $\Lambda^* E_k$ reduces that of θ_n^k as discussed above. This constradicts the fact that $\Lambda^* E_k$ is an ergodic state.

Thus $\Lambda^* E_k$ is a pure state. From this result we have

$$I(\rho; \Lambda^*) = \sup_E S(\sigma_E, \sigma_0) = S(\Lambda^* \rho)$$

because of $S(\Lambda^* E_k) = 0$.   □

The mutual entropy has been studied by several authors like Holevo, Ingarden, Levitin for a measure of information transmission in classical → quantum → classical processes. The above definition of the mutual entropy is that for quantum → quantum processes, so that it can be used in a direct study of quantum channel (transformation) and it contains other definitions as special cases [264]. This mutual entropy is a fundamental quantity to study the efficiency of information transmission, namely, the capacity of channels [265].

3.6. Applications

As an application of the mutual entropy, we can compute this entropy in *quantum Markov chain* (QMC for short) [3] by Accardi.

In statistical mechanics, a state of interacting system is changed under some actions from the outside of the system (reservoir), that is, the interaction between two systems is considered and the reduced state after interaction is studied. Mathematical expression of this process is given by ⟨4⟩ of the §3.4.

Let a system Σ_0 be described by a Hilbert space $\mathcal{H}_0$, which interacts with an external system Σ_1 described by another Hilbert space $\mathcal{H}_1$, and let the initial states of Σ_0, Σ_1 be $\rho \in \mathfrak{S}_0$, $\omega \in \mathfrak{S}_1$ ($\mathfrak{S}_i$ is the set of all states on $\mathcal{H}_i$ ($i = 0, 1$)), respectively. Then the combined state $\tilde{\rho} \in \mathfrak{S}_0 \otimes \mathfrak{S}_1$ after the interaction between the two systems is given by

$$\tilde{\rho} = U_t (\rho \otimes \omega) U_t^*, \tag{3.30}$$

where $U_t = \exp(-itH)$ with a total Hamiltonian H on $\mathcal{H}_0 \otimes \mathcal{H}_1$. Here we took $\hbar = 1$.

The above total Hamiltonian H for two weakly coupled oscillators in quantum optics is given by [221]:

$$H = H_0 + H_1 + H_{01}, \tag{3.31}$$

$$H_0 = a^* a, \quad H_1 = \sum_j b_j^* b_j, \quad H_{01} = \sum_j \left(\varepsilon_j b_j a^* + \varepsilon_j^* b_j^* a \right) \tag{3.32}$$

where H_0, H_1, H_{01} are the Hamiltonians for the obeserved system Σ_0, the external system (reservoir) Σ_1 and the interaction between Σ_0 and Σ_1, respectively. a, a^* on $\mathcal{H}_0$ and b_j, b_j^* on $\mathcal{H}_1$ are pairs of annihilation and creation operators, respectively, and ε_j ($j \in \mathbb{N}$) are the coupling constants.

For simplicity we assume that the reservoir has a single mode, that is, the total hamiltonian H is given by

$$\begin{aligned} H &= H_0 + H_1 + H_{01} \\ &= a^*a + b^*b + \varepsilon\left(a^*b + ab^*\right) \end{aligned} \tag{3.33}$$

By using a channel and a lifting shown in the previous section, the evolution of the initial state ρ after the interaction is mathematically expressed as

$$\rho \mapsto \bar{\rho}_t = \Lambda_t^* \rho = \mathrm{tr}_{\mathcal{H}_1} \mathcal{E}_t^* \rho = \mathrm{tr}_{\mathcal{H}_1} U_t \left(\rho \otimes \omega\right) U_t^*. \tag{3.34}$$

This model can be applied to QMC as follows.

In order to formulate QMC, we need the concept of a transition expectation [3].

Let $\mathcal{B}_0, \mathcal{B}_1$ be the algebras of all bounded operators on Hilbert spaces $\mathcal{H}_0, \mathcal{H}_1$ respectively, and let $\mathcal{B}_0 \otimes \mathcal{B}_1$ be a fixed tensor product of $\mathcal{B}_0$ and $\mathcal{B}_1$. A *transition expectation* from $\mathcal{B}_0 \otimes \mathcal{B}_1$ to $\mathcal{B}_0$ is a completely positive linear map $\mathcal{E} : \mathcal{B}_0 \otimes \mathcal{B}_1 \to \mathcal{B}_0$ satisfying

$$\mathcal{E}\left(1 \otimes 1\right) = 1. \tag{3.35}$$

When $\mathcal{B} = \mathcal{B}_0 = \mathcal{B}_1$ we simply say that $\mathcal{E}$ is a transition expectation on $\mathcal{B}$.

By using the transition expectation, a generalized QMC [3] is defined as follows: Let $\{\mathcal{E}_n\}_{n \geq 0}$ be a sequence of transition expectations on $\mathcal{B}$. Then there exists a unique completely positive identity preserving map $E_{0]} : \otimes^n \mathcal{B} \to \mathcal{B}$ such that for each integer n and for each $A_0, A_1, \cdots, A_n \in \mathcal{B}$, one has

$$\begin{aligned} E_{0]}\left(A_0 \otimes A_1 \otimes \cdots \otimes A_n \otimes 1 \otimes \cdots\right) = \\ \mathcal{E}_0\left(A_0 \otimes \mathcal{E}_1\left(A_1 \otimes \cdots \otimes \mathcal{E}_n\left(A_n \otimes 1\right) \cdots\right)\right). \end{aligned} \tag{3.36}$$

Let φ_0 be a state on $\mathcal{B}$ and define a state φ on $\otimes^n \mathcal{B}$ by

$$\varphi = \varphi_0 \circ E_{0]}. \tag{3.37}$$

Then φ satisfies

$$\begin{aligned} \varphi\left(A_0 \otimes A_1 \otimes \cdots \otimes A_n \otimes 1 \otimes \cdots\right) = \\ \varphi_0\left(\mathcal{E}_0\left(A_0 \otimes \mathcal{E}_1\left(A_1 \otimes \cdots \otimes \mathcal{E}_n\left(A_n \otimes 1\right) \cdots\right)\right)\right). \end{aligned} \tag{3.38}$$

The state φ, characterized by (3.37) is called a *generalized Markov chain* associated to the pair $(\varphi_0, \{\mathcal{E}_n\})$. If for each n

$$\mathcal{E} \equiv \mathcal{E}_n = \mathcal{E}_0, \tag{3.39}$$

then the state φ is called *homogeneous*.

Clearly, the dual of a "linear" lifting is a transition expectation, therefore to any linear lifting one can associate a QMC in the standard way [6].

In order to compute the mutual entropy in QMS rigorously, the Stinespring expression is concretely obtained in the above model [7].

The Stinespring-Kraus' theorem [203] is well known in the following form.

THEOREM 3.14. *Let $\mathcal{H}$ be a Hilbert space. A linear map $\Lambda : B(\mathcal{H}) \to B(\mathcal{H})$ is completely positive if and only if it has the form*

$$\Lambda\,(A) = \sum_{i=1}^{n} V_i^* A V_i \qquad A \in \mathcal{A}, \tag{3.40}$$

where $V_i : \mathcal{H} \to \mathcal{H}$ are partial isometries and $n \leq \dim(\mathcal{H})$.

We shall give a concrete expression for partial isometries V_i in our model.

From the form (3.33) of the Hamiltonian H of the total system $\mathcal{H}_0 \otimes \mathcal{H}_1$, it is easy to see that

$$[H_0 + H_1,\, H_{01}] = 0. \tag{3.41}$$

By using (3.41) and some calculations [7], we get the following formula:

$$\Lambda^* \rho = \sum_{\nu=0}^{\infty} T_\nu \,(\rho \otimes \omega)\, T_\nu^*, \tag{3.42}$$

where

$$T_\nu \equiv \sum_{n=\nu}^{\infty} |n - \nu\rangle \left\langle \varphi_{n-\nu}^{(n)} \right|,$$

$$\left\langle \varphi_{n-\nu}^{(n)} \right| \equiv \sum_{\beta=0}^{n} d_{n-\nu,\beta}^{(n)} \langle \beta| \otimes \langle n - \beta| \in \mathcal{K}_n \tag{3.43}$$

$$d_{\alpha,\beta}^{(n)} \equiv \sum_{j=0}^{n} e^{-it\epsilon \lambda_j^{(n)}} C_\alpha^{(n,j)} \overline{C_\beta^{(n,j)}}. \tag{3.44}$$

Since $\left\{ \left| \varphi_{n-\nu}^{(n)} \right\rangle;\ n \geq \nu,\ n, \nu \in N \right\}$ is othogonal [7], the above T_ν is a partial isometry. Therefore, (3.42) and (3.43) give a concrete expression for Stinespring-Kraus' theorem.

These formulae are useful for the computation of several entropies. In the following, we compute the mutual entropy.

From (3.42) and (3.43), we have

$$\Lambda^* \rho = \sum_i \left(\sum_{\substack{l,\nu \\ (\nu-l)+i \geq 0}} \lambda_l^\rho \mu_{(\nu-l)+i}^\omega \left| d_{i,l}^{(\nu+i)} \right|^2 \right) |i\rangle \langle i| . \qquad (3.45)$$

Here, we take the following Schatten decompositions:

$$\rho = \sum_l \lambda_l^\rho |l\rangle \langle l|, \quad \omega = \sum_\gamma \mu_\gamma^\omega |\gamma\rangle \langle \gamma|. \qquad (3.46)$$

Clearly, (3.45) is also a Schatten decomposition. Suppose that the decomposition (3.46) of ρ is unique. From (3.45) we obtain the mutual entropy as

$$I(\rho; \Lambda^*) = \qquad\qquad\qquad\qquad\qquad\qquad\qquad\qquad\qquad (3.47)$$

$$\sum_{k,t} \lambda_k^\rho \sum_{\substack{\nu \\ (\nu-k)+t \geq 0}} \mu_{(\nu-k)+t}^\omega \left| d_{t,k}^{(\nu+t)} \right|^2 \log \frac{\displaystyle\sum_{\substack{\nu_1 \\ (\nu_1-k)+t \geq 0}} \mu_{(\nu_1-k)+t}^\omega \left| d_{t,k}^{(\nu_1+t)} \right|^2}{\displaystyle\sum_{\substack{l,\nu \\ (\nu_2-l)+t \geq 0}} \lambda_l^\rho \mu_{(\nu_2-l)+t}^\omega \left| d_{t,l}^{(\nu_2+t)} \right|^2}.$$

The typical cases are:

(i) The state $\omega \in \mathfrak{S}_1$ is a vacuum state, $\omega = |0\rangle \langle 0|$. Then

$$I(\rho; \Lambda^*) = \sum_{k,t} \lambda_k^\rho \left| d_{t,k}^{(k)} \right|^2 \log \frac{\left| d_{t,k}^{(k)} \right|^2}{\sum_l \lambda_l^\rho \left| d_{t,l}^{(l)} \right|^2} \qquad (3.48)$$

(ii) The state $\omega \in \mathfrak{S}_1$ is a Gibbs state,

$$\omega_\beta = \frac{e^{-\beta H_1}}{\mathrm{tr} e^{-\beta H_1}} = \left(1 - e^{-\beta}\right) \sum_n e^{-\beta n} |n\rangle \langle n|.$$

Then

$$I(\rho; \Lambda^*) = \left(1 - e^{-\beta}\right) \sum_{k,t} \lambda_k^\rho D_{k,t}^\beta \log \frac{D_{k,t}^\beta}{\sum_l \lambda_l^\rho D_{l,t}^\beta}, \qquad (3.49)$$

where

$$D_{k,t}^\beta = \sum_{\substack{\nu \\ (\nu-k)+t \geq 0}} e^{-\beta((\nu-k)+t)} \left| d_{t,k}^{(\nu+t)} \right|^2. \qquad (3.50)$$

We discuss another application of the mutual entropy, which solves the difficulties encountered in the classical Gaussian communication processes. That is, we can solve the difficulty appeared in classical system by using quantum setting.

Let $(\Omega, \mathcal{F}, \mu)$ be a probability space and $F(\mathcal{G})$ be the set of all *finite partitions* of Ω for a σ-subfield $\mathcal{G}$ of $\mathcal{F}$ (i.e., $\tilde{A} = \{A_k; k = 1, \cdots, n < +\infty\} \in F(\mathcal{G})$ iff $A_k \in \mathcal{G}, A_k \cap A_j = \emptyset$ $(k \neq j)$ and $\cup_k A_k = \Omega$). We like to characterize the entropy of this state (measure) μ. Along the line of Shannon's philosophy for a discrete system, it is natural to define the entropy of μ as

$$S(\mu) = \sup\{S(\mu : \tilde{A}); \tilde{A} \in F(\mathcal{F})\} \tag{3.51}$$

with

$$S(\mu : \tilde{A}) = -\sum_k \mu(A_k) \log \mu(A_k). \tag{3.52}$$

This definition is of course mathematically consistent, but $S(\mu)$ takes the value $+\infty$ for almost all states of continuous systems, for instance, for every Gaussian measure μ in any real Hilbert space.

There exists another definition of the entropy for some continuous systems: Let f be a random variable and $F(t)$ be the probability distribution associated to f, $F(t) = \mu(\{\omega \in \Omega; f(\omega) \leq t\})$. We only treat the case when there exists the density distribution $\rho_f(t)$ such that

$$F(t) = \int_{-\infty}^t \rho_f(x) dx.$$

This density distribution $\rho_f(t)$ corresponds to the distribution p for a discrete system. The probability that f takes the value between x and $x + \delta x$ is given by $\rho_f(a)\delta x$ according to the mean value theorem, where a is a proper value between x and $x + \delta x$. Therefore, dividing $\mathbb{R}$ as $\mathbb{R} = \cup_k I_k$, $I_k = [x_k, x_{k+1})$, $x_{k+1} = x_k + \delta x$, $I_k \cap I_j = \emptyset$ $(k \neq j)$, the probability p_k in I_k is given by $\rho_f(a_k)\delta x$ by a certain constant $a_k \in I_k$. The Shannon entropy of $\{p_k\}$ becomes

$$\begin{aligned} S &= -\sum_k \rho_f(a_k)\delta x \log \rho_f(a_k)\delta x \\ &= -\sum_k \rho_f(a_k)\delta x \log \rho_f(a_k) - \sum_k \rho_f(a_k)\delta x \log \delta x, \end{aligned}$$

which becomes, as $\delta x \to 0$,

$$\begin{aligned} S &= -\int \rho_f(x) \log \rho_f(x) dx - \int \rho_f(x) dx \cdot (\lim_{\delta x \to 0} \delta x \log \delta x) \\ &= +\infty. \end{aligned}$$

After subtracting the infinity from the above (a kind of renormalization), the entropy with respect to (μ, f) is defined as

$$S(\mu : f) = -\int_{\mathbb{R}} \rho_f(x) \log \rho_f(x) dx. \qquad (3.53)$$

More generally, the entropy for a density distribution ρ is given by

$$S(\rho) = -\int_{\mathbb{R}} \rho(x) \log \rho(x) dx, \qquad (3.54)$$

which is usually called the differential entropy of ρ and was first introduced by Gibbs and Boltzmann independently, cf. Chap. 2. As an example, for the Gaussian distribution ρ in n-dimensional Hilbert space $\mathbb{R}^n$ with a covariance matrix A, we have

$$S(\rho) = \log(2\pi e)^{\frac{n}{2}} \mid \det A^{-1} \mid^{-\frac{1}{2}} .$$

However, this differential entropy has some inconvenient properties as a measure of information or uncertainty, for instance, it does not have the positivity and the scaling invariance [322].

The Gaussian communication processes have been beautifully treated by several authors [29], [104], [135], [349]. However most of these discussions have two drawbacks still unsolved; namely (1) Shannon's type entropy $S(\mu)$ defined by (3.51) is always infinite for any Gaussian measure μ; (2) If we take the differential entropy $S(\mu)$ by (3.53) as information of an input state, then this entropy becomes less than the mutual entropy $I(\mu; \lambda)$. These points are not well-matched with Shannon's communication theory. In order to avoid these difficulties and discuss Gaussian communication processes on a Hilbert space consistently, we need the quantum entropy and the quantum mutual entropy [251].

Let $\mathcal{B}_k$ be the Borel σ-field of a real separable Hilbert space $\mathcal{H}_k$ $(k = 1, 2)$ and let $(\mathcal{H}_1, \mathcal{B}_1)$ and $(\mathcal{H}_2, \mathcal{B}_2)$ be an input space and an output space, respectively. We denote the set of all Gaussian probability measures on $(\mathcal{H}_k, \mathcal{B}_k)$ by $P_G^{(k)}$ $(k = 1, 2)$. That is, the Gaussian measure μ in $\mathcal{H}_k$ is a Borel measure in $\mathcal{H}_k$ such that for each $x \in \mathcal{H}_k$, there exist real numbers m_x and $\sigma_x(> 0)$ satisfying

$$\mu\{y \in \mathcal{H}_k; \langle y, x \rangle \leq a\} = \int_{-\infty}^{a} \frac{1}{\sqrt{2\pi\sigma_x}} \exp\left(-\frac{(t - m_x)^2}{2\sigma_x}\right) dt,$$

where $\langle , \rangle$ is the inner product in $\mathcal{H}_k$. The notation $\mu_k = [m_k, R_k]$ means that μ_k is a Gaussian measure with a mean vector m_k and a covariance operator R_k, where the mean vector $m_k \in \mathcal{H}_k$ and the covariance operator R_k are defined by

$$\langle x_k, m_k \rangle = \int_{\mathcal{H}_k} \langle x_k, z_k \rangle d\mu_k(z_k),$$

$$\langle x_k, R_k y_k \rangle = \int_{\mathcal{H}_k} \langle x_k, z_k - m_k \rangle \langle z_k - m_k, y_k \rangle d\mu_k(z_k),$$

for any $x_k, y_k, \in \mathcal{H}_k$.

Moreover, let $\mu_0 \in P_G^{(2)}$ be a Gaussian measure indicating noise of the channel. A mapping $\lambda : \mathcal{H}_1 \times \mathcal{B}_2 \to [0,1]$ is called a Gaussian channel if λ satisfies the following conditions: (1) $\lambda(x, \cdot) \in P_G^{(2)}$ for each fixed $x \in \mathcal{H}_1$, (2) $\lambda(\cdot, Q)$ is a measurable function on $(\mathcal{H}_1, \mathcal{B}_1)$ for each fixed $Q \in \mathcal{B}_2$. Then the Gaussian measure $\mu_2 \in P_G^{(2)}$ is expressed as follows:

$$\mu_2(Q) \equiv \int_{\mathcal{H}_1} \lambda(x, Q) d\mu_1(x), \tag{3.55}$$

$$\lambda(x, Q) \equiv \mu_0(Q^x), \tag{3.56}$$

$$Q^x \equiv \{y \in \mathcal{H}_2 ; Ax + y \in Q\}, \quad x \in \mathcal{H}_1, \quad Q \in \mathcal{B}_2, \tag{3.57}$$

where A is a linear transformation from $\mathcal{H}_1$ to $\mathcal{H}_2$. The (classical) mutual entropy with respect to μ_1 and λ is defined through the Kullback-Leibler information such as

$$I(\mu_1; \lambda) \equiv S(\mu_{12}, \mu_1 \otimes \mu_2) \tag{3.58}$$

$$= \begin{cases} \int_{\mathcal{H}_1 \otimes \mathcal{H}_2} \frac{d\mu_{12}}{d\mu_1 \otimes \mu_2} \log \frac{d\mu_{12}}{d\mu_1 \otimes \mu_2} d\mu_1 \otimes \mu_2 & \mu_{12} \ll \mu_1 \otimes \mu_2 \\ +\infty & \text{otherwise,} \end{cases} \tag{3.59}$$

where $\frac{d\mu_{12}}{d\mu_1 \otimes \mu_2}$ is the Radon-Nikodym derivative of $d\mu_{12}$ with respect to $d\mu_1 \otimes \mu_2$ and $\mu_{12} \ll \mu_1 \otimes \mu_2$ means that μ_{12} is absolutely continuous with respect to $\mu_1 \otimes \mu_2$. Furthermore, $\mu_1 \otimes \mu_2$ is a product measure of μ_1 and μ_2, and μ_{12} is a compound measure given by

$$\mu_{12}(Q_1 \times Q_2) = \int_{Q_1} \lambda(x, Q_2) d\mu_1(x), \tag{3.60}$$

for any $Q_1 \in \mathcal{B}_1$ and $Q_2 \in \mathcal{B}_2$.

The above conventional treatment is not suitable for the Gaussian communication by constructing a simple model as shown below.

For simplicity, we put $\mathcal{H}_1 = \mathcal{H}_2 = \mathbb{R}^2$. Let two Gaussian measures μ_1 and μ_0 be given by $\mu_1 = [0, R_1]$, $\mu_0 = [0, R_0]$ with the mean 0 and the covariance operators R_1 and R_0, respectively:

$$R_1 = \begin{pmatrix} \frac{1}{2} & 0 \\ 0 & \frac{1}{2} \end{pmatrix}, \; R_0 = \begin{pmatrix} \frac{1}{48} & 0 \\ 0 & \frac{1}{48} \end{pmatrix}.$$

Note that we have many other choices for the covariance operators R_1 and R_0. We take the linear transformation A as

$$A = \begin{pmatrix} \sqrt{\frac{23}{24}} & 0 \\ 0 & \sqrt{\frac{23}{24}} \end{pmatrix}.$$

Based on the above settings, the covariance operator R_2 of the output measure μ_2 becomes

$$R_2 = \begin{pmatrix} \frac{1}{2} & 0 \\ 0 & \frac{1}{2} \end{pmatrix}.$$

If we take

$$S(\mu_1) \equiv \sup\left\{ - \sum_{A \in \tilde{A}} \mu_1(A) \log \mu_1(A); \; \tilde{A} \in \mathfrak{S}(\mathcal{B}_1) \right\}$$

as the definition of entropy for an input Gaussian measure as a straight extension of the Shannon entropy for a discrete probability distribution (where $\mathfrak{S}(\mathcal{B}_1)$ is the set of all finite partitions of $\mathcal{B}_2$), then this entropy is always infinite for any Gaussian measure $\mu_1 \in P_G^{(1)}$. The mutual entropy with respect to μ_1 and the channel λ determined by μ_0 is calculated as

$$I(\mu_1; \lambda) = S(\mu_{12}, \mu_1 \otimes \mu_2) = \log 24$$

In this case, the mutual entropy is smaller than $S(\mu_1)$, but it is difficult to comprehend the physical meaning of the fact that every input Gaussian measure carries infinite information. Moreover, it is impossible to distinguish an input Gaussian state from other Gaussian states only by using the entropy $S(\mu_1)$, because it is always infinite even on finite dimensional space. This fact is not convenient.

If we use the differential entropy as the definition of entropy for an input Gaussian measure (this definition is often used in literature dealing with communication theory), then

$$S_d(\mu_1)(\equiv S(\frac{d\mu_1}{dm})) = - \int_{\mathbb{R}^2} \frac{d\mu_1}{dm} \log \frac{d\mu_1}{dm} dm = \log \pi e$$

where m is a Lebesgue measure of $\mathbb{R}^2$. This entropy becomes less than the mutual entropy $I(\mu_1; \lambda)$. This fact is a contradiction of Shannon's usual theory.

Therefore, we should find some other expressions (quantities) characterizing a Gaussian state and a Gaussian channel so that we can discuss the Gaussian communication processes consistently. We solve this problem by using the formulation of quantum entropies [251].

Let $P_{G,1}^{(k)}$ be the set $\{\mu = [0, R] \in P_G^{(k)}; \operatorname{tr}R = 1\}$ $(k = 1, 2)$. We assume that $A^*A = (1 - \operatorname{tr}R_0)I_1$ holds for the covariance operator R_0 of μ_0. We define a transformation Γ from $P_{G,1}^{(1)}$ to $P_{G,1}^{(2)}$ associated with the Gaussian channel λ:

$$(\Gamma\mu_1)(Q) = \int_{\mathcal{H}_1} \lambda(x, Q) d\mu_1(x)$$

for any $\mu_1 \in P_{G,1}^{(1)}$ and any $Q \in \mathcal{B}_2$. This equation can be expressed as $\Gamma(\mu_1) = [0, A\rho_1 A^* + R_0]$ for any $\mu_1 = [0, \rho_1] \in P_{G,1}^{(1)}$. There exists a bijection Ξ_k from $P_{G,1}^{(k)}$ to $\mathfrak{S}(\mathcal{H}_k)$ given by

$$\operatorname{tr}\Xi_k(\mu_k)A_k = \int_{\mathcal{H}_k} \langle \xi, A_k\xi \rangle d\mu_k(\xi)$$

for any $A_k \in B(\mathcal{H}_k)$ and any $\mu_k \in P_{G,1}^{(k)} (k = 1, 2)$. We further define a map from $\mathfrak{S}(\mathcal{H}_1)$ to $\mathfrak{S}(\mathcal{H}_2)$ such as

$$\Lambda^*(\rho_1) \equiv \Xi_2 \circ \Gamma \circ \Xi_1^{-1}\rho_1 = A\rho_1 A^* + R_0 \tag{3.61}$$

for any $\rho_1 \in \mathfrak{S}(\mathcal{H}_1)$,

$$
\begin{array}{ccc}
P_{G,1}^{(1)} \ni \mu_1 & \xrightarrow{\ \Gamma\ } & \Gamma\mu_1 \in P_{G,1}^{(2)} \\
\big\downarrow{\scriptstyle \Xi_1} & & \big\downarrow{\scriptstyle \Xi_2} \\
\mathfrak{S}(\mathcal{H}_1) & \xrightarrow[\ \Lambda^*\ ]{} & \mathfrak{S}(\mathcal{H}_2).
\end{array}
$$

Λ is the dual map of Λ^* from $B(\mathcal{H}_2)$ to $B(\mathcal{H}_1)$ given by

$$\Lambda(Q) = A^*QA + (\operatorname{tr}R_0 Q)I_1$$

for any $Q \in B(\mathcal{H}_2)$.

THEOREM 3.15. *The map Λ^* is a quantum mechanical complete positive normal channel from $\mathfrak{S}(\mathcal{H}_1)$ to $\mathfrak{S}(\mathcal{H}_2)$.*

Proof. From the definition of quantum mechanical complete positive channel, we have only to show the following three properties of the map Λ from $B(\mathcal{H}_2)$ to $B(\mathcal{H}_1)$: (1) Complete positivity of Λ; (2) $\Lambda(I_2) = I_1$; (3) Normality of Λ.

(1) For any $\{Q_i\}_{i=1}^n \subset B(\mathcal{H}_1)$ and $\{R_j\}_{j=1}^n \subset B(\mathcal{H}_2)$ with $n \in \mathbb{N}$, we have

$$\sum_{i,j=1}^n Q_i^* \Lambda(R_i^* R_j) Q_j$$

$$= \sum_{i,j=1}^n Q_i^* (A^* R_i^* R_j A + (\mathrm{tr} R_0 R_i^* R_j) I_1) Q_j$$

$$= \sum_{i,j=1}^n Q_i^* A^* R_i^* R_j A Q_j + \sum_{i,j=1}^n (\mathrm{tr} R_0 R_i^* R_j) Q_i^* Q_j$$

$$= \Big(\sum_{i=1}^n R_i A Q_i\Big)^* \Big(\sum_{j=1}^n R_j A Q_j\Big) + \sum_{i,j=1}^n \sum_k \langle x_k,\ R_0 R_i^* R_j x_k\rangle Q_i^* Q_j.$$

Let $\{y_m\}$ be any CONS in $\mathcal{H}_1$ and put $C = \sum_{i=1}^n R_i A Q_i$. Then the above equality is identical to

$$C^* C + \sum_{i,j=1}^n \sum_{k,m} \langle x_k,\ R_j R_0^{1/2} y_m\rangle \langle y_m R_0^{1/2} R_i^* x_k\rangle Q_i^* Q_j$$

$$= C^* C + \sum_{k,m} \Big(\sum_{i=1}^n \langle y_m,\ R_0^{1/2} R_i^* x_k\rangle Q_i^*\Big) \Big(\sum_{j=1}^n \langle x_k,\ R_j R_0^{1/2} y_m\rangle Q_j\Big)$$

$$= C^* C + \Big(\overline{\sum_{i=1}^n \langle x_k,\ R_i R_0^{1/2} y_m\rangle Q_i^*}\Big) \Big(\sum_{j=1}^n \langle x_k,\ R_j R_0^{1/2} y_m\rangle Q_j\Big)$$

$$= C^* C + \sum_{k,m} \Big(\sum_{i=1}^n \Big(\sum_{j=1}^n \langle x_k,\ R_j R_0^{1/2} y_m\rangle Q_j\Big)\Big)$$

$$= C^* C + \sum_{k,m} \Big(\sum_{i=1}^n \langle x_k,\ R_i R_0^{1/2} y_m\rangle Q_i\Big)^* \Big(\sum_{j=1}^n \langle x_k,\ R_j R_0^{1/2} y_m\rangle Q_j\Big) \geq 0.$$

This inequality holds for any $n \in \mathbb{N}$, so that Λ is a completely positive map from $B(\mathcal{H}_2)$ to $B(\mathcal{H}_1)$.

(2) $\Lambda(I_2) = A^* I_2 A + (\mathrm{tr} R_0 I_2) I_1 = A^* A + (\mathrm{tr} R_0) I_1 = (1 - \mathrm{tr} R_0) I_1 + (\mathrm{tr} R_0) I_1 = I_1$.

(3) For any increasing net $\{B_\alpha\}$ $(\subset B(\mathcal{H}_2))$ converging to $B \in B(\mathcal{H}_2)$ and for any sequence $\{x_n\}$ $(\subset \mathcal{H}_1)$ satisfying $\sum_n \|x_n\|^2 < \infty$, we have

$$\sum_n \|(\Lambda(B) - \Lambda(B_\alpha)) x_n\|^2$$

$$= \sum_n \|\Lambda(B - B_\alpha) x_n\|^2$$

$$= \sum_n \|\{A^*(B - B_\alpha) A + (\mathrm{tr} R_0 (B - B_\alpha)) I_1\} x_n\|^2$$

$$\leq \sum_n \|A^*(B - B_\alpha)Ax_n\|^2 + \sum_n |\mathrm{tr}R_0(B - B_\alpha)|^2\|x_n\|^2$$

$$\leq \|A^*\|^2 \sum_n \|(B - B_\alpha)Ax_n\|^2 + |\mathrm{tr}R_0(B - B_\alpha)|^2 \sum_n \|x_n\|^2$$

Since $\sum_n \|Ax_n\|^2 \leq \|A\|^2 \sum_n \|x_n\|^2 < \infty$ is satisfied and R_0 is a trace class operator, then $\Lambda(B_\alpha)$ (ultrastrongly) converges to $\Lambda(B)$.

$\square$

We define the mutual entropy with respect to the input Gaussian measure μ_1 and the Gaussian channel λ as

$$I(\mu_1; \lambda) \equiv I(\rho; \Lambda^*) \tag{3.62}$$

by using the mutual entropy (3.28), where $\rho = \Xi_1(\mu_1)$. The entropy of an input Gaussian measure $\mu_1 = [0, \Xi_1(\mu_1)]$ expressing certain "information" of μ_1 is given by

$$S(\mu_1) \equiv -\mathrm{tr}\Xi_1(\mu_1) \log \Xi_1(\mu_1). \tag{3.63}$$

Then the above entropies satisfy Shannon's fundamental inequality by a direct application of Theorem 3.13.

THEOREM 3.16. *For any* $\mu_1 \in P_{G,1}^{(1)}$ *and for some Gaussian channel* λ, *the following inequality holds:*

$$0 \leq I(\mu_1; \lambda) \leq S(\mu_1).$$

For the model discussed in this section, we calculate the entropy functional $S(\mu_1)$ and the mutual entropy functional $I(\mu_1; \lambda)$ concretely:

$$I(\mu_1; \lambda) = \frac{47}{48} \log 47 - \log 24,$$
$$S(\mu_1) = \log 2 > I(\mu_1; \lambda).$$

Consequently, the difficulty appearing in the model is resolved by our

formulation. Besides the mathematical formulation, our functionals classify the Gaussian inputs and are useful to analyse the Gaussian communication processes [251]. Here we only considered the case $\mathrm{tr}R = 1$ briefly. The case $\mathrm{tr}R \neq 1$ can be also treated under the same conditions, but this case without any condition has not completed yet. Once this question can be solved, all studies on Gaussian communication processes can be consistently discussed in the above framework.

3.7. Dynamical change of entropy

In this section, we discuss various dynamical changes of entropies in both CDS and QDS. The rigorous study of the dynamical change of entropy was started by Boltzmann, but it was too difficult to obtain a satisfactory answer. This original problem is obviously related to the irreversibility of nature, and many physicists have tried to explain the irreversibility from the first principle of dynamics such as the Newton equations or the Schrödinger equation. However, the first principle, e.g., the fundamental equation of motion, is reversible under time reflection, so that we have to find some extra conditions hidden in nature for the interpretation of irreversibility.

We consider the dynamical change of the entropy and the mutual entropy. Concerning the change of entropy, we address ourselves to the following questions:

(E1) For a dynamical change of a state φ to $\bar{\varphi}$ under some external actions, under what conditions does the entropy $S(\varphi)$ increase: $S(\varphi) \leq S(\bar{\varphi})$?

(E2) When a state depends on time, for which time evolution does the entropy $S(\varphi_t)$ converge to some definite value: does there exist a state ψ and the limit: $\lim_{t \to \infty} S(\varphi_t) = S(\psi)$?

(E3) Under what conditions are both (E1) and (E2) satisfied: $S(\varphi_t) \uparrow S(\psi)$?

We study these problems in this section, but it is not so easy to get complete answers to them, so we also consider the following converse question to the above:

(E4) Under what conditions is the entropy invariant in the course of dynamics?

Up to now, little mathematically rigorous studies exist for such questions.

First, we consider the entropy change in classical systems. Let Ω be a finite set and $p(\omega, t_0)$ be the occurence probability of an event $\omega \in \Omega$ at time t_0. Moreover, let $p(\omega', t_{i+1}|\omega, t_i)$ be a transition probability from $\omega \in \Omega$ at t_i to $\omega' \in \Omega$ at t_{i+1}.

Suppose the probability distribution $\{p(\omega, t)\}$ satisfies the following equality of Chapman-Kolmogorov type:

$$[\text{C.K}] \qquad p(\omega', t_{n+1}) = \sum_{\omega \in \Omega} p(\omega', t_{n+1}|\omega, t_n) p(\omega, t_n) \qquad (3.64)$$

Then we have a theorem for the entropy at time t_n given as

$$S(t_n) = -\sum_{\omega \in \Omega} p(\omega, t_n) \log p(\omega, t_n)$$

THEOREM 3.17. *If (3.64) is satisfied and $\{p(\omega', t_{n+1}|\omega, t_n)\}$ is a doubly stochastic matrix (i.e., $\sum_{\omega \in \Omega} p(\omega', t_{n+1}|\omega, t_n) = \sum_{\omega' \in \Omega} p(\omega', t_{n+1}|\omega, t_n) = 1$), then we have*

$$S(t_{n+1}) \geq S(t_n).$$

Proof. For any two probability distibutions $\{p_i; i = 1, \ldots, n\}, \{q_j; j = 1, \ldots, n\}$, applying Jensen's inequality to a concave function $t \mapsto \log t$, one obtains

$$\sum_{j=1}^{n} q_j \log p_j \leq \log \left(\sum_{j=1}^{n} q_j p_j \right),$$

which implies

$$\prod_{j=1}^{n} (p_j)^{q_j} \leq \sum_{j=1}^{n} q_j p_j.$$

Using this inequality,

$$
\begin{aligned}
S(t_{n+1}) &= -\sum_{\omega'} p(\omega', t_{n+1}) \log p(\omega', t_{n+1}) \\
&= -\sum_{\omega'} \sum_{\omega} p(\omega', t_{n+1}|\omega, t_n) p(\omega, t_n) \log p(\omega', t_{n+1}) \\
&= -\sum_{\omega} p(\omega, t_n) \log \prod_{\omega'} p(\omega', t_{n+1})^{p(\omega', t_{n+1}|\omega, t_n)} \\
&\geq -\sum_{\omega} p(\omega, t_n) \log \left\{ \sum_{\omega'} p(\omega', t_{n+1}) p(\omega', t_{n+1}|\omega, t_n) \right\} \\
&\geq -\sum_{\omega} p(\omega, t_n) \log p(\omega, t_n) = S(t_n).
\end{aligned}
$$

$\square$

THEOREM 3.18. *When Ω is finite, then for any t_n*

$$S(t_n) \leq \log |\Omega|,$$

where $|\Omega|$ is the cardinal of Ω.

Proof. Take $p(\omega', t_{n+1}|\omega, t_n) = \frac{1}{|\Omega|}$ and apply Theorem 3.18. $\square$

Let us consider the case when the state change is caused by a time independent or time dependent channel; namely, $\varphi \to \Lambda^* \varphi$ or $\Lambda_t^* \varphi$. The following theorem might be one of the most general results concerning the question (E1).

Let $\mathcal{H}$ and $\overline{\mathcal{H}}$ be two Hilbert spaces and α be a positive linear map from $B(\mathcal{H})$ to $B(\overline{\mathcal{H}})$ preserving the unity.

THEOREM 3.19. *If the dual map Λ of a channel $\Lambda^* : \mathfrak{S}(\mathcal{H}) \to \mathfrak{S}\left(\overline{\mathcal{H}}\right)$ satisfies the equality* $\mathrm{tr}\Lambda(\rho) = \mathrm{tr}\rho$ *for any* $\rho \in \mathfrak{S}(\mathcal{H})$, *then* $S(\Lambda^*\rho) \geq S(\rho)$.

The proof of this theorem is a consequence of the following two propositions.

Let $B(I; \mathcal{H})$ be the set of all operators on Hilbert space $\mathcal{H}$, whose spectrum lays in the interval I.

PROPOSITION 3.20. *A positive linear map α preserving the unity from $B(\mathcal{H})$ to $B(\overline{\mathcal{H}})$ has the following properties:*
(1) $\Lambda : B((a, b); \mathcal{H}) \to B((a, b); \overline{\mathcal{H}})$;
(2) $\Lambda(A^2) \geq \Lambda(A)^2$ *for any* $A \in B((-\infty, \infty); \mathcal{H})$;
(3) $\Lambda(A^{-1}) \geq \Lambda(A)^{-1}$ *for any* $A \in B((0, \infty); \mathcal{H})$.

Proof. (1): Since $\Lambda(I_{\mathcal{H}}) = I_{\overline{\mathcal{H}}}$ ($I_{\mathcal{H}}$ is the identity operator on $\mathcal{H}$.), $\|\Lambda\| = 1$, so that $\|\Lambda(A)\| \leq \|A\|$, which implies (1).
(2): It is sufficient to show that the matrix

$$\begin{pmatrix} \Lambda(A^2) & \Lambda(A) \\ \Lambda(A) & \Lambda(I_{\mathcal{H}}) \end{pmatrix}$$

is positive definite. Let the spectral decomposition of A be

$$A = \int_R \lambda E(d\lambda).$$

Then

$$A^2 = \int \lambda^2 E(d\lambda), \quad I_{\mathcal{H}} = \int E(d\lambda),$$

which implies

$$\begin{pmatrix} \Lambda(A^2) & \Lambda(A) \\ \Lambda(A) & \Lambda(I_{\mathcal{H}}) \end{pmatrix} = \int_R \begin{pmatrix} \lambda^2 & \lambda \\ \lambda & 1 \end{pmatrix} \otimes \Lambda(E(d\lambda))$$

The matrix $\begin{pmatrix} \lambda^2 & \lambda \\ \lambda & 1 \end{pmatrix}$ is positive definite, from which it follows (2).
(3): We can show this equality similarly as (2) with $0 < \lambda < \infty$.

□

PROPOSITION 3.21. *For* $\eta(t) = -t \log t$ ($t \in [0, 1)$), *and the above mapping Λ, we have*
$$\Lambda(\eta(A)) \leq \eta(\Lambda(A))$$
for any $A \in B(\mathcal{H})_+$.

Proof. The function $\eta(t) = -t\log t$ $(t \in [0,1])$ can be written as

$$\eta(t) = \int_0^\infty \left\{1 - \frac{u}{t+u} - \frac{t}{1+u}\right\} du.$$

Thus,

$$\Lambda(\eta(A)) = \int_0^\infty \left\{I_{\mathcal{H}} - \Lambda(\frac{u}{A+u}) - \Lambda(\frac{A}{1+u})\right\} du$$
$$= \int_0^\infty \left\{I_{\mathcal{H}} - \Lambda((1 + \frac{A}{u})^{-1}) - \frac{1}{1+u}\Lambda(A)\right\} du.$$

By the way, $\eta(\Lambda(A))$ can be written

$$\eta(\Lambda(A)) = \int_0^\infty \left\{I_{\mathcal{H}} - \frac{u}{\Lambda(A)+u} - \frac{\Lambda(A)}{1+u}\right\} du$$
$$= \int_0^\infty \left\{I_{\mathcal{H}} - (\Lambda(\frac{A}{u} + I_{\mathcal{H}}))^{-1} - \frac{1}{1+u}\Lambda(A)\right\} du.$$

According to (3) of Proposition 3.21, we have

$$\Lambda\left((1 + \frac{A}{u})^{-1}\right) \geq \left(\Lambda(\frac{A}{u} + I_{\mathcal{H}})\right)^{-1},$$

which implies

$$\Lambda(\eta(A)) \leq \eta(\Lambda(A)).$$

$\square$

Some special cases of the above theorem have been discussed by several authors such as Nakamura and Umegaki [239], Davies [64], Ingarden and Urbanik [137]. When we measure an operator Q having a discrete spectral decomposition such as $Q = \sum_n q_n P_n$, the state ρ describing the system suffers a change after the measurement of Q, viz. $\rho \to \bar{\rho} \equiv \Gamma_Q \rho \equiv \sum_n P_n \rho P_n \mathrm{tr}(\sum_n P_n q_n P_n)$. The transformation Γ_Q is a channel and satisfies all conditions from $B(\mathcal{H})$ to the Von Neumann algebra $\mathfrak{M}$ generated by the operators Q and $I : \mathfrak{M} = \{Q\}''$. For this Γ_Q, the inequality $S(\Gamma_Q \rho) \geq S(\rho)$ holds and it is easily seen that the necessary and sufficient condition satisfying the equality $S(\Gamma_Q \rho) = S(\rho)$ is $[S(\Gamma_Q \rho) = S(\rho)$ iff $\rho \in \{Q\}']$.

We here give a simpler proof of this fact: If $\rho \in \{Q\}'$, then $\Gamma_Q \rho = \rho$, hence the equality holds. Conversely, if the equality $S(\Gamma_Q \rho) = S(\rho)$ holds, then

$$
\begin{aligned}
0 &= S(\Gamma_Q \rho) - S(\rho) = -\mathrm{tr}\Gamma_Q \rho \log \Gamma_Q \rho + \mathrm{tr}\rho \log \rho \\
&= -\mathrm{tr}\Gamma_Q (\rho \log \Gamma_Q \rho) + \mathrm{tr}\rho \log \rho = -\mathrm{tr}\rho \log \Gamma_Q \rho + \mathrm{tr}\rho \log \rho \\
&= S(\rho, \Gamma_Q \rho).
\end{aligned}
$$

According to the inequality $\|\varphi - \phi\|^2/4 \le S(\varphi, \phi)$ of the relative entropy, we have $\Gamma_Q \rho = \rho$, so that $\rho \in \{Q\}'$. Ingarden and Urbanik called the entropy $S(\Gamma_Q \rho)$ the Q-entropy and showed that it has similar properties as the Von Neumann entropy, such as positivity, concavity, additivity.

From Theorem 3.17 we can conclude that a time-dependent channel Λ_t^* satisfying all conditions of the theorem yields a monotone increase of the entropy $S(\Lambda_t^* \rho)$. Thus, it is interesting to find physical systems having these properties or to find some conditions under which a suitable time-dependent channel can be constructed from the first principles of quantum mechanics. However, this question is not so easy to answer.

Now, we discuss the problem (E2) in a more concrete dynamics, having irreversibility in itself, namely, in the linear response dynamics. A mathematically rigorous treatment of the linear response theory (Kubo theory) has been discussed by several authors. Here we introduce the entropy for the linear response theory and study the dynamical change of the entropy in the linear response dynamics. Let a physical system be described by a triple $(A, \mathfrak{S}, \alpha(\mathbb{R}))$, where $A = B(\mathcal{H})$, $\mathfrak{S}$ is the set of all density operators on $\mathcal{H}$ and $\alpha_t(\cdot) = U_t \cdot U_t^* = \exp(itH)$ with a Hamiltonian H. Take a state $\varphi \in \mathfrak{S}$ such that $\varphi(\cdot) = \mathrm{tr}\rho\cdot$. We suppose that φ is a *faithful* (i.e., $\varphi(A^*A) = 0 \Rightarrow A = 0$) KMS state at $\beta = 1$ w.r.t. α_t and the KMS Hamiltonian H is bounded from below.

When the physical system described by the above state φ is perturbed by an α-analytic self-adjoint operator λV ($\lambda \in (0,1)$) in A, the perturbed time evolution α_t^V and the perturbed state φ^V are expressed as follows:

$$
\alpha_t^V(A) = \sum_{n \ge 0}(i\lambda)^n \int \cdots \int_{0 \le t_1 \le t_2 \le \cdots t_n \le t} [\alpha_{t_1}(V), [\cdots[\alpha_{t_n}(V), \alpha_t(A)]\cdots]]dt_1 \cdots dt_n
$$

(3.65)

for $t \ge 0$ (the case $t < 0$ is due to the exchange of 0 and t in the integral domain) and

$$
\varphi^V(A) = \frac{\varphi(W^*AW)}{\varphi(W^*W)}, \tag{3.66}
$$

$$
W = \sum_{n \ge 0}(-\lambda)^n \int \cdots \int_{0 \le t_1 \le t_2 \le \cdots t_n \le \frac{1}{2}} \alpha_{it_1}(V) \cdots \alpha_{it_n}(V)dt_1 \cdots dt_n. \tag{3.67}
$$

Taking the first order approximation w.r.t. λ in (3.65) and (3.66), we have

$$\alpha_t^{V,1}(A) \; = \; \alpha_t(A) + i\lambda \int_0^t ds[\alpha_s(V), \alpha_t(V)], \qquad (3.68)$$

$$\varphi^{V,1}(A) \; = \; \varphi(A) - \lambda \int_0^1 ds\varphi(A\alpha_{is}(V)) + \lambda\varphi(A)\varphi(V), \qquad (3.69)$$

where we need some computation to obtain (3.69) from (3.67). In the sequel we call $\varphi^{V,1}$ and $(\alpha_t^{V,1})^*\varphi$ a *linear response perturbed state* and a *linear response state*, respectively, although they do not exhibit the positivity of a state and the linear response time evolution $\alpha_t^{V,1}$ is not an automorphism of $\mathcal{A}$.

It is known that φ^V satisfies the KMS condition at $\beta = 1$ w.r.t. α_t^V, hence φ^V is an equilibrium state for the Hamiltonian $H + \lambda V$. Therefore, it is natural to expect the state change under α_t^V such that

$$w^* - \lim_{t\to\infty} (\alpha_t^V)^*\varphi = \varphi^V. \qquad (3.70)$$

That is, when the perturbation λV acts on the system in equilibrium described by φ and H, we should observe that the state φ gradually approaches new equilibrium state for the Hamiltonian $H + \lambda V$. Regardless of (3.70), the entropy of the time-dependent state $(\alpha_t^V)^*\varphi$ is always equal to that of φ (i.e., $S((\alpha_t^V)^*\varphi) = S(\varphi)$) because α_t^V is implemented by a unitary operator generated by $H + \lambda V$. This means that the entropy of a state is a dynamical invariant even when the state changes to a different equilibrium state like (3.70), which contradicts the usual macroscopic behaviour of physical systems in our real world.

This result is not surprising because quantum mechanics does not contain the irreversibility in itself, as discussed before. The linear response theory suceeded to explain some irreversible phenomena, so we hope that the entropy of the linear response state might change in time. From the assumptions imposed on our dynamical system and φ, we obtain, after a simple calculation,

$$\varphi(\alpha_t^{V,1}(A)) \; = \; \mathrm{tr}\rho^{V,1}(t)A,$$
$$\varphi^{V,1}(A) \; = \; \mathrm{tr}\rho^{V,1}A,$$

with

$$\rho^{V,1} \; = \; (I - \lambda \int_0^1 \alpha_{is}(V)ds + \mathrm{tr}\rho V)\rho,$$

$$\rho^{V,1}(t) \; = \; (I - i\lambda \int_0^t \alpha_{-s}(V)ds + i\lambda \int_0^t \alpha_{-s+i}(V)ds)\rho.$$

Since the set of all trace class operators is an ideal for $\mathcal{A}$, both $\rho^{V,1}$ and $\rho^{V,1}(t)$ are trace class operators, so we define normal states $\theta(t)$ and θ by

$$\theta(t) = \frac{|\rho^{V,1}(t)|}{\mathrm{tr}|\rho^{V,1}(t)|},$$

$$\theta = \frac{|\rho^{V,1}|}{\mathrm{tr}|\rho^{V,1}|},$$

where $|A| \equiv (A^*A)^{1/2}$. The entropy $S(\rho^{V,1}(t))$ of the linear response state $\rho^{V,1}(t)$ and the entropy $S(\rho^{V,1})$ of the linear response perturbed state $\rho^{V,1}$ are defined as

$$S(\rho^{V,1}(t)) = -\mathrm{tr}\theta(t)\log\theta(t),$$
$$S(\rho^{V,1}) = -\mathrm{tr}\theta\log\theta.$$

If $\rho^{V,1}(t) \geq 0$, then our entropy is identical to that of Von Neumann because of $\mathrm{tr}\rho^{V,1}(t) = 1$.

Now, our problem is as follows: "When the linear response state $\rho^{V,1}(t)$ approaches the linear response perturbed state $\rho^{V,1}$ in the weak topology (i.e., $\lim_{t\to\infty}\mathrm{tr}\rho^{V,1}(t)A = \mathrm{tr}\rho^{V,1}A$ for any A in $\mathcal{A}$), under what conditions does the entropy $S(\rho^{V,1}(t))$ dynamically change to $S(\rho^{V,1})$? " Here, we assume the existence of the limit $w\text{--}\lim_{t\to\infty}\rho^{V,1}(t) = \rho^{V,1}$, but this existence is realized under some ergodic conditions. Then we can prove the following lemma.

LEMMA 3.2. If $\rho^{V,1}(t)$ converges weakly to $\rho^{V,1}$ as $t \to \infty$, then we have (1) $\|\theta(t) - \theta\|_1 \to 0$ as $t \to \infty$, where $\|\cdot\|_1 \equiv \mathrm{tr}|\cdot|$, (2) there exist positive constants a, b such that $a\rho \leq |\rho^{V,1}(t)| \leq b\rho$ for a sufficiently small λ.

  Proof. (1) Since $\mathrm{tr}\rho^{V,1}(t) = \mathrm{tr}\rho^{V,1} = 1$, the following implications holds: $\rho^{V,1}(t) \to \rho^{V,1}(\text{weak}) \Rightarrow \|\rho^{V,1}(t) - \rho^{V,1}\|_1 \to 0 \Rightarrow \|\rho^{V,1}(t) - \rho^{V,1}\| \to 0 \Rightarrow \||\rho^{V,1}(t)| - |\rho^{V,1}|\| \to 0$ and $|\mathrm{tr}|\rho^{V,1}(t)| - \mathrm{tr}|\rho^{V,1}|| \to 0 \Rightarrow \theta(t) \to \theta\ (\text{weak}) \Rightarrow \|\theta(t) - \theta\|_1 \to 0.$
(2) Since φ is faithful and normal, $\varphi(\alpha_t^{V,1}(A))$ can be expressed as

$$\varphi(\alpha_t^{V,1}(A)) = \varphi(A) - \langle TA^*x,\ T(I - \exp(-itH))Vx \rangle,$$

where $T = \{1 - \exp(-H)\}/H$. This fact with the boundedness from below of H implies the conclusion.

$\square$

We immediately get

THEOREM 3.22. *If $\rho^{V,1}(t)$ weakly converges to $\rho^{V,1}$ as $t \to \infty$ and $S(\rho)$ is finite, then $S(\rho^{V,1}(t))$ converges to $S(\rho^{V,1})$ as $t \to \infty$.*

This theorem might provide one reason why the linear response theory has been useful for the description of some irreversible phenomena.

Next, we discuss the problem (E5). For two dynamical systems $(\mathcal{A}, \mathfrak{S}, \alpha)$ and $(\bar{\mathcal{A}}, \bar{\mathfrak{S}}, \bar{\alpha})$ and a channel Λ^* from $\mathfrak{S}$ to $\bar{\mathfrak{S}}$, let S be a weak* compact convex subset of $\mathfrak{S}$ and $\bar{S}$ be that of $\bar{\mathfrak{S}}$. If we have a certain one-to-one correspondence between S and $\bar{S}$ by the channel Λ^*, then the equality $S^S(\varphi) = S^{\bar{S}}(\Lambda^*\varphi)$ is expected. We can prove the following intuitively trivial facts about the entropy invariance [249].

THEOREM 3.23. *For a stationary channel Λ^*, the following statesments hold.*
(1) If Λ^* is surjective and a *-homomorphism, then $S^{K(\alpha)}(\varphi) = S^{K(\bar{\alpha})}(\Lambda^*\varphi)$ for any $\varphi \in K(\alpha)$.
(2) If Λ^* is a bijection from $exI(\alpha)$ to $exI(\bar{\alpha})$, then $S^{I(\alpha)}(\varphi) = S^{I(\bar{\alpha})}(\Lambda^*\varphi)$ for any $\varphi \in I(\alpha)$.
(3) If Λ^* is a bijection, then $S(\varphi) = S(\Lambda^*\varphi)$ for any $\varphi \in \mathfrak{S}$.

Finally, we mention the time development of the mutual entropy when the state is changed by a time-dependent channel Λ_t^*. Here we assume that $\mathcal{A}$ is a Von Neumann algebra acting on a Hilbert space $\mathcal{H}$ and $\Lambda(\mathbb{R}^+) = \{\Lambda_t ; t \in \mathbb{R}^+\}$ is a dynamical semigroup on a von Neumann algebra $\mathcal{A}$ (i.e., $\Lambda(\mathbb{R}^+)$ is a weak* continuous semigroup from $\mathcal{A}$ to $\bar{\mathcal{A}}$ and Λ_t^* is a normal channel for each $t \in \mathbb{R}^+$) and there exists at least one faithful normal stationary state ψ w.r.t. Λ_t^* (i.e., $\Lambda_t^*\psi = \psi$ for all $t \in \mathbb{R}^+$). Define two subsets $\mathcal{A}(\alpha)$ and $\mathcal{A}(C)$ of $\mathcal{A}$ such as

$$\mathcal{A}(\Lambda) = \{A \in \mathcal{A} : \Lambda_t(A) = A, t \in \mathbb{R}^+\},$$
$$\mathcal{A}(C) = \{A \in \mathcal{A} : \Lambda_t(A^*A) = \Lambda_t(A^*)\Lambda_t(A), t \in \mathbb{R}^+\}.$$

Then $\mathcal{A}(\Lambda)$ is a Von Neumann subalgebra of $\mathcal{A}$ and there exists a unique conditional expectation $\mathcal{E}$ from $\mathcal{A}$ to $\mathcal{A}(\Lambda)$ [101].

THEOREM 3.24. *If μ is orthogonal and $\mathcal{A}$ is type I with $\mathcal{A}(\Lambda) = \mathcal{A}(C)$, then $I_\mu(\varphi; \Lambda_t^*)$ decreses in time and approaches $I_\mu(\varphi; \mathcal{E}^*)$ as $t \to \infty$.*

Proof. Since $\mathcal{A}(\Lambda) = \mathcal{A}(C)$, it is easily seen that $\Lambda_t^*\omega$ weakly converges to $\mathcal{E}^*\omega$ for any normal state ω. The assumption that $\mathcal{A}$ is α-finite and type I implies the norm convergence of $\Lambda_t^*\omega$:

$$\|\Lambda_t^*\varphi - \mathcal{E}^*\varphi\| \to 0 \quad \text{and} \quad \|\Lambda_t^*\varphi_n - \mathcal{E}^*\varphi_n\| \to 0, \quad t \to \infty,$$

where $\varphi = \sum_n \mu_n \varphi_n$ with $\varphi_n \perp \varphi_m (n \neq m)$. As there exists a constant $\lambda_n \in \mathbb{R}^+$ satisfying $\varphi < \lambda_n \varphi$ for each n, the inequality $\Lambda_t^* \varphi_n \leq \lambda_n \Lambda_t^* \varphi$ holds for all $t \in \mathbb{R}^+$. Therefore, we obtain

$$\lim_{t \to \infty} S(\Lambda_t^* \varphi_n, \Lambda_t^* \varphi) = S(\mathcal{E}^* \varphi_n, \mathcal{E}^* \varphi).$$

This equality proves the existence of $\lim_{t \to \infty} I_\mu(\varphi; \Lambda_t^*)$. This limit is decreasing in time because of $S(\Lambda_{t+s}^* \varphi_n, \Lambda_{t+s}^* \varphi) \leq S(\Lambda_t^* \varphi_n, \Lambda_t^* \varphi)$ for all $s \in \mathbb{R}^+$. $\qquad\qquad\qquad\qquad\qquad\qquad\qquad\qquad\qquad\qquad\qquad\qquad\qquad\square$

3.8. Entropy production

Before closing this chapter we briefly discuss the *entropy production*.

As is seen in the preceding section, there are many approaches to explain the nonequilibrium irreversible processes. The entropy production is one of such approaches. Mathematical formulation of this concept was started by Spohn [307], although there were many other studies of the entropy production.

Here we briefly explain Spohn's formulation [307] and Ojima's more recent formulations [266], [267] of the entropy production.

In nonequilibrium balance equations, the balance equation for the entropy is

$$\frac{\partial S}{\partial t} = -\mathrm{div}\mathbf{J}_S + \sigma,$$

where S is local entropy, $\mathbf{J}_S$ is the vector of the entropy flow per unit area and unit time, and σ is the (local) entropy production associated with an irreversible change of the system ($\sigma = 0$ for a reversible process).

In quantum statistical mechanics, Spohn discussed the entropy production as follows, [307]: Let Λ_t^* ($t \in \mathbb{R}^+$) be a dynamical semigroup for the set of all trace operators $\mathfrak{S}(\mathcal{H})$ to $\mathfrak{S}(\mathcal{H})$; namely, Λ_t^* is a CP (completely positive) channel satisfying $\Lambda_{t+s}^* = \Lambda_t^* \Lambda_s^*$ for any $t, s \in \mathbb{R}^+$ and $\lim_{t \to 0} \| \Lambda_t^* \rho - \rho \| = 0$ for all $\rho \in \mathfrak{S}(\mathcal{H})$ (strong continuity). The strong continuity of $\{\Lambda_t^*; t \in \mathbb{R}^+\}$ implies the existence of the generator L [219] such that $\lim_{t \to 0} \| L\rho - \frac{1}{t}(\Lambda_t^* - \rho) \|_1 = 0$. If one assumes

$$S(\rho) = -\mathrm{tr}\rho \log \rho,$$

$$\mathrm{div}\mathbf{J}_S = \frac{d}{dt}\mathrm{tr}\Lambda_t^* \rho \log \rho_\beta \Big|_{t=0},$$

where β is the inverse temperature of the system, then the entropy production is

$$\sigma(\rho) = \frac{d}{dt}\{\mathrm{tr}\Lambda_t^* \rho \log \rho_\beta - \mathrm{tr}\Lambda_t^* \rho \log \Lambda_t^* \rho\}\Big|_{t=0}$$

$$= -\frac{d}{dt}S(\Lambda_t^* \rho, \rho_\beta)|_{t=0}$$

Spohn defined the *entropy production* $\sigma(\rho)$ in relation to the Λ_t^*-invariant state ρ_0 by

$$\sigma(\rho) = -\frac{d}{dt}S(\Lambda_t^* \rho, \rho_0)|_{t=0}.$$

It is easily shown that $S(\Lambda_t^* \rho, \rho_0)$ is decreasing and continuous in t from the right. Then we have the following results.

THEOREM 3.25. *Let* $\dim \mathcal{H} < \infty$ *and the range* $\operatorname{ran} \rho_0 = \mathcal{H}$. *Then the entropy production* σ *is defined on* $\mathfrak{S}(\mathcal{H})$ *and is given by*

$$\sigma(\rho) = \operatorname{tr}\{L(\rho)(\log \rho_0 - \log \rho)\}, \quad \rho \in \mathfrak{S}(\mathcal{H}),$$

where $\Lambda_t^* = e^{tL}$, $t \in \mathbb{R}^+$. *Moreover,* σ *is convex and it takes values in* $[0, \infty]$.

THEOREM 3.26. *(1)* $\{\rho \in \mathfrak{S}(\mathcal{H}); L(\rho) = 0\} \subset \{\rho \in \mathfrak{S}(\mathcal{H}); \sigma(\rho) = 0\}$. *If* $\sigma(\rho) = 0$ *only for* $\rho = \rho_0$, *then* $\lim_{t \to \infty} \Lambda_t \rho = \rho_0$ *for all* $\rho \in \mathfrak{S}(\mathcal{H})$.
(2) $\sigma = 0$ *iff* $L = -i[H, \cdot]$.

This theorem is related to the theorem 3.19.

Ojima discussed the entropy production in terms of Kubo's linear response theory.

Let $\mathcal{A}$ be a C^*-algebra, α_t $(t \in \mathbb{R})$ be a one-parameter group of automorphisms of $\mathcal{A}$ and ω_β be a KMS state with respoect to α_t and inverse temperature β. The KMS state ω_β is an equilibrium state prepared at $t = -\infty$ and an external perturbation $\mathbf{V}(t) = (V_1(t), ..., V_n(t))$ is adiabatically effected to the system. The interaction energy of this perturbation with the system is expressed by an observable vector $\mathbf{Q} = (Q_1, ..., Q_n)$, $Q_k \in \mathcal{A}$ as follows

$$H_{\mathrm{int}}(t) = -\mathbf{Q} \cdot \mathbf{V}(t) = -\sum_{i=1}^{n} Q_i V_i(t).$$

Using the derivation $\delta(\cdot)$ of Sakai obtained by

$$\frac{d\alpha_t(A)}{dt} = \alpha_t(\delta(A)), \qquad A \in \mathcal{A},$$

the time evolution $\{\alpha_{t,s;\mathbf{V}}; t, s \in \mathbb{R}\}$ with the perturbation $\mathbf{V}$ is determined by

$$\frac{d}{dt}\alpha_{t,s;\mathbf{V}}(A) = \alpha_{t,s;\mathbf{V}}(\delta(A) + [H_{\mathrm{int}}(t), A])$$

and

$$\alpha_{t=s;\mathbf{V}}(A) = A, \qquad A \in \mathcal{A}.$$

Put $U(t,s; \mathbf{V}) = T\exp\{i\int_s^t \alpha_\tau(Q)\cdot\mathbf{V}(\tau)d\tau\}$ with the time ordering operator T. Then

$$\alpha_{t,s;\mathbf{V}}(A) = \alpha_{-s}[U(t,s;\mathbf{V})^*\alpha_t(A)U(t,s;\mathbf{V})]. \quad A \in \mathcal{A}.$$

When the input state $\omega_\beta(= \varphi_{t=t_0=-\infty})$ changes to $\varphi_t \equiv \omega_\beta\circ\alpha_{t_0,t}; \mathbf{V}$ at time t, the relative entropy between ω_β and φ_t is evaluated as

$$S(\varphi_t, \varphi_{t_0} = \omega_\beta) = \beta\int_{t_0}^t \varphi_s(\delta(\mathbf{Q}))\mathbf{V}(s)ds \geq 0. \tag{3.71}$$

The current operator $\mathbf{J}$ conjugated to the external force $\mathbf{V}(t)$ is defined by

$$\mathbf{J} = \delta(\mathbf{Q}).$$

Then the time-dependent entropy production $\sigma(t, t_0; \mathbf{V})$ becomes

$$\sigma(t, t_0; \mathbf{V}) = \frac{d}{dt}S(\varphi_t, \varphi_{t_0} = \omega_\beta) = \beta\varphi_t(\delta(\mathbf{Q}))\mathbf{V}(t)$$

which is not always positive because ω_β is not $\alpha_{t,t_0;\mathbf{V}}$-invariant, so that Ojima, Hasegawa and Ichiyanagi [266] considered the mean entropy production as

$$\bar{\alpha} = \lim_{T\to\infty}\frac{1}{T}\int_0^T dt \lim_{T_0\to\infty}\frac{1}{T_0}\int_{-T_0}^0 dt_0\sigma(t+t_0, t_0; \mathbf{V}) \tag{3.72}$$

$$= \lim_{T\to\infty, T_0\to\infty}\frac{1}{TT_0}\int_{-T_0}^0 dt_0 S(\varphi_{T+t_0}, \varphi_{t_0} = \omega_\beta) \geq 0. \tag{3.73}$$

In order for these two limiting procedures to be meaningful, Ojima [267] gave certain characterization to the external force $\mathbf{V}(t)$, namely, he assumed that the continuous function $\mathbf{V}(t)$ is almost periodic, i.e., it is uniformly approximated by a linear combination of periodic functions. Here "uniformly" means the convergence in the uniform topology with norm $\|\mathbf{V}\| = \sup\{|\mathbf{V}(t)|; t \in \mathbb{R}\}$.

By Bochner's theorem, this is equivalent to the condition that the orbit $\{\mathbf{V}(t-\lambda); \lambda \in \mathbb{R}\}$ is precompact in the uniform topology whose completion $M_{\mathbf{V}} = \overline{\{\mathbf{V}(t-\lambda); \lambda \in \mathbb{R}\}}$ becomes an Abelian compact group. Therefore, the external force $\mathbf{V}(t)$ can be expressed as an element in the set $C(M_{\mathbf{V}})$ of all continuous functions on $M_{\mathbf{V}}$, that is,

$$\exists\tilde{\mathbf{V}} \in C(M_{\mathbf{V}}), \quad \xi_0 \in M_{\mathbf{V}} \quad \text{such that} \quad \mathbf{V}(t) = \tilde{\mathbf{V}}(\lambda_t(\xi_0)),$$

where λ_t is the time flow on $M_{\mathbf{V}}$ defined by

$$\lambda_t\xi = \xi_t, \quad \xi \in M_{\mathbf{V}}.$$

We construct a compound system of $(\mathcal{A}, \mathbb{R}, \alpha)$ and $(C(M_V), \mathbb{R}, \lambda)$, which is denoted by $(\mathcal{B}, \mathbb{R}, \gamma)$:

$$\mathcal{B} = \mathcal{A} \otimes C(M_V) = C(M_V, \mathcal{A})$$
$$\gamma_t(\tilde{B})(\xi) = \alpha_{0,t;\xi}(\tilde{B}(\lambda_{-t}\xi)), \quad \tilde{B} \in \mathcal{B}.$$

Since the Haar measure μ on M_V is ergodic, we have

$$(\varphi \otimes \mu)(\gamma_t(\tilde{B})) = \int_{M_V} \varphi(\gamma_t(\tilde{B})(\xi)) d\mu(\xi)$$
$$= \lim_{T_0 \to \infty} \frac{1}{T_0} \int_{-T_0}^{0} \varphi(\alpha_{t_0, t+t_0;\xi}(\tilde{B}(\lambda_{-t-t_0}\xi))) dt_0$$

for any state φ on $\mathcal{A}$. Then the relative entropy and the mean entropy production becomes

$$S(\gamma_t^*(\omega_\beta \otimes \mu), \omega_\beta \otimes \mu) = \lim_{T_0 \to \infty} \frac{1}{T_0} \int_{-T_0}^{0} S(\varphi_{t+t_0}, \varphi_{t_0} = \omega_\beta) dt_0, \quad (3.74)$$

$$\overline{\sigma} = \lim_{n \to \infty} \frac{1}{T_n} \int_{0}^{T_n} S(\gamma_t^*(\omega_\beta \otimes \mu), \omega_\beta \otimes \mu) dt, \quad (3.75)$$

where

$$\tilde{\varphi} \equiv \lim_{n \to \infty} \frac{1}{T_n} \int_{0}^{T_n} \gamma_t^*(\omega_\beta \otimes \mu) dt$$

with the proper sequence $\{T_n\}$.

Ojima discussed the relation between the above expression (3.73) and the linear response theory of Kubo [267].

Chapter 4

Information Dynamics

Various physical or nonphysical systems can be described by states, so that the dynamics of a system is described by the state change. One of essential characters of a state is expressed by its complexity. Complexity such as entropy is a key concept in information theory. We call the study of the state change together with such complexities *information dynamics (ID)*, which is a kind of synthesis of the dynamics of state change and complexity. Here we explain what information dynamics is and indicate how it can be used to define the dynamical entropy. Some of concrete applications of information dynamics are discussed in Chap. 8 of this volume.

4.1. Information Dynamics

Let an input dynamical system and an output dynamical system be described by $(\mathcal{A}, \mathfrak{S}, \alpha(G))$ and $(\overline{\mathcal{A}}, \overline{\mathfrak{S}}, \overline{\alpha}(\overline{G}))$, respectively. Here $\mathcal{A}$ is the set of all objects to be observed and $\mathfrak{S}$ is the set of all means getting the observed value for each element A in $\mathcal{A}$, and α describes inner evolution of the input system parameteraized by a group G. We call $\mathfrak{S}$ the "state space" here. Same for the output system $(\overline{\mathcal{A}}, \overline{\mathfrak{S}}, \overline{\alpha}(\overline{G}))$. Thus we may say

[Giving a mathematical structure to input and output triples
$\equiv$ Having a theory].

A map providing a bridge between the two systems is called a channel if it sends a state of the input system to that of the output system, $\Lambda^* : \mathfrak{S} \to \overline{\mathfrak{S}}$. There exist several channels whose properties specify the character of two systems. Examples of various channels are given in Chap. 3. Mathematical structure of almost all systems can be expressed by the following charts.

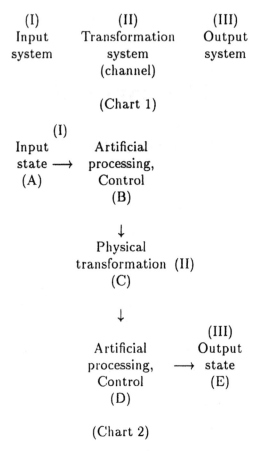

(Chart 1)

(Chart 2)

In chart 2 above, (I) (= (A) + (B)) corresponds to an *input system*, (II) (= (B) + (C) + (D)) to a *transformation (channel) system*, (III) (= (D) + (E)) to an *output system*. More complex systems are constructed from this first structure.

A *"naked" state* state (A) is artificially processed or controlled to a *"dressed" state* state, and it is suffered to change by a physical (natural) transformation, and it is again artificially processed and controlled. The fundamental part of the process for this state change is obviously "A → C → E".

Let us give some examples of an input system (I), a transformation system (channel) (II) and an output system (III).

(1) *Causal System*: (A) $x \in \mathbb{R}^n$; (B) $\dot{x} = f(x)$; (C) $x(t) = \Phi_t(x)$, where Φ_t is an evolution (semi)group generated by f. Remark that if $x(t)$ can not be obtained directly, then we have to examine the properties of f and finds some proper approximation to obtain Φ_t, which may cause chaos.

(2) *Signal Transmission*: (A) causal signal $x(t)$; (B) coding causal signal $x(t)$ to $\{x_n\}$ by e.g., sampling theorem with cut-off; (C) some transformation $y_n = f(x_n)$ for any n; (III) interpolating or decoding $\{y_n\}$ to $y(t)$.

(3) *Discrete system*: (A) input probability distribution $p = \{p_1, \ldots, p_n\}$ of events $X = \{x_1, \ldots, x_n\}$; (C) transition probability $(p(i|j))$; (E) output probability distribution $q = (q_i)$; $q_i = \sum_j p(i|j)p_j$.

(4) *Continuous system*: (A) probability measure μ on measurable space $(\Omega, \mathfrak{F})$; (C) Markov kernel $\lambda : \Omega \times \bar{\mathfrak{F}} \to [0, 1]$ s.t. $\lambda(x, \cdot) \in P(\bar{\Omega})$, where $(\bar{\Omega}, \bar{\mathfrak{F}})$ is output measureable space and $P(\bar{\Omega})$ is the set of all probability measures on Ω, and $\lambda(\cdot, A) \in M(\Omega)$, the set of all random variables; (E) $\bar{\mu} \equiv \int_\Omega \lambda(x, \cdot)d\mu$, $\mu \in P(\Omega)$, the set of all probability measures on Ω.

(5) *Quantum system 1*: (A) $x \in \mathcal{H}$ (Hilbert space); (C) unitary operator U or isometric operator V; (E) $y = Ux$ or $Vx \in \mathcal{H}$.

(6) *Quantum system 2*: (A) density operator $\rho \in \mathfrak{S}(\mathcal{H})$; (C) $\Lambda^* \equiv \mathrm{Ad}U$ (i.e, $\mathrm{Ad}U(\rho) \equiv U\rho U^*$) or $\mathrm{Ad}V$. (E) $\bar{\rho} = \Lambda^*\rho$.

(7) *C*-system containing all above*: (A) $(\mathcal{A}, \mathfrak{S}, \alpha(G))$ C*-triple; (C) Λ^* : $\mathfrak{S}(\mathcal{A}) \to \mathfrak{S}(\bar{\mathcal{A}})$ a dual map of a completely positive map $\Lambda : \bar{\mathcal{A}} \to \mathcal{A}$, where $(\bar{\mathcal{A}}, \bar{\mathfrak{S}}, \alpha(\bar{G}))$ another C*-triple; (E) $\bar{\varphi} = \Lambda^*\varphi$.

Once input and output systems are mathematically fixed and a transformation rule (channel) is given, we next consider some complexities of the state associated with the systems, which are a corner stone of information dynamics. The first complexity is one for a state itself : For a state φ, the complexity seen from a reference system S, a certain subset of $\mathfrak{S}$, is denoted by $C^S(\varphi)$. The second complexity is determined by both input and output states φ, $\bar{\varphi}$ or an input state φ and a channel Λ^*, so that it is denoted by $T^S(\varphi; \bar{\varphi})$ or $T^S(\varphi; \Lambda^*)$, which is called the transmitted complexity from φ to $\bar{\varphi}$ or $\Lambda^*\varphi$. Typical examples of these complexities are the entropy and mutual entropy, which play essential roles in several fields as discussed in Chap. 3.

Let $(\mathcal{A}_t, \mathfrak{S}_t, \alpha^t(G^t))$ be the total system of $(\mathcal{A}, \mathfrak{S}, \alpha(G))$ and $(\bar{\mathcal{A}}, \bar{\mathfrak{S}}, \bar{\alpha}(\bar{G}))$, and S be a subset of $\mathfrak{S}$ in which we are measuring observables (e.g., $S = I(\alpha)$, $K(\alpha)$ in C*-system the set of all α-invariant states and the set of all KMS states, respectively).

C is the *complexity* of a state φ with respect to S, and T is the *transmitted complexity* associated with the state change $\varphi \to \Lambda^*\varphi$, both of which should satisfy the following properties:

(i) For any $\varphi \in S \subset \mathfrak{S}$,

$$C^S(\varphi) \geq 0, \quad T^S(\varphi; \Lambda^*) \geq 0.$$

(ii) For any orthogonal bijection $j : ex\mathfrak{S} \to ex\mathfrak{S}$ (the set of all extreme points in $\mathfrak{S}$),

$$C^{j(S)}(j(\varphi)) = C^{\mathcal{S}}(\varphi),$$

$$T^{j(S)}(j(\varphi);\Lambda^*) = T^{\mathcal{S}}(\varphi;\Lambda^*).$$

(iii) For $\Phi \equiv \varphi \otimes \psi \in \mathcal{S}_t \subset \mathfrak{S}_t,$

$$C^{\mathcal{S}_t}(\Phi) = C^{\mathcal{S}}(\varphi) + C^{\overline{\mathcal{S}}}(\psi).$$

(iv) $0 \leq T^{\mathcal{S}}(\varphi;\Lambda^*) \leq C^{\mathcal{S}}(\varphi).$
(v) $T^{\mathcal{S}}(\varphi; id) = C^{\mathcal{S}}(\varphi)$, where "id" is an identity map from $\mathfrak{S}$ to $\mathfrak{S}$.

Instead of (iii), when "(iii')" $\Phi \in \mathcal{S}_t \subset \mathfrak{S}_t$, put $\varphi \equiv \Phi \mid_{\mathcal{A}}$ (i.e., the restriction of Φ to $\mathcal{A}$), $\psi \equiv \Phi \mid_{\overline{\mathcal{A}}}$, then $C^{\mathcal{S}_t}(\Phi) \leq C^{\mathcal{S}}(\varphi) + C^{\overline{\mathcal{S}}}(\psi)$ " is satisfied, C and T is called the *pair of strong complexity*.

Therefore ID can be defined as follows:

DEFINITION 4.1. *Information Dynamics (ID)* is defined by

$$\left(\mathcal{A}, \mathfrak{S}, \alpha(G); \overline{\mathcal{A}}, \overline{\mathfrak{S}}, \overline{\alpha}(\overline{G}); \ \Lambda^*; \ C^{\mathcal{S}}(\varphi), T^{\mathcal{S}}(\varphi; \ \Lambda^*)\right)$$
and some relations R among them.

Thus, in the framework of ID, we have to
(i) determine mathematically

$$\mathcal{A}, \mathfrak{S}, \alpha(G); \overline{\mathcal{A}}, \overline{\mathfrak{S}}, \overline{\alpha}(\overline{G}),$$

(ii) choose Λ^* and R, and
(iii) define $C^{\mathcal{S}}(\varphi), T^{\mathcal{S}}(\varphi; \Lambda^*)$.

Information Dynamics can be applied to the study of chaos in the following ways:
(a) ψ is more chaotic than φ as seen from the reference system $\mathcal{S}$ if $C^{\mathcal{S}}(\psi) \geq C^{\mathcal{S}}(\varphi)$.
(b) When φ changes to $\Lambda^*\varphi$, a *degree of chaos* associated to this state change is given by

$$D^{\mathcal{S}}(\varphi;\Lambda^*) = C^{\overline{\mathcal{S}}}(\Lambda^*\varphi) - T^{\mathcal{S}}(\varphi;\Lambda^*).$$

In ID, several different topics can be treaded from the same standpoint, so that we can find a new clue bridging several different fields. For examples, we may have the following applications, cf. [259], [8], [256], [254], [252], [257], [260]:
(1) The study of optical communication processes.

(2) Definition of fractal dimensions of states, and the study of complexity for some systems.

(3) Definition of genetic matrix for genome sequences and construction of phylogenetic tree for evolution of species.

(4) Entropic complexities $\Longrightarrow$ KS type complexities (entropy) $\Longrightarrow$ classification of dynamical systems.

(5) The study of optical illusion (psychology)

(6) The study of some economic models.

In this chapter, we only discuss (4).

4.2. Example of Complexities

Though the complexity T depends on a channel Λ^*, severel examples of the channel were given in Section 3.4 of Chap. 3, so that we only give a few examples of complexities C and T related to information.

$\langle 1 \rangle$ The first examples of C and T are the entropy S and the mutual entropy I, respectively. Both classical and quantum S and I satisfied the conditions of the complexities, which were shown in Chap. 1 for the classical case and Chap. 3 for the quantum case. Here we only repeat the quantum cases. For a density operator ρ in a Hilbert space and a channel Λ^*, the entropy $S(\rho)$ and the mutual entropy $I(\rho; \Lambda^*)$ are defined as

$$S(\rho) \;\; = \;\; -\mathrm{tr}\rho \log \rho,$$

$$I(\rho; \Lambda^*) \;\; = \;\; \sup\left\{\sum_k \lambda_k S(\Lambda^* E_k, \Lambda^* \rho); \; \{E_k\}\right\},$$

where the supremum is taken over all Schatten decompositions $\{E_k\}$ of ρ; $\rho = \sum_k \lambda_k E_k$. From Theorem 3.1 of Chap. 3, the entropy $S(\rho)$ satisfies (i) $S(\rho) \geq 0$, (ii) $S(j(\rho)) = S(\rho)$ for an orthogonal bijection j, that is, it is a map from a set of orthogonal pure states to another set of orthogonal pure states, (iii) $S(\rho_1 \otimes \rho_2) = S(\rho_1) + S(\rho_2)$, so that $S(\rho)$ is a complexity C of ID.

The mutual entropy $I(\rho; \Lambda^*)$ satisfies the conditions (i), (ii), (iv) from the fundamental inequality of mutual entropy discussed in Section 3.5. Further, for the identity channel, $\Lambda^* = id$,

$$I(\rho; id) \;\; = \;\; \sup\left\{\sum_k S(E_k; \rho); \; \{E_k\}\right\}$$

$$= \;\; \sup\left\{\sum_k \lambda_k \mathrm{tr} E_k (\log E_k - \log \rho); \; \{E_k\}\right\}$$

$$= \;\; -\mathrm{tr}\rho \log \rho$$

because of $S(E_k) = 0$, hence it satisfies the condition (v). Thus S and I become a pair of the complexities. Moreover, S satisfies the condition of the strong complexity (subadditivity).

⟨2⟩ Fuzzy entropy has been defined by several auhtors like Zadeh [351], DeLuca and Termini [73], and Ebanks [77]. Here we take Ebanks's fuzzy entropy and we show that we can use it to construct the complexity C. Let X (this is $\mathcal{A}$ of ID) be a countable set $\{x_1, \cdots, x_n\}$ and f_A be membership function from X to $[0,1]$ associated with a subset $A \subset X$. If $f_A = 1_A$, then A is a usual set, which is called a sharp set, and if $f_A \neq 1_A$, then A is called a fuzzy set. Therefore the correspondence between a fuzzy set and a membership function is one to one. Take a membership function f and let us denote $f_i = f(x_i)$ for each $x_i \in X$. Then Ebanks' fuzzy entropy $S(f)$ for a membership function f is defined by

$$S(f) = - \sum_{i=1}^{n} f_i^\nu \log f_i, \qquad (\nu = \log_2 e).$$

When f is sharp, that is, $f_i = 0$ or 1 for any $x_i \in X$, $S(f) = 0$. When $f_i = \frac{1}{2}$ for any i, $S(f)$ attains the maximum value. Moreover, we have the following order $\prec$ on membership functions (or equivalently on fuzzy sets):

$$f \prec f' \equiv \begin{cases} f(x) \geq f'(x) & \left(\text{when } f'(x) \geq \frac{1}{2}\right) \\ f(x) \leq f'(x) & \left(\text{when } f'(x) \leq \frac{1}{2}\right) \end{cases}$$

Then $f \prec f' \Leftrightarrow \left| f_i - \frac{1}{2} \right| \geq \left| f_i' - \frac{1}{2} \right|$, which implies

$$S(f) \leq S(f').$$

This fuzzy entropy $S(f)$ defines a *complexity* $C(f)$ by

$$C(f) = S\left(\frac{f}{n}\right).$$

The positivity of $C(f)$ is proved as follows: From Klein's inequlity, $\log \frac{1}{x} \geq 1 - x$ for any $x > 0$, we have

$$\begin{aligned} C(f) &= -\sum_{i=1}^{n} \left(\frac{f_i}{n}\right)^\nu \log \frac{f_i}{n} \\ &\geq -\sum_{i=1}^{n} \left(\frac{f_i}{n}\right)^\nu \left(1 - \frac{n}{f_i}\right) \\ &= \sum_{i=1}^{n} \left(\frac{f_i}{n}\right)^{\nu-1} \left(1 - \frac{f_i}{n}\right) \geq 0 \end{aligned}$$

The invariance under a permutation π of indicies i of x_i (i.e, $i \to \pi\,(i)$), comes directly from the invariance of S under π.

This $C\,(f)$ satisfies not only the additivity but also the subadditivity. Let Y be another set $\{y_1, \ldots, y_m\}$ and g be a membership function from Y to $[0,1]$. Moreover, let h be a membership function on $X \times Y$ to $[0,1]$ satisfying

$$\sum_{j=1}^{m} h\,(x_i, y_j) = f\,(x_i), \qquad \sum_{i=1}^{n} h\,(x_i, y_j) = g\,(y_j)$$

What we have to show is the inequality

$$C\,(h) \leq C\,(f) + C\,(g).$$

Without loss of generality, we assume $n \geq m \geq 2$. Put

$$\eta\,(t) = -t^\nu \log t \qquad (\nu = \log_2 e).$$

$\eta\,(t)$ is monotone increasing in $0 \leq t \leq \frac{1}{2}$, so that we have

$$\eta\left(\frac{f_i}{nm}\right) \geq \eta\left(\frac{h_{ij}}{nm}\right) \qquad (n \geq m \geq 2)$$

because of $h_{ij} \equiv h\,(x_i, y_j) \leq f_i \equiv f\,(x_i)$ and hence $0 \leq h_{ij}/nm \leq f_i/nm \leq \frac{1}{2}$ for any i, j.

Thus we have

$$m\eta\left(\frac{f_i}{nm}\right) \geq \sum_{j=1}^{m} \eta\left(\frac{h_{ij}}{nm}\right).$$

Now

$$\eta\left(\frac{f_i}{n}\right) - \frac{1}{2}m\eta\left(\frac{f_i}{nm}\right)$$

$$= \left(\frac{f_i}{n}\right)^\nu \left\{\left(\frac{1}{2m^{\nu-1}} - 1\right) \log f_i + \left(1 - \frac{1}{2m^{\nu-1}}\right) \log n - \frac{1}{2m^{\nu-1}} \log m\right\}$$

$$\geq \left(\frac{f_i}{n}\right)^\nu \left\{\left(\frac{1}{2m^{\nu-1}} - 1\right) \log f_i + \left(1 - \frac{1}{m^{\nu-1}}\right) \log n\right\}$$

which is positive since $n \geq m \geq 2$. Hence

$$\eta\left(\frac{f_i}{n}\right) \geq \frac{1}{2}\sum_{j=1}^{m} \eta\left(\frac{h_{ij}}{nm}\right)$$

which implies

$$C(f) \geq \frac{1}{2}C(h).$$

Similarly we can prove

$$C\left(g\right) \geq \frac{1}{2}C\left(h\right) .$$

Therefore, we have the subadditivity

$$C\left(h\right) \leq C\left(f\right) + C\left(g\right) .$$

⟨3⟩ Kolmogorov [190] and Chaitin [52] discussed the complexity of sequences. Consider e.g. the following two sequences a and b composed of 0-s and 1-s:

$$a \ : \ 010101010101$$
$$b \ : \ 011010000110$$

In both a and b, the occurrence probabilities $p(0)$ and $p(1)$ are the same, $p(0) = p(1) = 1/2$. However, the sequence b seems to be more complicated than a. It suffices to know the first two letters to guess the whole a, but one may need the whole sequence of letters to know b. In general, we consider a computer (an automaton) transforming binary input sequences into output ones. Formalizing the description of such a machine it is possible to introduce the notion of the *minimum programme* as the "simplest" algorithm which produces a given output sequence. The amount of information contained in the specification of the minimum programme (measured e.g. in "bits") is called the complexity of a given (output) sequence.

Let $\mathcal{A}$ be the set of all finite sequences over an alphabet, say $\{0, 1\}$, and $\bar{\mathcal{A}}$ be another set. Further, let f be a *partial function* from $\mathcal{A}$ to $\bar{\mathcal{A}}$ (i.e., f is not necessarily defined on the whole $\mathcal{A}$). The triple $(\mathcal{A}, \bar{\mathcal{A}}, f)$ can be regarded as a language describing certain objects. For an element $a \in \mathcal{A}$, the length of a is denoted by $\ell(a)$. The minimum length of $a \in \mathcal{A}$ describing $\bar{a} \in \bar{\mathcal{A}}$ (that is $f(a) = \bar{a}$) is called the *complexity of description*. If there is no $a \in \mathcal{A}$ such that $f(a) = (a)$, then we set the complexity of $\bar{a}$ to ∞.

When both $\mathcal{A}$ and $\bar{\mathcal{A}}$ are sets of binary sequences, we consider partial *computable* functions $f : \mathcal{A} \to \bar{\mathcal{A}}$ (i.e., such that there exists a programme which, given an input $a \in \mathcal{A}$, terminates (halts) with the output $f(a) \in \bar{\mathcal{A}}$ whenever $a \in \text{Domain}(f)$, but it need not halt at all for inputs $a \notin \text{Domain}(f)$). The *complexity* $H_f(\bar{a})$ determined by $(\mathcal{A}, \bar{\mathcal{A}}, f)$ is defined as

$$H_f(\bar{a}) = \begin{cases} \min\{\ell(a); a \in \mathcal{A}, f(a) = \bar{a}\} & \text{(when } \exists a \in \mathcal{A} \text{ s.t. } f(a) = \bar{a}) \\ \infty & \text{(otherwise)} \end{cases}$$

For $a \in \mathcal{A}$ let $0^k 1a$ denote the sequence obtained from a by appending k symbols 0 and a single 1 on the left, i.e. $0 \ldots 01a$. Then it is shown

[52] that there exist $k \in \mathbb{N}$ and a computable partial function f_U such that for any f, $f_U(0^k 1a) = f(a)$. This f_U is called the universal partial function. With this function certain *universal computer* U is associated. In particular, U is capable of computing $f(a)$ for $a \in$ Domain (f). Some important consequences of the above are: (1) there exists a constant ε such that $H_{f_U}(\bar{a}) \leq H_f(\bar{a}) + \varepsilon$ for any f, and (2) there exists a constant ε' for two universal partial functions $f_U, f_{U'}$ satisfying $|H_{f_U}(\bar{a}) - H_{f_{U'}}(\bar{a})| \leq \varepsilon'$.

The above facts imply that H_{f_U} gives the minimum value for H_f if we neglect the constant ε. Kolmogorov and Chaitin introduced the following complexity

$$H(\bar{a}) = H_{f_U}(\bar{a})$$

which does not depend on a choice of f_U because of the statement (2). Moreover, Chaitin introduced the mutual entropy type complexity in the same framework as above. This complexity and mutual entropy type complexity can be associated with our complexities C and T, respectively.

It is possible to construct other complexities, not of the entropy type, in several fields like genetics, economics and computer sciences.

The details of all topics mentioned in this section are discussed in the recent paper [263].

4.3. Entropic complexity in GQS

Generalizing the entropy S and the mutual entropy I, we have several *complexities of entropy type*:

Let $(\mathcal{A}, \mathfrak{S}(\mathcal{A}), \alpha(G))$, $(\overline{\mathcal{A}}, \overline{\mathfrak{S}}(\overline{\mathcal{A}}), \overline{\alpha}(\overline{G}))$ be C^* systems as before. Let $\mathcal{S}$ be a weak $*$-compact convex subset of $\mathfrak{S}(\mathcal{A})$ and $M_\varphi(\mathcal{S})$ be the set of all maximal measures μ on $\mathcal{S}$ with the fixed barycenter φ

$$\varphi = \int_{\mathcal{S}} \omega d\mu .$$

Moreover let $F_\varphi(\mathcal{S})$ be the set of all measures of finite support with the fixed barycenter φ. Three pairs of complexity are:

$$T^{\mathcal{S}}(\varphi; \Lambda^*) \equiv \sup \left\{ \int_{\mathcal{S}} S(\Lambda^* \omega, \Lambda^* \varphi) d\mu; \mu \in M_\varphi(\mathcal{S}) \right\}$$

$$C_T^{\mathcal{S}}(\varphi) \equiv T^{\mathcal{S}}(\varphi; id)$$

$$I^{\mathcal{S}}(\varphi; \Lambda^*) \equiv \sup \left\{ S \left(\int_{\mathcal{S}} \omega \otimes \Lambda^* \omega d\mu, \varphi \otimes \Lambda^* \varphi \right) \mu \in M_\varphi(\mathcal{S}) \right\}$$

$$C_I^{\mathcal{S}}(\varphi) \equiv I^{\mathcal{S}}(\varphi; id)$$

$$J^{\mathcal{S}}(\varphi; \Lambda^*) \equiv \sup \left\{ \int_{\mathcal{S}} S(\Lambda^* \omega, \Lambda^* \varphi) d\mu; \mu \in F_\varphi(\mathcal{S}) \right\}$$

$$C_j^S(\varphi) \equiv J^S(\varphi; id)$$

These complexities and the mixing S-entropy $S^S(\varphi)$, the CNT (Connes-Narnhofer-Thirring) entropy $H_\varphi(\mathcal{A})$, discussed in Chap. 3, satisfy the following relations, cf. [234], [257]:

THEOREM 4.2. *(1)* $0 \leq I^S(\varphi; \Lambda^*) \leq T^S(\varphi; \Lambda^*) \leq J^S(\varphi; \Lambda^*)$.
(2) $C_I^{\mathfrak{S}}(\varphi) = C_T^{\mathfrak{S}}(\varphi) = C_J^{\mathfrak{S}}(\varphi) = S^{\mathfrak{S}}(\varphi) = H_\varphi(\mathcal{A})$.
(3) When $\mathcal{A} = \bar{\mathcal{A}} = B(\mathcal{H})$, *for any density operator* ρ

$$0 \leq I^S(\rho; \Lambda^*) = T^S(\rho; \Lambda^*) \leq J^S(\rho; \Lambda^*).$$

4.4. KS-type complexities

The Kolmogorov-Sinai (KS) entropy is used in CDS to compute the mean information and the degree of chaos for dynamical systems. The notion of quantum dynamical entropy has been formulated by several authors. In this section, we discuss the dynamical entropy in the context of ID.

Let us remind the definition of the classical KS-entropy cf. [188, 303] and the quantum dynamical entropy of Connes-Narnhofer-Thirring [59]. Let $(\Omega, \mathcal{F}, \mu)$ be a probability space, and $\mathcal{P}$ be the set of all finite partitions of Ω for σ-field $\mathcal{F}$. Then we introduce a *measure-preserving transformation* θ of Ω, i.e., $\theta^{-1}\mathcal{F} \subset \mathcal{F}$ and $\mu \circ \theta^{-1} = \mu$. For any finite partitions (experiments) $\check{A}$, $\check{B} \in \mathcal{P}$, we define the following operations on finite partitions:

$$\check{A} \vee \check{B} = \{A_i \cap B_j; A_i \in \check{A}, B_j \in \check{B}\},$$

$$\theta^{-1}\check{A} = \{\theta^{-1}A_i; A_i \in \check{A}\}.$$

It can be proved that the following limit exists:

$$\check{S}_\mu(\theta, \check{A}) = \lim_{n \to \infty} \sup \frac{1}{n} S\left(\mu : \bigvee_{k=0}^{n-1} \theta^{-k}\check{A}\right),$$

where the entropy $S(\mu : \check{A})$ is defined in Chap. 3, Eq. (3.52). Finally, we define the *classical KS-entropy* of θ as follows

$$\check{S}_\mu(\theta) = \sup\{\check{S}_\mu(\theta, \check{A}); \check{A} \in \mathcal{P}\}.$$

The quantum dynamical entropy was first introduced by Connes and Størmer, and its general version is written by Connes-Narnhofer-Thirring [59]: Let $H_\varphi(\mathfrak{N})$ be CNT entropy for a subalgebra $\mathfrak{N}$ of a Von Neumann algebra $\mathcal{A}$ and θ be an automorphism of $\mathcal{A}$. Put $\bigvee_{k=1}^{n} \theta^{k-1}\mathfrak{M}$ be a Von

Neumann algebra generated by $\{\theta^{k-1}\mathfrak{M}; k = 1, \ldots, n\}$. Then the *dynamical entropy* with respect to θ and φ is given by

$$\tilde{H}_\varphi(\theta; \mathfrak{M}) \equiv \limsup_{n \to \infty} \frac{1}{n} H_\varphi \left(\bigvee_{k=1}^n \theta^{k-1}\mathfrak{M} \right),$$

$$\tilde{H}_\varphi(\theta) \equiv \sup\{\tilde{H}_\varphi(\theta; \mathfrak{M}); \mathfrak{M} \subset \mathcal{A}\}.$$

We now define the *dynamical entropy by the complexities*. Let θ (resp. $\overline{\theta}$) be an automorphism of $\mathcal{A}$ (resp. $\overline{\mathcal{A}}$) such that $\varphi \circ \theta = \varphi$ and Λ be a covariant (i.e., $\Lambda \circ \theta = \overline{\theta} \circ \Lambda$) CP (completely positive) map from $\overline{\mathcal{A}}$ to $\mathcal{A}$. Take $\mathcal{A}_k$ (resp. $\overline{\mathcal{A}}_k$) as a finite subalgebra of $\mathcal{A}$ (resp. $\overline{\mathcal{A}}$) and α_k (resp. $\bar{\alpha}_k$) as a unital map from $\mathcal{A}_k$ (resp. $\overline{\mathcal{A}}_k$) to $\mathcal{A}$ (resp. $\overline{\mathcal{A}}$). Put $\alpha^M \equiv (\alpha_1, \alpha_2, \cdots, \alpha_M)$, $\bar{\alpha}_\Lambda^N \equiv (\Lambda \circ \bar{\alpha}_1, \Lambda \circ \bar{\alpha}_2, \cdots, \Lambda \circ \bar{\alpha}_N)$. Then two compound states for α^M and $\bar{\alpha}_\Lambda^N$ w.r.t $\mu \in M_\varphi(\mathcal{S})$ are

$$\Phi_\mu^S \left(\alpha^M \right) = \int_S \bigotimes_{m=1}^M \alpha_m^* \omega \, d\mu$$

$$\Phi_\mu^S \left(\alpha^M \cup \bar{\alpha}_\Lambda^N \right) = \int_S \bigotimes_{m=1}^M \alpha_m^* \omega \bigotimes_{n=1}^N \bar{\alpha}_n^* \Lambda^* \omega \, d\mu$$

The transmitted complexities are written as

$$T^S(\varphi; \ \alpha^M, \bar{\alpha}_\Lambda^N)$$
$$\equiv \sup \left\{ \int_S S \left(\bigotimes_{m=1}^M \alpha_m^* \omega \bigotimes_{n=1}^N \bar{\alpha}_n^* \Lambda^* \omega, \Phi_\mu^S(\alpha^M) \otimes \Phi_\mu^S(\bar{\alpha}_\Lambda^N) \right) d\mu \ ; \right.$$
$$\left. \mu \in M_\varphi(\mathcal{S}) \right\}$$

$$I^S(\varphi; \ \alpha^M, \bar{\alpha}_\Lambda^N)$$
$$\equiv \sup \left\{ S \left(\Phi_\mu^S(\alpha^M \cup \bar{\alpha}_\Lambda^N), \Phi_\mu^S(\alpha^M) \otimes \Phi_\mu^S(\bar{\alpha}_\Lambda^N) \right) ; \mu \in M_\varphi(\mathcal{S}) \right\}$$

$$J^S(\varphi; \ \alpha^M, \bar{\alpha}_\Lambda^N)$$
$$\equiv \sup \left\{ \int_S S \left(\bigotimes_{m=1}^M \alpha_m^* \omega \bigotimes_{n=1}^N \bar{\alpha}_n^* \Lambda^* \omega, \Phi_\mu^S(\alpha^M) \otimes \Phi_\mu^S(\bar{\alpha}_\Lambda^N) \right) d\mu_f \ ; \right.$$
$$\left. \mu_f \in F_\varphi(\mathcal{S}) \right\}$$

When $\mathcal{A}_k = \mathcal{A}_0 = \overline{\mathcal{A}}_k$, $\mathcal{A} = \overline{\mathcal{A}}$, $\theta = \overline{\theta}$ and $\alpha_k = \theta^{k-1} \circ \alpha = \overline{\alpha}_k$ with a unital CP map $\alpha : \mathcal{A}_0 \to \mathcal{A}$,

$$\tilde{T}_\varphi^S(\theta, \alpha, \Lambda^*) \equiv \limsup_{N \to \infty} \frac{1}{N} T^S(\varphi; \ \alpha^N, \overline{\alpha}_\Lambda^N)$$

$$\tilde{T}^S_\varphi(\theta, \Lambda^*) \equiv \sup_\alpha \tilde{T}^S_\varphi(\theta, \alpha, \Lambda^*)$$

$\tilde{T}^S_\varphi(\theta, \Lambda^*)$ is the mean transmitted complexity w.r.t θ, Λ^*. We similarly define $\tilde{I}^S_\varphi, \tilde{J}^S_\varphi$. Then the CNT-type theorem holds for these complexties [234]:

THEOREM 4.3. *If there exist* $\alpha'_m : \mathcal{A} \to \mathcal{A}_m$ *with such that* $\alpha_m \circ \alpha'_m = id$ *as* $m \to \infty$, *then*

$$\tilde{T}^S_\varphi(\theta, \Lambda^*) = \lim_{m \to \infty} \tilde{T}^S_\varphi(\theta, \alpha_m, \overline{\alpha}, \Lambda^*)$$

The same for $\tilde{I}^S_\varphi, \tilde{J}^S_\varphi$.

Our complexties generalize usual KS entropy in the following senses.

(1) When $\mathcal{A}_n, \mathcal{A}$ are abelian C*-algebra and α_k embedding,

$$T^S(\mu; \alpha^M) = S^{classical}_\mu \left(\bigvee_{m=1}^M \tilde{A}_m \right) ,$$

$$I^S(\mu; \alpha^M, \overline{\alpha}^N) = I^{classical}_\mu \left(\bigvee_{n=1}^M \tilde{A}_n, \bigvee_{n=1}^N \tilde{B}_n \right)$$

for any finite partitions $\tilde{A}_n, \tilde{B}_n$ on the probability space $(\Omega = \text{spec}(\mathcal{A}), \mathcal{F}, \mu)$.

(2) When Λ is the restriction from Von Neumann algebra $\mathcal{A}$ to its subalgebra $\mathcal{B}$ (i.e., $\Lambda = |_\mathcal{B}$) and id is an identity map from $\mathcal{A}$ to $\mathcal{A}$,

$$H_\varphi(\mathcal{B}) = J^S(\varphi; |_\mathcal{B}) = J^S(\varphi; id, |_\mathcal{B})$$

Moreover, let $\mathfrak{M}$ be a subalgebra of a Von Neumann algebra $\mathcal{A}_0$ and $\mathcal{A} = \otimes^N \mathcal{A}_0$, $\theta \in \text{Aut}(\mathcal{A})$, $\alpha = \overline{\alpha}$ be an embedding from $\mathcal{A}_0$ to $\mathcal{A}$, and $\mathfrak{M}_N = \otimes^N_1 \mathfrak{M}$. Then

$$\tilde{J}^S_\varphi(\theta; \mathfrak{M}) \equiv \limsup_{N \to \infty} \frac{1}{N} J^S(\varphi; \alpha^N, |_{\mathfrak{M}_N})$$

and

$$\tilde{J}^S_\varphi(\theta) = \sup\{\tilde{J}^S_\varphi(\theta; \mathfrak{M}); \mathfrak{M} \subset \mathcal{A}_0\} .$$

which becomes $\tilde{H}_\varphi(\theta)$.

Finally, we note that the quantum KS dynamical entropy can be formulated in several different ways [36], [20], [8], [59], [58], [84]. Their formulations have their own merits to compute the dynamical entropy in each model, but the dynamical entropy stated above contains them as special cases [10]. We have briefly review other two definitions of the dynamical entropy.

A construction of the *dynamical entropy through quantum Markov chain* is as follows [9]: Let $\mathcal{A}$ be a Von Neumann algebra acting on a Hilbert space $\mathcal{H}$ and let φ be a state on $\mathcal{A}$ and $A_0 = M_d$ ($d \times d$ matrix algebra). Take the transition expectation $\mathcal{E}_\gamma : A_0 \otimes \mathcal{A} \to \mathcal{A}$ of Accardi [3, 5] such that

$$\mathcal{E}_\gamma(\tilde{A}) = \sum_i \gamma_i A_{ii} \gamma_i,$$

where $\tilde{A} = \sum_{i,j} e_{ij} \otimes A_{ij} \in A_0 \otimes \mathcal{A}$ and $\gamma = \{\gamma_j\}$ is a finite partition of unity $I \in \mathcal{A}$. Quantum Markov chain is defined by $\psi \equiv \{\varphi, \mathcal{E}_{\gamma,\theta}\} \in \mathfrak{S}(\overset{\infty}{\underset{1}{\otimes}} A_0)$ such that

$$\psi(j_1(A_1) \cdots j_n(A_n)) \equiv \varphi \left(\mathcal{E}_{\gamma,\theta}(A_1 \otimes \mathcal{E}_{\gamma,\theta}(A_2 \otimes \cdots \otimes A_{n-1} \mathcal{E}_{\gamma,\theta}(A_n \otimes I) \cdots))) \right),$$

where $\mathcal{E}_{\gamma,\theta} = \theta \circ \mathcal{E}_\gamma$, $\theta \in \mathrm{Aut}(\mathcal{A})$ and j_k is an embeding of A_0 into $\overset{\infty}{\underset{1}{\otimes}} A_0$, namely, $j_k(A) = I \otimes \cdots \otimes I \otimes \underset{k-th}{A} \otimes I \cdots$.

Suppose that for φ there exists a unique density operator ρ such that $\varphi(A) = \mathrm{tr}\,\rho A$ for any $A \in \mathcal{A}$. Let us define a state ψ_n on $\overset{n}{\underset{1}{\otimes}} A_0$ by

$$\psi_n(A_1 \otimes \cdots \otimes A_n) = \psi(j_1(A_1) \cdots j_n(A_n)).$$

Then the density operator ξ_n for ψ_n is given by

$$\xi_n \equiv \sum_{i_1} \cdots \sum_{i_n} \mathrm{tr}_{\mathcal{A}}(\theta^n(\gamma_{i_n}) \cdots \gamma_{i_1} \rho \gamma_{i_1} \cdots \theta^n(\gamma_{i_n}))) e_{i_1 i_1} \otimes \cdots \otimes e_{i_n i_n}.$$

Put

$$P_{i_n \cdots i_1} = \mathrm{tr}_{\mathcal{A}}(\theta^n(\gamma_{i_n}) \cdots \gamma_{i_1} \rho \gamma_{i_1} \cdots \theta^n(\gamma_{i_n}))).$$

The mean dynamical entropy through QMC is defined by Von Neumann entropy $S(\cdot)$:

$$\tilde{S}_\varphi(\theta; \gamma) \equiv \limsup_{n \to \infty} \frac{1}{n} S(\xi_n) = \limsup_{n \to \infty} \frac{1}{n} \left(-\sum_{i_1, \cdots, i_n} P_{i_n \cdots i_1} \log P_{i_n \cdots i_1} \right).$$

If $P_{i_n \cdots i_1}$ satisfies the Markov property, then the above equality is written as

$$\tilde{S}_\varphi(\theta; \gamma) = -\sum_{i_1, i_2} P(i_2|i_1) P(i_1) \log P(i_2|i_1).$$

The dynamical entropy through QMC with respect to θ is

$$\tilde{S}_\varphi(\theta) \equiv \sup\left\{\tilde{S}_\varphi(\theta;\gamma);\quad \gamma \subset \mathcal{A}\right\}.$$

Another formulation has been given by Alicki and Fannes [20]. Let $\mathcal{A}$ be a unital C^*-algebra, θ be an automorphism on $\mathcal{A}$ and φ be a stationary state with respect to θ. A set $\gamma = \{\gamma_1, \gamma_2, \ldots, \gamma_k\}$ of elements of $\mathcal{A}$ is called a finite operational partition of unity of size k if γ satisfies the following condition:

$$\sum_{i=1}^{k} \gamma_i^* \gamma_i = I. \tag{4.1}$$

Let the operation $\circ$ be defined by

$$\gamma \circ \xi \equiv \{\gamma_i \xi_j;\quad i = 1, 2, \ldots, k,\quad j = 1, 2, \ldots, l\}$$

for any partitions $\gamma = \{\gamma_1, \gamma_2, \ldots, \gamma_k\}$ and $\xi = \{\xi_1, \xi_2, \ldots, \xi_l\}$. For any partition γ of size k, a $k \times k$ density matrix $\rho[\gamma] = (\rho[\gamma]_{i,j})$ is given by

$$\rho[\gamma]_{i,j} = \varphi(\gamma_j^* \gamma_i).$$

Then the dynamical entropy $\tilde{S}_\varphi(\theta, \gamma)$ with respect to the partition γ and shift θ is defined by

$$\tilde{S}_\varphi(\theta, \gamma) = \limsup_{n\to\infty} \frac{1}{n} S(\rho[\theta^{n-1}(\gamma) \circ \cdots \circ \theta(\gamma) \circ \gamma]). \tag{4.2}$$

The dynamical entropy is obtained by taking the supremum over operational partitions of unity in $\mathcal{A}$ as

$$\tilde{S}_\varphi(\theta) = \sup\left\{\tilde{S}_\varphi(\theta, \gamma); \gamma \subset \mathcal{A}\right\}. \tag{4.3}$$

4.5. Some model computations

We compute the dynamical entropy by complexity in order to see which modulation is effective for *optical fiber communication* [260]. Let $\mathcal{H}_0$ and $\overline{\mathcal{H}}_0$ be input and output Hilbert spaces, respectively. In order to send a state carrying information to the output system, we might need to modulate the state in proper way. The modulation is defined as a map (denoted by $\Gamma^*_{(\mathcal{M})}$) from $\mathfrak{S}(\mathcal{H}_0)$ (the set of all density operators on $\mathcal{H}_0$) to a certain state space $\mathfrak{S}(\mathcal{H}_{(\mathcal{M})})$ on a proper Hilbert space $\mathcal{H}_{(\mathcal{M})}$. Take

$$\mathcal{A} \equiv \mathop{\otimes}_{i=-\infty}^{\infty} B(\mathcal{H}_0),\quad \overline{\mathcal{A}} \equiv \mathop{\otimes}_{i=-\infty}^{\infty} B(\overline{\mathcal{H}}_0).$$

Let $\mathfrak{S}$ (resp. $\overline{\mathfrak{S}}$) be the set of all density operators in $\mathcal{A}$ (resp. $\overline{\mathcal{A}}$). Let θ (resp. $\overline{\theta}$) be a shift on $\mathcal{A}$ (resp. $\overline{\mathcal{A}}$) and α (resp. $\overline{\alpha}$) be an embedding map from $B(\mathcal{H}_0)$ to $\mathcal{A}$, (resp. $B(\overline{\mathcal{H}}_0)$ to $\overline{\mathcal{A}}$) as before. Let Λ^* be an attenuation channel (i.e., $\Lambda^*_{\sqrt{\eta},\sqrt{1-\eta}}$ discussed in Chap. 3 with the transmission rate η).

Put $\tilde{\Lambda} \equiv \overset{\infty}{\underset{i=-\infty}{\otimes}} \Lambda$ and $\tilde{\Gamma}_{(\mathcal{M})} \equiv \overset{\infty}{\underset{i=-\infty}{\otimes}} \Gamma_{(\mathcal{M})}$ and define

$$\alpha^N_{(\mathcal{M})} \equiv (\alpha \circ \tilde{\Gamma}_{(\mathcal{M})}, \cdots, \theta^{N-1} \circ \alpha \circ \tilde{\Gamma}_{(\mathcal{M})}),$$

$$\overline{\alpha}^N_{\Lambda(\mathcal{M})} \equiv (\tilde{\Gamma}_{(\mathcal{M})} \circ \tilde{\Lambda} \circ \overline{\alpha}, \cdots, \tilde{\Gamma}_{(\mathcal{M})} \circ \tilde{\Lambda} \circ \overline{\theta}^{N-1} \circ \overline{\alpha}).$$

Here we consider only two modulations PAM *(pulse amplitude modulation)* and PPM *(pulse position modulation)*. The modulations $\Gamma^*_{(PAM)}$ for PAM and $\Gamma^*_{(PPM)}$ for PPM are written as

$$\Gamma^*_{(PAM)}(E_n) = |n\rangle\langle n|$$

$$\Gamma^*_{(PPM)}(E_n) = \underbrace{|0\rangle\langle 0| \otimes \cdots \otimes |0\rangle\langle 0| \otimes \overbrace{|d\rangle\langle d|}^{n-th} \otimes |0\rangle\langle 0| \otimes \cdots \otimes |0\rangle\langle 0|}_{M}$$

for any $E_n \in \mathfrak{S}(\mathcal{H}_0)$, where $|k\rangle\langle k|$ is a k-photon number state. For a stationary initial state $\rho = \sum_m \mu_m \overset{\infty}{\underset{i=-\infty}{\otimes}} \rho^{(i)}_m \in \mathfrak{S}$, a density operator, and an attenuation channel Λ^*, the transmitted complexities of the mutual entropy type for two modulations PAM and PPM are calculated as

$I^{\mathfrak{S}}(\rho;\ \alpha^N_{(PAM)}, \overline{\alpha}^N_{\Lambda(PAM)})$

$$= \sum_{j_0=0}^{M} \cdots \sum_{j_{N-1}=0}^{M} \sum_{n_0=J_0}^{M} \cdots \sum_{n_{N-1}=J_{N-1}}^{M} \left(\sum_m \mu_m \prod_{k=0}^{N-1} \lambda^{(m)}_{n_k}\right) \left(\prod_{k'=0}^{N-1} |C^{n_{k'}}_{j_{k'}}|^2\right)$$

$$\times \left\{ \log \prod_{k=0}^{N-1} |C^{n_k}_{j_k}|^2 - \log \left(\sum_{n'_0=J_0}^{M} \cdots \sum_{n'_{N-1}=J_{N-1}}^{M} \left(\sum_{m'} \mu_{m'} \prod_{k'=0}^{N-1} \lambda^{(m')}_{n'_{k'}}\right) \right. \right.$$

$$\times \left. \left. \left(\prod_{k''=0}^{N-1} |C^{n'_{k''}}_{j_{k''}}|^2\right)\right)\right\}.$$

$I^{\mathfrak{S}}(\rho;\ \alpha^N_{(PPM)}, \overline{\alpha}^N_{\Lambda(PPM)})$

$$= -\sum_{n_0=1}^{M} \cdots \sum_{n_{N-1}=1}^{M} \left(\sum_m \mu_m \prod_{k=0}^{N-1} \lambda^{(m)}_{n_k}\right) \left\{ \sum_{p=1}^{N} \sum_{\{q_1,\cdots,q_p\}\subset\{1,2,\cdots,N\}} \right.$$

$$\times \left. \left(\sum_{\ell_1=1}^{d} \cdots \sum_{\ell_p=1}^{d} |C^d_{\ell_1}|^2 \cdots |C^d_{\ell_p}|^2 (1-\eta)^{N-p}\eta^p \log\left(\sum_{m'} \mu_{m'} \prod_{k'=0}^{p} \lambda^{(m')}_{n_{q_{k'}}}\right)\right)\right\}.$$

where

$$|C_{j_i}^{n_i}|^2 = \frac{n_i!}{j_i!(n_i - j_i)!}\eta^{j_i}(1 - \eta)^{(n_i - j_i)}.$$

Then we have

THEOREM 4.4. *If* $\mathcal{A} = \overline{\mathcal{A}}$, $\theta = \overline{\theta}$, $\alpha = \overline{\alpha}$ *and* $d \geq N$, *then*

$$\tilde{I}_\rho^{\mathfrak{S}}(\theta, \alpha_{(PPM)}, \Lambda^*) \geq \tilde{I}_\rho^{\mathfrak{S}}(\theta, \alpha_{(PAM)}, \Lambda^*).$$

This theorem tells us that PPM sends more information, hence PPM is more effective than PAM in optical communication.

Chapter 5

Information Thermodynamics I

5.1. Complete and incomplete measurement

Since in the problem of maximization of entropy the property of convexity is the most important, we first assume for simplicity that our Hilbert space (we consider the quantum case) is finite dimensional, $n = \dim \mathcal{H} < \infty$, cf. [341], [196].

We denote the set of all states (density operators) by

$$\mathfrak{S} = \{\rho : \mathcal{H} \to \mathcal{H}, \text{linear}; \ \rho \geq 0, \quad \text{tr}\rho = 1\}, \tag{5.1}$$

and by $\mathfrak{S}_0$ the set of all idempotent operators (pure states) in $\mathfrak{S}$.

Now we have

THEOREM 5.1. $\mathfrak{S}$ is a closed (n^2-1)-dimensional convex set. $\mathfrak{S}_0$ is the set of all extremal elements of $\mathfrak{S}$. An element $\rho \in \mathfrak{S}$ is a border point (interior point) of $\mathfrak{S}$ iff $\det(\rho) = 0$ ($\det(\rho) > 0$).

The proof follows immediately from the definitions of $\mathfrak{S}, \mathfrak{S}_0$ and the properties of convex sets, cf. e.g. [80].

The set of all observables (selfadjoint linear operators in $\mathcal{H}$) will be denoted by $\mathcal{A}$. A *spectral measure* is a function E defined on the Borel sets of the real axis $\mathbb{R}^1$ with values among the projection operators in $\mathcal{H}$ such that

$$E(\mathbb{R}^1) = I, \quad E(\bigcup_{i=1}^{\infty} \epsilon_i) = \sum_{i=1}^{\infty} E(\epsilon_i), \tag{5.2}$$

where I is the identity operator in $\mathcal{H}$, and ϵ_i, $(i = 1, 2, \ldots)$ are disjoint Borel sets of $\mathbb{R}^1$.

It is well-known, cf. e.g. [295], that for any observable $A \in \mathcal{A}$ there exists one and only one spectral measure E_A such that

$$A = \int_{-\infty}^{+\infty} \lambda E_A(d\lambda). \tag{5.3}$$

(This is also true for any infinite dimensional separable $\mathcal{H}$; in our case of finite n this reduces to

$$A = \sum_{i=1}^{n} \lambda_i E_i, \tag{5.4}$$

where E^i is a projection on the eigenspace belonging to eigenvalue λ_i.)

According to the probabilistic interpretation of quantum mechanics, cf. [333], the probability $p_A^\rho(\epsilon)$ of the event that a measurement of A on a system in the state ρ gives a result contained in the Borel set ϵ is

$$p_A^\rho(\epsilon) = \text{tr}(\rho E_A(\epsilon)). \tag{5.5}$$

The mean (expectation) value of an observable A in a state ρ

$$\int_{-\infty}^{+\infty} \lambda p_A^\rho(d\lambda) = \text{tr}(\rho A) \tag{5.6}$$

will be denoted by $m_A(\rho)$.

Assuming that the hamiltonian H of the system is independent of time t and the system is isolated (closed), we have the usual Von Neumann equation of evolution

$$i\hbar \frac{d}{dt}\rho(t) = [H, \rho(t)], \tag{5.7}$$

with the initial condition $\rho(0) = \rho_0$. As is well-known, the solution of this equation can be written in the form

$$\rho(t) = U_t^* \rho_0 U_t \qquad (t \geq 0), \tag{5.8}$$

where U_t is a unitary operator depending on t

$$U_t = \exp(\frac{i}{\hbar}Ht). \tag{5.9}$$

If the initial condition ρ_0 is known, we can calculate from (5.7) and (5.8) $\rho(t)$ for any $t \geq 0$, and consequently, we can calculate mean values $m_A(\rho(t))$ for any observable $A \in \mathcal{A}$. The state ρ_0 can be fixed by the measurement of $n^2 - 1$ mean values of $n^2 - 1$ linearly independent observables, cf. below (5.18). If n is large, such a measurement will be difficult or impossible. Therefore, in the stationary case below we shall discuss a method of the estimation of state ρ_0 by means of a smaller number of measurements. It is hoped that in the thermodynamical equilibrium the number of independent parameters is rather small, and the experimental evidence seems to support this hope.

To show a method of fixing ρ_0, let us consider as $\mathcal{L}^2_{\mathcal{H}}$ a real Hilbert space of operators of Hilbert-Schmidt type, cf. [228], defined in $\mathcal{H}$. Elements of $\mathcal{L}^2_{\mathcal{H}}$ are selfadjoint operators on $\mathcal{H}$ such that

$$||A||^2_{HS} = (A, A) < \infty \qquad (A \in \mathcal{L}^2_{\mathcal{H}}), \tag{5.10}$$

where $(\cdot, \cdot)$ is a scalar product in $\mathcal{L}^2_{\mathcal{H}}$ defined by

$$(A, B) = \text{tr}(AB) \qquad (A, B \in \mathcal{L}^2_H). \tag{5.11}$$

If operators $E_k \in \mathcal{L}^2_H$ $(k = 1, \ldots, n^2)$ form a complete orthonormal base in $\mathcal{L}^2_{\mathcal{H}}$, then for any $X \in \mathcal{L}^2_{\mathcal{H}}$ we have

$$X = \sum_{k=1}^{n^2} \text{tr}(X E_k) E_k. \tag{5.12}$$

For any $\rho \in \mathfrak{S}$ we have

$$||\rho||^2_{HS} = \text{tr}(\rho^2) \leq 1, \tag{5.13}$$

so all states ρ are the elements of $\mathcal{L}^2_{\mathcal{H}}$. Putting in (5.12) $X = \rho$, we obtain

$$\rho = \sum_{k=1}^{n^2} \text{tr}(\rho E_k) E_k. \tag{5.14}$$

Since $\text{tr}(\rho I) = 1$ for any $\rho \in \mathfrak{S}$, among n^2 quantities $\text{tr}(\rho E_k)$ $(k = 1, \ldots, n^2)$ being mean values of the observables $E_1, \ldots, E_{n^2}$, only $n^2 - 1$ are independent (as we mentioned above).

If we choose a base in $\mathcal{L}^2_{\mathcal{H}}$ such that

$$E_0 = \frac{I}{n}, E_1, \ldots, E_{n^2-1}, \tag{5.15}$$

then from the orthonormal conditions it follows

$$(E_0, E_0) = 1, \quad (E_k, E_l) = \delta_{kl} \quad (k, l = 1, \ldots, n^2 - 1) \tag{5.16}$$

and

$$(E_0, E_k) = \frac{1}{\sqrt{n}} \text{tr}(I E_k) = \frac{1}{\sqrt{n}} \text{tr} E_k = 0 \quad (k = 1, \ldots, n^2 - 1). \tag{5.17}$$

In this base every state ρ has the form

$$\rho = \frac{I}{n} + \sum_{k=1}^{n^2-1} \text{tr}(\rho E_k) E_k. \tag{5.18}$$

The above result can be generalized by the introduction, instead of observables E^k $(k = 1, \ldots, n^2 - 1)$ fulfilling conditions (5.16) and (5.17), the observables $A_k \in \mathcal{L}_{\mathcal{H}}^2$ $(k = 1, \ldots, n^2 - 1)$ such that the operators $I, A_1, \ldots, A_{n^2-1}$ are linearly independent. An example of such a system is for an observable $A \neq I$ the set of operators $I, A, A^2, \ldots, A^{n^2-1}$.

We shall call a set $\mathbf{A}^{(p)} = \{A_1, A_2, \ldots, A_p\}$ $(p = 1, 2, \ldots, n^2 - 1)$ such that operators $I, A_1, \ldots, A_p$ are linearly independent a *complete set of observables* if $p = n^2 - 1$, and an incomplete set of observables if $p < n^2 - 1$.

Recapitulating, we may say that a measurement of a complete set of observables fixes uniquely the state of a system. Saying that the state of the system is known we always mean that at time $t = 0$ a complete measurement has been performed, while for $t \geq 0$ it is fixed by (5.8) with ρ_0 determined by the measurement.

5.2. Convex properties of entropy

Let us now discuss the convex properties of von Neumann entropy (for its axiomatic definition cf. [147])

$$s(\rho) = -\mathrm{tr}(\rho \ln \rho), \qquad (\rho \in \mathfrak{S}). \tag{5.19}$$

Using the results of Wichmann [341], we shall quote some theorems about the properties of Von Neumann entropy.

THEOREM 5.2. *Von Neumann entropy $s(\rho)$ (5.19) in a Hilbert space $\mathcal{H}$ with $n = \dim \mathcal{H} < \infty$ satisfies the inequalities*

$$0 \leq s(\rho) \leq \ln n \qquad (\rho \in \mathfrak{S}). \tag{5.20}$$

The smallest value, $s = 0$, is taken by $s(\rho)$ for pure states $\rho \in \mathfrak{S}_0$, while the largest value, $s = \ln n$, is taken for the "middle state" $\rho = \frac{I}{n}$.

THEOREM 5.3. *The entropy $s(\rho)$ is a convex function on $\mathfrak{S}$, i.e., for arbitrary real numbers λ_i $(i = 1, \ldots, m)$, $\sum_{i=1}^m \lambda_i = 1$, and $\rho_i \in \mathfrak{S}$ $(i = 1, \ldots, m)$ we have*

$$s\left(\sum_{i=1}^m \lambda_i \rho_i\right) \geq \sum_{i=1}^m \lambda_i s(\rho_i), \tag{5.21}$$

where the equality sign occurs iff there exists $\rho \in \mathfrak{S}$ such that $\rho_i = \rho$ for all $i = 1, \ldots, m$ (all ρ_i are equal).

THEOREM 5.4. *If Φ is an arbitrary closed and convex subset of set $\mathfrak{S}$, then there exists one and only one state $\rho^* \in \Phi$ such that*

$$s(\rho^*) = S(\Phi) := \sup_{\rho \in \Phi} s(\rho). \tag{5.22}$$

THEOREM 5.5. *Let* $\mathbf{A}^{(p)} = \{A_1, \ldots, A_p\}$ *be the set of* $p \leq n^2 - 1$ *observables such that* $I, A_1, \ldots, A_p$ *are linearly independent. To each* $\rho \in \mathfrak{S}$ *we assign a vector* $\mathbf{a} = (a_1, \ldots, a_p)$ *in a* p-*dimensional real linear space* $\mathbb{R}^p$ *by means of the relation*

$$a_k = \mathrm{Tr}\,(\rho A_k) \qquad (k = 1, \ldots, p). \tag{5.23}$$

We denote the linear representation (5.23) of $\mathfrak{S}$ *in* $\mathbb{R}^p$ *by* $\pi_{\mathbf{A}} : \mathfrak{S} \to \mathbb{R}^p$. *For each* $\mathbf{a} \in \pi(\mathfrak{S})$ *the set* $\pi_{\mathbf{A}}^{-1}(\mathbf{a})$ *is a closed and convex subset of* $\mathfrak{S}$.

THEOREM 5.6. *Let* $\mathbf{A}^{(p)} = \{A_1, \ldots, A_p\}$ *be the set of* $p \leq n^2 - 1$ *observables such that operators* $I, A_1, \ldots, A_p$ *are linearly independent and* $\mathbf{a} \in \pi_{\mathbf{A}}(\mathfrak{S})$. *Then there exists one and only one state* $\rho^\star \in \pi_{\mathbf{A}}^{-1}(\mathbf{a})$ *such that*

$$s(\rho^\star) = S(\pi_{\mathbf{A}}^{-1}(\mathbf{a})) = \sup_{\rho \in \pi_{\mathbf{A}}^{-1}(\mathbf{a})} s(\rho) \tag{5.24}$$

of the form

$$\rho^\star = \rho^\star(\mathbf{a}) = Z^{-1}(\beta_1, \ldots, \beta_p) \exp(-\sum_{i=1}^{p} \beta_i A_i) \tag{5.25}$$

where

$$Z(\beta) = Z(\beta_1, \ldots, \beta_p) = \mathrm{Tr}\,\exp(-\sum_{i=1}^{p} \beta_i A_i). \tag{5.26}$$

The vector $\beta = (\beta_1, \ldots, \beta_p)$ *is a unique function of the vector* $\mathbf{a} = (a_1, \ldots, a_p) \in \pi_{\mathbf{A}}(\mathfrak{S})$ *defined by means of the equation*

$$-\frac{\partial \ln Z(\beta)}{\partial \beta_k} = a_k \qquad (k = 1, \ldots, p). \tag{5.27}$$

Using (5.19), (5.24) and (5.25) we obtain

$$S(\pi_{\mathbf{A}}^{-1}(\mathbf{a})) = \ln Z(\beta) + \sum_{i=1}^{p} \beta_i a_i. \tag{5.28}$$

If we interpret $s(\rho)$ as a measure of uncertainty of the result of a measurement of ρ, or, using more traditional terminology, as a measure of chaoticity (disorderliness) of the statistical ensemble described by the density operator ρ, then the entropy $S(\pi_{\mathbf{A}}^{-1}(\mathbf{a}))$ is a measure of disorderliness of the statistical ensembles described by set $\pi_{\mathbf{A}}^{-1}(\mathbf{a})$ of density operators determined by mean values $a_1, \ldots, a_p$ of observables $A_1, \ldots, A_p$ ($p \leq n^2 - 1$). Therefore, the state $\rho^\star$ characterizes the most disorderly statistical ensemble in the class of the statistical ensembles described by elements of set π_A^{-1}. (Wichmann, [341] p. 888, calls it "the most chaotic ensemble determined by p ensemble averages".) Parameters β_i ($i = 1, \ldots, p$) are Lagrange multipliers (Lagrange coefficients) analoguous to the inverse temperature $\beta = 1/kT$ in the canonical distribution of statistical thermodynamics (Gibbs state). We obtain the latter in our formalism if $p = 1$ and $A = H = $ hamiltonian of the system. Therefore, we shall call the former parameters *generalized (inverse) temperatures* of the A_i-type or A_i-*temperature coefficients*. Etymologically, such a terminology is correct since the word "temperature" is derived from the Latin word *tempero, -are* which means "to abstain, be moderate, be indulgent; mix properly, regulate, govern" [344], p. 119, and not from the word "energy". This corresponds well to our case, where the maximum entropy gives the "proper mixing", "moderation" etc.

We remark that when $p = n^2 - 1$, i.e., when $\mathbf{A}$ is a complete set of observables, for any $\mathbf{a} \in \pi_A(\mathfrak{S})$ the sets $\pi_A^{-1}(\mathbf{a})$ are one-element sets, or $\rho^\star = \pi_A^{-1}(\mathbf{a})$. In this case, any state $\rho \in \mathfrak{S}$ can be represented in the form (5.25), together with (5.26) and (5.27), as a unique operator function of $n^2 - 1$ independent mean values $a_1, \ldots, a_{n^2-1}$ of observables $A_1, \ldots, A_{n^2-1}$ from the set $\mathbf{A}$.

From a property of trace, the definition of entropy as a trace (5.19), and the form of state evolution in a closed system, (5.8) and (5.9), it follows that

THEOREM 5.7. *In an isolated system, we have for $t \geq 0$*

$$s(\rho(t)) = s(\rho(0)), \tag{5.29}$$

where $\rho(t)$ is defined by (5.8).

Thus in isolated (closed, Hamiltonian) systems the information (entropy) of a statistical state (statistical ensemble) is preserved in time, as a simple consequence of the fact that the entropy is a trace of some operator. This theorem is, in general, no more true for open (non-Hamiltonian) systems which will be considered in the next chapter.

5.3. The principle of maximum entropy

Theorem 5.6 is a base for the *principle of maximum entropy* which, in quantum statistical physics (actually, so far only for $n < \infty$, i.e. for spin systems), is an information-theoretical estimation principle (decision rule). This principle was formulated independently in three papers: classically or implicitly for finite n by E. T. Jaynes [173], explicitly for infinite n by R. S. Ingarden and K. Urbanik [138], and finally explicitly for finite n by

E. H. Wichmann [341] (and later for infinite n by W. Ochs and W. Bayer [31], [243]). (In the second case, the paper by Jaynes was found by the authors after their paper was already written, but before the publication, so Jaynes' paper has been mentioned. We have to remark also that the mere principle that "the entropy is maximized under the subsidiary conditions that a set of ensemble averages $\langle A_i \rangle_{Av}$ assumes specified values" has been mentioned, without larger elaboration, by Peter Bergman yet in 1951 [38], p.1032, in connection with the problem of relativistic thermodynamics. The latter problem, however, is too complicated and too vague to be discussed here.)

The condition introduced in [137] which has to be added for the case of infinite dimensional separable Hilbert space is as follows: the set of observables $\mathbf{A} = (A_1, \ldots, A_p)$ has to be *thermodynamically regular* , i.e., such that there exist positive numbers $\beta = (\beta_1, \ldots, \beta_p)$ such that

$$Z(\beta) = Z(\beta_1, \ldots, \beta_p) = \operatorname{tr} \exp(-\beta \mathbf{A}) < \infty. \qquad (5.30)$$

In [137] this condition was formulated for $p = 1$ and $A = H$. Introduction of many observables, actually the powers of $H : H, H^2, \ldots, H^p$ (a generalization of the classical problem of moments, cf. [302]), has been done in the paper [140]. In general, operators $I, A_1, \ldots, A_p$ should be linearly independent, as we mentioned above, but not necessarily commuting among themselves. Only for the stationary case (which will be discussed in Sec. 5.5 below) we shall assume that all A_i commute with the Hamiltonian H

$$[A_i, H] = 0, \qquad (i = 1, \ldots, p), \qquad (5.31)$$

i.e., that they should be constants of motion, as $H, H^2, \ldots, H^p$.

Now we define some macroscopic concepts introduced first in [138]. Let $\mathbf{A}$ be an incomplete set of observables $A_1, \ldots, A_p$. We say that states $\rho_1, \rho_2 \in \mathfrak{S}$ are *equivalent with respect to* $\mathbf{A}$, and write $\rho_1 \sim_{\mathbf{A}} \rho_2$, if

$$\operatorname{tr}(\rho_1 A_j) = \operatorname{tr}(\rho_2 A_j) \qquad (j = 1, \ldots, p). \qquad (5.32)$$

Relation $\sim_{\mathbf{A}}$ is an equivalence relation and it divides $\mathfrak{S}$ into disjoint classes being closed and convex subsets of $\mathfrak{S}$. We shall call these classes $\mathbf{A}$-*macrostates* and denote them by $\Phi_{\mathbf{A}}, \Psi_{\mathbf{A}}$. The $\mathbf{A}$-macrostate containing state ρ will be denoted by $[\rho]_{\mathbf{A}}$

In the case when $\mathbf{A}$ is a complete set of observables (for $n < \infty$), all $\mathbf{A}$-macrostates are one-element sets, so they correspond bijectively to states.

The joint value of $\operatorname{tr}(\rho A_j)$ $(j = 1, \ldots, p)$ for all $\rho \in \Phi_{\mathbf{A}}$ is said to be the *value* $M_{A_j}(\Phi_{\mathbf{A}})$ *of observable* $A_j \in \mathbf{A}$ in $\mathbf{A}$-macrostate $\Phi_{\mathbf{A}}$. Putting

$a_j = M_{A_j}(\Phi_{\mathbf{A}})$ $(j = 1, \ldots, p)$, it is easy to see that $\Phi_{\mathbf{A}} = \pi^{-1}(\mathbf{a})$, where $\mathbf{a} = (a_1, \ldots, a_p)$, cf. the definition of π after formula (5.23).

The results of Sec. 5.2, in particular Theorem 5.6, can be formulated using the concepts of macrostate $\Phi_{\mathbf{A}}$ and the value $M_{A_j}(\Phi_{\mathbf{A}})$ of observable A_j in macrostate $\Phi_{\mathbf{A}}$ $(j = 1, \ldots, p)$.

The entropy $S(\pi_{\mathbf{A}}^{-1}(\mathbf{a})) = S(\Phi_{\mathbf{A}})$ is said to be the *entropy of macrostate* $\Phi_{\mathbf{A}}$, cf. (5.24), (5.28).

Now, let $\mathbf{A} = \{A_1, \ldots, A_p\}$ be an incomplete set of observables and $\Phi_{\mathbf{A}}$ — a given macrostate. The problem arises how to define the mean value of an arbitrary observable X in macrostate $\Phi_{\mathbf{A}}$? Otherwise, we have to define the mean value of X when we know the mean values $M_{A_1}(\Phi_{\mathbf{A}}), \ldots, M_{A_p}(\Phi_{\mathbf{A}})$ and nothing else. If

$$X = x_0 I + \mathbf{x}\mathbf{A} \tag{5.33}$$

with real constants x_0, $\mathbf{x} = (x_1, \ldots, x_p)$, then of course

$$M_X(\Phi_{\mathbf{A}}) = \mathrm{tr}(\rho X) = x_0 + \mathbf{x} M_{\mathbf{A}}(\Phi_{\mathbf{A}}), \tag{5.34}$$

where $\rho \in \Phi_{\mathbf{A}}$.

If, however, X is not a linear combination of $I, A_1, \ldots, A_p$, then the mean value of X in $\Phi_{\mathbf{A}}$ is not uniquely defined in this way since for each $\rho \in \Phi_{\mathbf{A}}$ we obtain, in general, a different value of $\mathrm{tr}(\rho X)$. In order to define in a unique way the (estimated) mean value of observable X in macrostate $\Phi_{\mathbf{A}}$ it is necessary to choose (decide) a unique state from the macrostate $\Phi_{\mathbf{A}}$. Theorem 5.4 gives us a constructive criterion for the unique choice of the state $\rho^\star \in \Phi_{\mathbf{A}}$ which describes the most chaotic (disorderly) state in this class of states (realizing the maximum of entropy under the given conditions). This state eliminates any bias (prejudice) towards any of the given observables. Such a method of choice will be called the *information-theoretic decision rule*.

We shall call the *(estimated) value* of X in $\Phi_{\mathbf{A}}$ the trace

$$M_X(\Phi_{\mathbf{A}}) = \mathrm{tr}(\rho^\star X), \tag{5.35}$$

where

$$\rho^\star = \rho^\star[M_{A_1}(\Phi_{\mathbf{A}}), \ldots, M_{A_p}(\Phi_{\mathbf{A}})] = Z^{-1}(\beta)\exp(-\beta\mathbf{A}). \tag{5.36}$$

The temperature vector $\beta = (\beta_1, \ldots, \beta p)$ is a unique function of mean values $M_{A_k}(\Phi_{\mathbf{A}})$ $(k = 1, \ldots, p)$ of observables $A_1, \ldots, A_p$ defined by the unique solution of the system of equations

$$-\frac{\partial \ln Z(\beta)}{\partial \beta_k} = M_{A_k}(\Phi_{\mathbf{A}}) \qquad (k = 1, \ldots, p), \tag{5.37}$$

where $Z(\beta)$ is defined by (5.26). Our formulae are valid both for finite n and for infinite-dimensional separable Hilbert space when $\mathbf{A}$ is thermodynamically regular (notice: p is always finite).

Since $\rho^* \in \Phi_{\mathbf{A}}$, in the case (5.33) the value $M_X(\Phi_{\mathbf{A}})$ of the operator X given by (5.35) coincides with (5.34).

A representative state ρ^* is an operator function of mean values $a_k = M_{A_k}(\Phi_{\mathbf{A}})$ $(k = 1, \ldots, p)$ of observables $A_1, \ldots, A_p$ of set $\mathbf{A}$. It follows that also the estimated value $M_X(\Phi_{\mathbf{A}})$ (5.35) of an arbitrary observable X is a function of mean values a_k $(k = 1, \ldots, p)$.

If $\mathbf{A}$ is a complete set of observables (for $n < \infty$), then the representative state ρ^* is identical with a unique element of macrostate $\Phi_{\mathbf{A}}$. In this case, the estimated value $M_X(\Phi_{\mathbf{A}})$ is the usual mean value of X in state ρ^* which is uniquely fixed by the mean values of the observables of the set $\mathbf{A}$.

5.4. Foundations of information thermodynamics

Let $\mathbf{A}^{(p)}$ and $\mathbf{B}^{(q)}$ be complete or incomplete sets of observables with p and q elements, respectively. Sets $\mathbf{A}^{(p)}$ and $\mathbf{B}^{(q)}$ are said to be *equivalent* if $p = q$ and there exists a linear nonsingular transformation such that

$$A_i = \sum_{k=1}^{p} a_{ik} B_k \qquad (i = 1, \ldots, p). \qquad (5.38)$$

We remark that all complete sets (for $n < \infty$) are equivalent.

We call a measurement of mean values of a complete (incomplete) set of observables a *microscopic (macroscopic) measurement*. If for $n < \infty$ we take a microscopic measurement as a base of quantum statistical mechanics, then it is independent of the choice of a complete set of observables. The trouble is, however, that usually $n = \infty = d$ or very large and a complete measurement is practically unattainable. Then we have to use a description relative to a macroscopic measurement based on incomplete set $\mathbf{A}$ or incomplete $\mathbf{A}$-*measurement*. This gives us *information thermodynamics based on* $\mathbf{A}$, or $\mathbf{A}$-*thermodynamics*, which essentially depends on $\mathbf{A}$ and is invariant only with respect to the transformation of equivalence (5.38). We have to resign then from a unique universal description independent of the kind of measurement. This is the key of the estimation method which is physically reliable if we fix some accuracy of measurement. Namely, if we measure mean values of more and more observables (e.g., higher and higher powers of the hamiltonian which gives statistical moments of energy) until we obtain no appreciable differences to the given accuracy between predictions by this method and direct measurements of the mean value of arbitrary observables, then our choice can be considered as sufficient. Thus practicability of the method can be established only by trial and error in each case.

Let $\mathbf{A} = \{A_1, \ldots, A_p\}$ be a fixed incomplete set of observables ($p < n^2 - 1$). The equivalence relation $\sim_\mathbf{A}$ divides $\mathfrak{S}$ into disjoint classes being $\mathbf{A}$-macrostates. By an $\mathbf{A}$-*macroscopic description* of a physical system we shall an assignment to each observable X its mean value in time $t \geq 0$ if for $t = 0$ the macrostate $\Phi_\mathbf{A}$ is known, i.e., the mean values $M_{A_1}(\Phi_\mathbf{A}), \ldots, M_{A_p}(\Phi_\mathbf{A})$ are given. In the frame of conventional statistical quantum mechanics it is impossible to solve this problem since for each $\rho \in \Phi_\mathbf{A}$ another, in general, mean value $\text{tr}(\rho(t)X)$ in state $\rho(t) = U_t^* \rho U_t$ for $t \geq 0$ is defined. This ambiguity can be, however, removed if we assume our decision rule assigning uniquely to every macrostate $\Phi_\mathbf{A}$ its reperesentative state $\rho^*(\Phi_\mathbf{A}) \in \Phi_\mathbf{A}$ which may be taken as the initial state of the system. Then we can uniquely define the estimated value of X for $t \geq 0$ by

$$\text{tr}(U_t^* \rho^*(\Phi_\mathbf{A}) U_t X).$$

Thus using as a decision rule the principle of maximum information explained in the previous section, we obtain a generalized thermodynamics called *information* $\mathbf{A}$-*thermodynamics*, cf. [138]. The generalization consists in a more general level of description than in the usual thermodynamics. The macroscopic point of view (measuring of mean values, not eigenvalues) is connected with the latter, but the number of the basic mean values is greater, in general. The maximization of information (entropy) gives the method the maximum possible objectivization under the existing conditions of measuring. Under some idealized conditions (e.g., when we discuss an ideal gas with infinite number of particles) we may have no real generalization because of the laws of large numbers (the thermodynamic limit). Then the problem reduces to that of only one temperature. Below we shall discuss the non-trivial examples when this idealization is not sufficient.

In closed (isolated) systems, for any observable X we can define its estimated value for every positive time t as

$$M_X^t(\Phi_\mathbf{A}) = \text{tr}(U_t^* \rho^* U_t X), \qquad (t \geq 0), \tag{5.39}$$

where ρ^* is the representative state for $\Phi_\mathbf{A}$ in $t = 0$, cf. (5.35), (5.37).

Of course, for $n < \infty$ information $\mathbf{A}$-thermodynamics goes over into conventional quantum statistical mechanics when $\mathbf{A}$ is a complete set of observables, since ρ^* is then the unique element of macrostate $\Phi_\mathbf{A}$.

5.5. Information thermodynamics of equilibrium

Now, we shall show that the Gibbs description of macroscopic equilibrium states is contained in information thermodynamics.

Using the notation of the previous sections, let **A** be an incomplete set of observables. We say that macrostate $\Phi_\mathbf{A}$ is *invariant in time* if for any $t \geq 0$ the following equality holds

$$\{\rho_t = U_t^* \rho U_t : \rho \in \Phi_\mathbf{A}\} = \Phi_\mathbf{A}. \tag{5.40}$$

Now, we have for closed systems

THEOREM 5.8. *All* **A**-*macrostates are invariant in time iff each operator* $A_i \in \mathbf{A}$ *(i = 1,..., p) commutes with the hamiltonian H of the system, cf. (5.31).*

The theorem is an extension of the analogous theorems of Urbanik [327] and Szafnicki [315] for macrostates composed of pure states, cf. [196].

Proof. We shall use the following definition of commutativity of observables: we say that observables A and B commute if their spectral measures E_A and E_B commute

$$[E_A(\epsilon), E_B(\epsilon')] = 0 \quad \forall \epsilon, \epsilon' \in B(\mathbb{R}^1), \tag{5.41}$$

where $B(\mathbb{R}^1)$ is the set of all Borel subsets of $\mathbb{R}^1$.

Let us assume that every operator $A_j \in \mathbf{A}$ $(j = 1,\ldots,p)$ commutes with hamiltonian H of the system. Then for every Borel set ϵ and any $t \geq 0$ spectral measure $E_{A_j}(\epsilon)$ $(j = 1, \ldots, p)$ commutes with operators U and U^*, cf. [295]. Hence it follows, cf. (5.4), (5.5),

$$\text{tr}(\rho(t)A_j) = \int_{-\infty}^{+\infty} \lambda \text{tr}(\rho(0)U_t E_{A_j}(d\lambda)U_t^*) \tag{5.42}$$

$$= \int_{-\infty}^{+\infty} \text{tr}(\rho(0)E_{A_j}(d\lambda)) = \text{tr}(\rho(0)A_j) \quad (j = 1,\ldots,p).$$

From (5.42) it follows that $\rho(t) \sim_\mathbf{A} \rho(0)$ for each $t \geq 0$ and $\rho(0) \in \Phi_\mathbf{A}$. Therefore, all **A**-macrostates are invariant in time.

Now let us assume that all **A**-macrostates are invariant in time. Let $\lambda_1 < \lambda_2 < \ldots < \lambda_m$ $(m \leq n)$ be a sequence of eigenvalues of observable A_1, and let $E_1, E_2, \ldots, E_m$ be a sequence of projection operators on the eigensubspaces $\mathcal{H}_1, \mathcal{H}_2, \ldots, \mathcal{H}_m$ corresponding to the particular eigenvalues. It is easy to check that equality $\text{tr}(\rho A_1) = \lambda_1$ is satisfied iff $\rho \in \mathfrak{S}_1$, where

$$\mathfrak{S}_1 = \{\rho \in \mathfrak{S} : E_1 \rho = \rho E_1 = \rho\}. \tag{5.43}$$

Since all **A**-macrostates are invariant in time, also the set-theoretical sum of all **A**-macrostates for which $M_{A_1}(\Phi_\mathbf{A}) = \lambda_1$, i.e. set $\mathfrak{S}_1$, is invariant in time. It follows that the operator E_1 commutes with U_t.

Let us consider operators A_1 and U_t in the subspace $\mathcal{H}_1^\perp$ (orthogonal to $\mathcal{H}_1$). Analogously we get $E_2 U_t = U_t E_2$ and, by iteration of this procedure,

$$[E_k, U_t] = 0 \qquad (k = 1, \ldots, m). \tag{5.44}$$

Therefore,

$$[A_1, H] = 0. \tag{5.45}$$

Repeating this reasoning for $i = 2, \ldots, p$ we obtain (5.31) which proves our theorem. $\square$

A-macrostates invariant in time will be called *equilibrium* A-*macrostates*.

Let $\mathbf{H} = \{H_1, \ldots, H_p\}$ be an incomplete set of observables such that

$$[H_j, H] = 0 \qquad (j = 1, \ldots, p). \tag{5.46}$$

Then each macrostate $\Phi_\mathbf{H}$ is invariant in time and the representative state $\rho^\star$ for the macrostate $\Phi_\mathbf{H}$ is a stationary state, i.e.,

$$\rho^\star(t) = U_t^* \rho^\star U_t = \rho^\star \qquad (t \geq 0), \tag{5.47}$$

where $\rho^\star$ has the form as in (5.25),

$$\rho^\star = \rho^\star(M_{H_1}(\Phi_\mathbf{H}), \ldots, M_{H_p}(\Phi_\mathbf{H})) = Z^{-1}(\beta_1, \ldots, \beta_p) \exp(-\sum_{j=1}^p \beta_j H_j),$$

$$\tag{5.48}$$

where

$$Z(\beta_1, \ldots, \beta_p) = \operatorname{tr} \exp(-\sum_{j=1}^p \beta_j H_j). \tag{5.49}$$

Lagrange parameters (*generalized inverse temperatures*) $(\beta_1, \ldots, \beta_p)$ are uniquely determined by the equations

$$-\frac{\partial \ln Z(\beta_1, \ldots, \beta_p)}{\partial \beta_j} = M_{H_j}(\Phi_\mathbf{H}) \qquad (j = 1, \ldots, p). \tag{5.50}$$

On the base of (5.39) the estimated value $M_X^t(\Phi_\mathbf{H})$ of an arbitrary observabe X in time $t \geq 0$ on macrostate $\Phi_\mathbf{H}$ does not depend on time and is given by

$$M_X^t(\Phi_\mathbf{H}) = \operatorname{tr}(\rho^\star X), \tag{5.51}$$

where $\rho^\star$ is given by (5.48), (5.49) and (5.50).

Since state $\rho^\star$ is an operator function of mean values of observables $H_1, \ldots, H_p$, also the estimated value $M_X^t(\Phi_\mathbf{H})$ of observable X in macrostate Φ_H and entropy $S(\Phi_H)$ of macrostate Φ_H are functions of mean values $M_{H_1}(\Phi_H), \ldots, M_{H_p}(\Phi_H)$.

The expression for $S(\Phi_H)$ has the form

$$S(\Phi_H) = -\mathrm{tr}(\rho^\star \ln \rho^\star) = \ln Z(\beta_1, \ldots, \beta_p) + \sum_{j=1}^{p} \beta_j M_{H_j}(\Phi_H), \qquad (5.52)$$

where $Z(\beta_1, \ldots, \beta_p)$ and β_j $(j = 1, \ldots, p)$ are defined by (5.49) and (5.50).

If H contains only one element, $p = 1$, $H_1 = H$, where H is the hamiltonian of the system, then the representative state $\rho^\star$ is identical with the well-known *canonical distribution (canonical state, Gibbs state)* and relations (5.49), (5.50), (5.51) and (5.52) are the well-known relations of conventional statistical thermodynamics of equilibrium states. Usually, the method of obtaining these formulae is more or less heuristic, as by using the symmetry arguments (Maxwell), etc., but the method of information theory proposed first by Jaynes seems to be the simplest and the most reasonable one. The subtle physical points of this reasoning will be discussed in the next section, together with the problem of the possibility of new physical applications of the case $p > 1$. Actually, two simple applications of the latter type $(p = 2)$ are already well-known in statistical physics. Namely, for $H_2 = N$ (number of particles) we obtain the *great canonical distribution (state, ensemble)*, and for $H_2 = V$ (volume) — the Boguslavski distribution of statistical thermodynamics (known also as (P, T)-distribution, where P is the pressure and T the temperature). The question is if also other cases with higher p, especially those proposed by Ingarden [140], [141], [142], [144], [148], i.e., $H_i = H^i$ $(i = 1, \ldots, p)$, have interesting physical applications. We will discuss this problem in the following sections.

5.6. Physical discussion and Q-entropy

When we imagine a macroscopic measurement, we actually have in mind a measurement of some mean values of microscopical quantities. Indeed, according to the Ehrenfest principle the macroscopic quantities are averages of the corresponding microscopical ones. Thinking about mean values we have to think about the probability distribution, classical or quantum, occurring in this experiment, let us call this distribution ρ. But which ρ has to be selected when we know only the mean values? It is clear that at first we can only fix the *set* of all possible distributions (statistical states) occurring in the given case, the set called by us a macrostate. The method of maximum entropy selects in this set a unique representative state $\rho^\star$ depending only on the given mean values. Thus, the difference between the unknown "true" ρ and the known, although only estimated, $\rho^\star$ is the clue of the new method. As we said, in older times ρ was heuristically postulated or guessed as the Gibbs state or the grand canonical state, and we now see that in most cases this guesses are correct from the point of view of the new method. The cases, rather unusual but important, in which the best estimated state can deviate from the mentioned traditional solutions will be discussed below.

Now we would like to discuss briefly the difference between the classical and the quantum approaches to statistics. Classical mechanics is a deterministic non-statistical (categorical or indicative) causal theory, while quantum mechanics is an indeterministic and statistical, therefore modal theory showing a more general type of causality called by M. Born and W. Pauli statistical causality (causality of the probability itself). Pauli in his letter to E. Schrödinger of 9 July 1935, cf. [47], p. 229, characterized the difference between the two theories in the following way:

> One cannot, however, — as the conservative, old gentlemen wish — declare the statistical results of quantum mechanics as being *correct* and *nevertheless* base this on a hidden variable causal mechanism. In this sense the system of quantum-mechanical laws appears to be logically closed (complete in the sense of axiomatics) — in contrast to kinetic gas theory.

In quantum mechanics even pure states are statistical with respect to any fixed representation Q, i.e., a Hilbert space base defined by a complete system of commuting observables. Since, contrary to the classical case, not all observables commute, we have for the noncommuting ones the phenomenon of uncertainty relations and complementarity unknown in classical physics. This is just the clue generalization: the possibility of rotation in the Hilbert space of an isolated system, and the fact that the states are then invariant with respect to this rotation (the group of unitary transformations). (In open systems the representation is fixed by the environment and the unitary invariance is cancelled.) In each fixed representation the probability refers to a definite point item given by the eigenvalues of the respective operators, e.g., positions (q) or momenta (p), but not both as in classical physics. Pauli wrote on this situation (in the letter to W. Heisenberg of October 9, 1926, cf. [211], p. 158),

> One can look at the world either with the p-eye or one can look at it with the q-eye, but if you will simultaneously open both eyes, you get lost.

Actually, a pure state has something much stronger than only a distribution of probability $\rho(Q)$ for any representation Q: it has a complex *amplitude of probability* $\Psi(Q) = |Q\rangle$ such that $\rho(Q) = |\Psi(Q)|^2$. (p and q are special cases of Q). As is well-known, $\Psi(Q)$ can describe interference and diffraction phenomena, i.e., wave properties of states while eigenvalues of Q, in particular, coordinates p and q describe particle properties (they can be arbitrarily sharply measured, although not simultaneously, by means of the so-called *squeezed states* which are the eigenstates of Q).

Each representation Q is equivalent with a maximal (most subtle) partition of unit I in Hilbert space of the system, so we can write

$$Q = \{P_i^Q\}, \tag{5.53}$$

where P_i^Q (in general, $i = 1, 2, \ldots$) are all one-dimensional (mutually orthogonal) projectors belonging to base Q. Then for any state ρ (pure or mixed) and any representation Q we can define their *Q-entropy* as

$$S_Q(\rho) = - \sum_{i=1}^{\infty} \mathrm{tr}(\rho P_i^Q) \ln(\mathrm{tr}(\rho P_i^Q)) = S(\sum_{i=1}^{\infty} P_i^Q \rho P_i^Q). \tag{5.54}$$

Analogously and more generally, for any (not necessarily nondegenerate), thermodynamically regular observable $A = \sum_{i=1}^{\infty} a_i P_i^A$ we have

$$S_A(\rho) = - \sum_{i=1}^{\infty} \mathrm{tr}(\rho P_i^A) \ln(\mathrm{tr}\rho P_i^A). \tag{5.55}$$

This concept has been first explicitly introduced for a pure state and a thermodynamically regular observable A under the name of "*A*-entropy" by R. S. Ingarden and K. Urbanik in [138], for mixed states cf. [156]. Wehrl called it *Ingarden-Urbanik entropy* or *IU-entropy* in [337] and denoted by S_{IU}. He remarked, however, that

This concept in fact appeared very early, namely in the papers of the Ehrenfests (1911), Pauli (1928), and Von Neumann (1929), but was intensively studied in the 1960s.

Direct inspection of the mentioned papers did not show any explicit definition of this concept, which is not strange when we recall that quantum mechanics was introduced only in 1925 and density matrix was defined by Dirac only at the end of 1928. It is possible, however, that some implicit ideas near to this concept can be found in these and other papers and books, cf. e.g. formula (*) in Von Neumann's book [333], p. 380 of Engl. transl., which is near to the so-called Klein's inequality [183] p. 773, called by Ruelle, the "convexity inequality" [296], Sec. 2.5.2, (in a generalized sense), cf. also Uhlmann [321]. Detail investigation of *Q*-entropy was done by Wehrl [336], Staszewski [308] and Grabowski [111], [112]. Wehrl showed that "most of the properties of the usual entropy remain valid *cum grano salis*" and in [337] he added:

IU-entropy has (of course, besides invariance) many properties in common with classical discrete entropy, for instance concavity, additivity, and subadditivity (the latter ones in some appropriate sense).

Staszewski (who introduced the term *entropy of measurement* for Q-entropy) used Klein's inequality (which in this formulation was contained earlier in [156])

$$S_Q(\rho) \geq S(\rho) \qquad (5.56)$$

and proved the "characterization" of Von Neumann's entropy by

$$S(\rho) = \liminf_{(Q,\rho) \in Q_\rho} S_Q(\rho), \qquad (5.57)$$

where Q_ρ is the set of measurements of all observables of finite degenaration (not necessarily equal to 1) in the state ρ, and the lim inf is understood in the sense of Staszewski's relation "more informative" generalizing the concepts of Hardy, Littlewood and Polya (classical case) [126] and Uhlmann (quantum case) [321]. On the other hand if Q is the eigen-representation of ρ, $Q(\rho)$, we may write

$$S(\rho) = S_{Q(\rho)}(\rho). \qquad (5.58)$$

Grabowski defined A-entropy for spectrally absolutely continuous observables. He proved that "this kind of entropy increases after measurement of an observable with a continous spectrum in the sense of Von Neumann and assumes its maximum on a Gaussian state."

Since $S_Q(\rho)$ depends in general on Q, it is not unitary invariant, Q-entropy is not a geometrical concept in Hilbert space and therefore it has no covariant physical meaning in the theory of isolated quantum systems (it is neither a Hilbert scalar nor a Hilbert vector or tensor). It expresses only a relation between Q and ρ connected with a particular type of measurement, a concept typical for quantum mechanics. For theory of isolated systems, Q-entropy is therefore of secondary importance with respect to the Von Neumann entropy which is unitary invariant and is a unitary scalar. For open systems, however, where the representation of the environment is distinguished, Q-entropy has the first-hand importance. In isolated systems, in problems of maximization of entropy with commuting observables it is irrelevant if we use Von Neumann's entropy or Q-entropy for the eigen-representation of our observables. When, however, the observables do not commute, we have to use by maximization only Von Neumann's entropy as we did in the above presentation in this chapter.

We have to explain yet how we can measure mean values of noncommuting observables. In this case we have in mind the so-called *generalized measurement*, i.e., such that we measure each of the two noncommuting observables on a different copy of the same system (i.e., a system of the same type prepared in the same way or being in the same state). In this case, the respective *generalized Hilbert space* is a tensor product of the Hilbert spaces of each copy (a tensor square of the original Hilbert space). This construction corresponds to the old idea of an "ensemble" of systems where the elements of the ensemble are similar but independent copies of the same system. In the generalized Hilbert space the corresponding operators commute, while they do not commute if treated as operators on the original Hilbert space.

Finally, we have to remark that in the two independent axiomatic approaches to entropy in isolated quantum systems, namely, by Ingarden and Kossakowski [147] and by Ochs [244], the result is Von Neumann entropy and not Q-entropy. This confirms our above statement that only Von Neumann entropy has an invariant physical meaning (independent of representation) in isolated quantum systems, while Q-entropy, as follows from its name, characterizes information in a measurement of state in a given representation Q (and only when Q is the eigen-representation of the state both entropies coincide).

5.7. Self-organization and biology

It was suspected for a long time that for the description of biological systems, regarded as highly structurized complex systems, some new kind of generalized thermodynamics, something as "information dynamics" was needed. Great progress of statistical and phenomenological thermodynamics in the last decades, especially due to synergetics of H. Haken, cf. [122], [123], [124], and the dissipative structures of I. Prigogine and his school, cf. [282] [283] [242], has opened some new ways in this direction, i.e., to the physical explanation of self-organization of complex systems. Especially, in the book of Haken [124], the connection of synergetics and self-organization with information-theoretical approach to thermodynamics has been explicitly shown and developed. It is out of the scope of the present book to present the ideas of Haken and Prigogine in detail, we only briefly mention that these investigations have shown that besides the usual thermodynamic equilibrium, characterized by uniform density, temperature, pressure, concentrations etc., there exists a vast domain of "higher order equilibria", where all these local properties may be nonuniform. There are many types of the phase transitions between the "higher phases" called by mathematicians "bifurcations" or "catastrophes". The phases have, in general, a very complicated structure, as the so-called Benard cells or the Belousov-Zhabotinsky structures, cf. [124] p. 9. The phenomena are connected with nonlinear interactions ("synergons" or "collective interactions" acting according to the "slaving principle" in Haken's terminology), so they may be also connected with nonlinear dynamics and chaos theory. From the point of view of Prigogine, the new phases are "very far from equilibrium" (and therefore are called dissipatative structures) and represent "order out of chaos". All macroscopic phenomena, however, which are not transient but stationary, can be called equilibria or "generalized equilibria", since thermodyamics has no means of distiguishing between "true equilibria" and stationary states. We have to remember that even the classical equilibrium of a gas is actually a very complicated mechanical chaotic process which only macroscopically and statistically is stationary, representing also an "order out of chaos". So the adjective "dissipative" in "dissipative structures" seems to be unnecessary.

It is clear that the mentioned "higher order" equilibria can be also viewed from the standpoint of *higher-order thermodynamics*, i.e., our method of maximization of entropy with respect to many observables, e.g., higher powers of energy. Of course, such a theory is half-phenomenological since it has empirical macroscopical constants as higher-order temperatures. The usual thermodynamics, as a special case, has also such empirical macroscopical constants of temperature, chemical potentials, etc. Our generalized temperatures are global, not local (sometimes they are intensive quantities, sometimes not, but always global), but in general they are multiple. The respective probability distribution may have many maxima and minima, so they may discribe, in principle, such phenomena as Benard's cells (in the case of appropriate pumping). The information-theoretic method, however, does not describe the detailed mechanism of the phenomenon, it gives only its macroscopic or mesoscopic effect, as modal estimation. The higher-order temperatures have to be measured by the appropriate *higher-order thermometers*, i.e., macroscopic or mesoscopic bodies in contact with the system, and changing (shape or other macroscopic property) in a measurable way in response to changing temperatures. In higher equilibria the temperatures of two contacting bodies are equal if we use cumulants as statistical moments, cf. [144] [145], but to the fourth order exclusively the cumulants are equal to the central moments. For the fourth order cumulant U_4 the connection with the central moments μ_i is yet simple: $U_4 = \mu_4 - 3\mu_2^2$, cf.[143], Eq. (16). We can use cumulants as the basic moments, but in the case of total energy, for physical reasons explained below, it will be better to start from central moments. The reason of the exceptional position of cumulants (called also semi-invariant moments) is that they are additive (extensive) quantities, so their temperatures are intensive ones, and then the extended so-called Zeroth Principle of Thermodynamics is valid, cf. [277] and [146]. For other moments the Zeroth Principle is not true, but always there exist definite functional relations between moments and temperatures of different types, cf. [146].

There are 5 papers from Toruń connected with the information-theoretical approach to self-organizing systems: [150], [151], [152], [158], and [160]. The first three papers concern a laser model, and the last two discuss biological models. Since in the laser papers the concept of an open system is implicitly used (optical pumping), and the topic is rather complicated, we resign from presentation them here. Here we shall briefly discuss only the results of the fourth paper about a biological model. (The fifth paper uses also some other methods, those of information nets, so it does not fit in with the present Section).

One of the most mysterious effects in living eucariothic cells (with a distinguished nucleus) of plants and animals is a macroscopic or mesoscopic circular motion (cyclosis) of cytoplasm along the walls of the cell (in plants) or along the membranes of endoplasmatic reticulum (in animals), cf., e.g., [21], especially papers by T. Hayashi and R.I. Goldacre. It seems that the direction of motion is randomly unstable (the flip-flop effect) and changes occasionally, especially in animals, where a kind of irregular pulse motion is therefore observed. It is apparent that such flows, being a means of transportation of molecules and atoms, are necessary for metabolic and reproductory chemical reactions in the cell. It can be easily seen that for this transport the diffusion phenomena are not sufficient, as is the case in the much smaller procariotic cells (bacteria without differentiated nucleus). E.g., the diffusion transport along a usual animal cell would last ca. 12 seconds, and in an especially large plant cell of algae *Chara brauni* (length ca. 2 cm) ca. 23 days, cf. [158]. Therefore, it seems that some special electro-molecular mechanism, maybe of similar type as muscle contraction [316], engaging microfilaments and energy produced by mitochondria, is responsible for the macroscopically or mesoscopically (in microscope) observed cyclosis. It seems that so far the phenomenon is not yet explained sufficiently by molecular biology. But for macroscopic (or mesoscopic) description we do not need necessarily the detailed microscopic mechanism (as detailed description of a collision processe is not contained in the derivation of statistical thermodynamics of ideal gases). It is sufficient to have macroscopic measurements of some mean values, at least theoretically. In [158], the probability distribution $\rho(v)$ of (scalar) velocity v ($-\infty < v < +\infty$) of a particle in the border layer of a cell (with respect to the cell wall) has been considered. Assuming that the 2nd and the 4th statistical initial (power) moments of v, U_2 and U_4, are given, the maximum entropy principle gives the result

$$\rho(v) = Z^{-1}(\beta_2, \beta_4) \exp(-\beta_2 v^2 - \beta_4 v^4), \qquad (5.59)$$

$$U_2 = -\frac{\partial \ln Z}{\partial \beta_2}, \quad U_4 = -\frac{\partial \ln Z}{\partial \beta_4}, \qquad (5.60)$$

$$Z(\beta_2, \beta_4) = \sqrt{\pi} \beta_4^{-1/4} H_{-1/2}\left(\frac{\beta_2}{2\sqrt{\beta_4}}\right), \qquad (5.61)$$

where either $\beta_4 > 0$, or $\beta_4 = 0$ and $\beta_2 > 0$. $H_p(z)$ is a Hermite function of the first kind which is connected with the Weber parabolic cylinder function

$$H_p(z) = 2^{p/2} \exp\left(\frac{z^2}{2}\right) D_p(\sqrt{2}z), \qquad (5.62)$$

and for real x and $p = -1/2$ can be presented by cylinder functions

$$H_{-1/2}(x) = \begin{cases} \sqrt{\frac{x}{2\pi}} \exp\left(\frac{x^2}{2}\right) K_{1/4}\left(\frac{x^2}{2}\right) & \text{if } x > 0, \\ \sqrt{\frac{-\pi x}{8}} \exp\left(\frac{x^2}{2}\right) T_{1/4}\left(\frac{x^2}{2}\right) & \text{if } x < 0, \end{cases} \qquad (5.63)$$

where in general

$$K_p(z) = \frac{\pi}{2} \frac{I_{-p}(z) - I_p(z)}{\sin(p\pi)}, \quad |\arg z| < \pi, \quad p \neq 0, \pm 1, \pm 2, \ldots, \quad (5.64)$$

$$T_p(z) = \frac{I_{-p} + I_p}{\cos(p\pi)}, \quad |\arg z| < \pi, \quad p \neq \pm\frac{1}{2}, \pm\frac{3}{2}, \pm\frac{5}{2}, \ldots, \quad (5.65)$$

where $I_p(z)$ is a modified Bessel function of the first kind, and $K_p(z)$ is a MacDonald or Kelvin function, well-known in physics. For us it is important that $H_{-1/2}(x)$ is assymmetric with respect to the change of sign of x (this asymmetry has been overlooked in some tables and papers), i.e., in our case with respect to β_2, which corresponds to the difference between one maximum or two maxima of $\rho(v)$. For $x = 0$ or $\beta_2 = 0$ we have just a bifurcation point or a phase transition, and therefore this value of x is omitted in the analytical formulae (5.63). By differentiating $\rho(v)$ (5.59) and equating the result to zero we can easily find the possible extrema (minima or maxima) of this function

$$v = 0, \pm\sqrt{\frac{-\beta_2}{2\beta_4}}, \qquad (5.66)$$

where one minimum (for $v = 0$) and two real maxima can occur only for $\beta_2 < 0$ and $\beta_4 > 0$, while the case $\beta_2 > 0$, $\beta_4 = 0$ gives only one maximum (for $v = 0$). The case $\beta_4 < 0$ is impossible (because of the non-normalizability of $\rho(v)$), while for $\beta_2 > 0$, $\beta_4 = 0$ we have only one maximum for $v = 0$ (the Gaussian distribution). The half-line $\beta_2 = 0$, $\beta_4 > 0$ on the "phase half-surface" $\{-\infty < \beta_2 < +\infty, \ \beta_4 > 0\} \cup \{\beta_2 > 0), \ \beta_4 = 0\}$ is the bifurcation line.

Thus we obtained a qualitative correspondence of our simplified model with the biological phenomenon of the motion bifurcation with two maximal velocities in opposite directions. The Gaussian motion and the one-maximum distribution with $\beta_2 > 0$, $\beta_4 > 0$ correspond to the "death case" of the diffusion-type motions, while the two-maxima cases with $\beta_2 < 0$, $\beta_4 > 0$ are the "living cases" of cyclosis.

We have seen that an explicit analytic calculation of the Z-function (the partition function or the statistical sum) is possible also for two temperature coefficients (in our notation they are the second and the fourth-order coefficients, from the point of view of energy which is proportional to v^2 they are of the first and of the second order).

Similar calculations can be done for studying the shape of spectral lines [279], the problem of polymolecularity of polymers [61], the conformations of polymers [223], the problem of pattern (shape) recognition [161], etc. In all these cases we have to be able to perform calculations of the partition function Z and the Lagrange (temperature) coefficients β_i. In the next section we shall discuss the problem of these calculations.

5.8. Evaluation of the partition function

The simplest systems for the calculation of partition function are classical systems since then infinite sums reduce to integrals (in infinite domains) which, as a rule, are much easier for evaluation. The examples of a complete classical calculation will be given in the next section. Now we discuss only the partition function. The quantum case will be discussed at the end of this section. We first discuss the approximation method of calculating classical partition function and temperatures due to Czajkowski [62]. He assumed that the system is one-dimensional (as in the previous section, so the number f of degrees of freedom is $1/2$, just as a mathematical example which, however, for the ideal gas and solid state can be easily generalized to any positive integer f), and the argument x of the probability distribution varies over a real nonempty interval $X = [a, b]$, (possibly $a = -\infty$ or/and $b = +\infty$). Then our problem reduces to solving the following system of equations with respect to $\beta_0 = \ln Z, \beta_1, ..., \beta_p$:

$$a_i = \int_a^b f_i(x) \exp(-\sum_{k=0}^p \beta_i f_i) dx \quad (i = 0, 1, ..., p), \qquad (5.67)$$

where $a_0 = 1$, $f_0(x) \equiv 1$ denotes the normalization of probability, and $f_1(x), ..., f_p(x)$ represent classical observables. Then we choose a set of orthonormal polynomials $Q_j(x)$, $j = 0, 1, 2, ...$, on the interval X with respect to the appropriate weight function $w(x)$. We introduce the polynomial

$$P(x) = x^{p+1} + c_1 x^p + ... + c_{p+1}, \qquad (5.68)$$

which differs from Q_{p+1} only by a multiplicative constant. Then we have the approximate quadrature formula, cf. [204] Chap.7,

$$\int_X g(x)w(x)\,dx = \sum_{k=1}^{p+1} C_k g(x_k) + R(g), \tag{5.69}$$

where x_k are zeros of the polynomial $Q_{p+1}(x)$, and

$$C_k = \frac{c_{p+1}}{c_p}\frac{1}{Q'_{p+1}(x_k)Q_p(x_{k-1})}, \tag{5.70}$$

$$R(g) = \frac{g^{(2p+2)}(y)}{(2p+2)!}\int_X w(x)P^2(x)dx, \quad y \in X. \tag{5.71}$$

Applying eq. (5.69) to all equations (5.67), neglecting the remainder $R(g)$, and assuming that $w(x) \neq 0$ for all $x \in X$, we obtain the approximate system of linear equations

$$a_i = \sum_{k=1}^{p+1} A_{ik}v_k, \quad i = 0, 1, \ldots, p, \tag{5.72}$$

where

$$A_{ik} = C_k\frac{f_i(x_k)}{w(x_k)}, \quad i = 0, 1, \ldots, p, \quad k = 1, \ldots, p+1, \tag{5.73}$$

$$v_k = \exp(-\sum_{i=0}^{p} \beta_i f_i(x_k)). \tag{5.74}$$

Assuming that $\det(A_{ik}) \neq 0$ (probably this follows from the linear independence of functions $1, f_1, \ldots, f_p$) we can solve this system of linear equations with respect to v_k. If all $v_k > 0$ (this is a condition for the values of a_i: they have to be mutually consistent as in the general problem of moments), we can define

$$b_k = -\ln v_k, \quad k = 1, \ldots, p+1 \tag{5.75}$$

as real quantities, and then solve the second system of linear equations

$$b_k = \sum_{i=0}^{p} \beta_i f_i(x_k), \quad k = 1, \ldots, p+1 \tag{5.76}$$

with respect to the generalized temperatures β_i. The condition of solubility $\det f_i(x_k) \neq 0$ follows from $\det A_{ik} \neq 0$. Thus, the problem of the approximate evaluation of partition function and temperatures is solved. The values of the zeros of the standard orthogonal polynomials and the coefficients $C_k = C_k^{p+1}$ are tabulated with high accuracy, and can be found, e.g., in [204].

In [62], Czajkowski checked the accuracy of his method on the example of a two-moment problem in which exact solution can be easily found. He assumed $X = \mathbb{R}^1$, and

$$\Phi = \Phi_{a_1, a_2} = \left\{ \rho \in \mathfrak{S}(\mathbb{R}^1) : \int_{-\infty}^{+\infty} x^j \rho(x)\, dx = a_j, \quad j = 1, 2 \right\}. \qquad (5.77)$$

Denoting

$$a_1 = a, \quad a_2 = a^2 + \sigma^2, \qquad (5.78)$$

we obtain for the exact solution the Gaussian distribution with the temperatures

$$\beta_0 = \ln Z = \frac{a^2}{2\sigma^2} + \frac{1}{2}\ln \pi + \frac{1}{2}\ln 2\sigma^2, \qquad (5.79)$$

$$\beta_1 = -\frac{a^2}{\sigma^2}, \qquad \beta_2 = \frac{1}{2\sigma^2}. \qquad (5.80)$$

Using the Hermite polynomial $H_3(x)$ Czajkowski found the following formulae for the approximate solution ($c = x_3 = -x_1 = 1.225$):

$$\beta_0 = 0.573 - \ln(c^2 - a^2 - \sigma^2), \qquad (5.81)$$

$$\beta_1 = \frac{1}{2c}\ln\frac{a^2 - ca + \sigma^2}{a^2 + ca + \sigma^2}, \qquad (5.82)$$

$$\beta_2 = \frac{1.61 + 2\ln(c^2 - a^2 - \sigma^2) - \ln(a^2 - ca + \sigma^2) - \ln(a^2 + ca + \sigma^2)}{2c^2} > 0, \qquad (5.83)$$

where a and σ should be such that all β_i are real and β^2 is positive. Czajkowski chose special values of this type for a and σ and computed numerically the values of β_i, $i = 0, 1, 2$. Then he compared the latter with the corresponding values for the exact solution. The results are given in the table on the following page.

We see that Czajkowski's approximate method is quite general, but the exactness of evaluation is not always high (in the given example the absolute value of the relative error varies between 0 and 56 percent). Therefore, it is interesting to study the possibility of analytical rigorous solutions at least for the partition function Z which is the most important function in statistical thermodynamics. Such studies have been performed for different, more or less general cases by: 1) R.S. Ingarden [149] (for special classical

	Exact solutions	Approximate solutions	Relative error in %
$a = 0,\ \sigma^2 = 0.5$			
β_0	0.573	0.572	-0.2
β_1	0	0	0.0
β_2	1	0.999	-0.1
$a = 0,\ \sigma^2 = 0.25$			
β_0	0.225	0.350	55.6
β_1	0	0	0.0
β_2	2	1.61	-19.5
$a = 0,\ \sigma^2 = 0.75$			
β_0	0.775	0.861	11.1
β_1	0	0	0.0
β_2	0.667	0.536	-19.6
$a = 0.5,\ \sigma^2 = 0.5$			
β_0	0.822	0.861	4.7
β_1	-1	-0.940	-6.0
β_2	1	0.904	-9.6

TABLE 5.1.

cases of the electromagnetic radiation of one mode), 2) G.Z. Czajkowski [61], [63] (for rather general classical cases occurring in polymolecularity of polymers), 3) R. Antoniewicz [23] (for the quantum case of the second order coherence of electromagnetic field), and 4) W. Jaworski and R.S. Ingarden [166] (for typical classical situations of rather general type).

Here we shall discuss Czajkowski's ideas from [62] and the results of [166] and [23]. It is clear that in the cases with many temperatures one can expect the partition function to be a special function of many variables. Since up to now, as a rule, only one-variable special functions are elaborated and tabulated, we need definitions of some new types of special functions. In [63], Czajkowski introduced the *generalized gamma functions* only in the real domain of the arguments and for parameter $s = 1, 2, \ldots$. In [166] these functions have been formulated (in slightly different notation) in a more general form as complex functions of complex arguments and with complex parameter s as follows:

$$G_{p,s}(z_1, \ldots, z_p) = \int_0^\infty t^{s-1} \exp\left(-\sum_{i=1}^p z_i t^i\right) dt,$$

$$\Re s > 0, \quad \Re z^p > 0, \quad p = 1, 2, \ldots \tag{5.84}$$

(This is done from the mathematical point of view. Actually, in our physical applications we are interested only in the real domain of all z_i and for $s = 1, 2, \ldots$.) In particular, we obtain Euler's gamma function

$$G_{1,s}(1) = \Gamma(s) = \int_0^\infty t^{s-1} e^{-t} dt = (s-1)! = \Pi(s-1), \quad \Re s > 0, \quad (5.85)$$

and

$$G_{1,s}(z) = \frac{\Gamma(s)}{z^s} = \int_0^\infty t^{s-1} e^{-zt} dt, \tag{5.86}$$

$$G_{2,1}(0, z) = \frac{1}{2}\sqrt{\frac{\pi}{z}} = \int_0^\infty e^{-zt^2} dt, \tag{5.87}$$

$$G_{2,2n+1}(0, z) = \frac{1 \cdot 3 \cdot 5 \cdots (2n-1)}{(2z)^n} \frac{1}{2}\sqrt{\frac{\pi}{z}}$$

$$= \int_0^\infty t^{2n} e^{-zt^2} dt, \quad n = 1, 2, \ldots, \tag{5.88}$$

$$G_{2,2n}(0, z) = \frac{(n-1)!}{2z^n} = \int_0^\infty t^{2n+1} e^{-zt^2} dt, \tag{5.89}$$

$$G_{p,s}(0, \ldots, 0, z_p = 1) = \frac{1}{p}\Gamma\left(\frac{s-1}{p}\right) = \int_0^\infty t^{s-1} e^{-t^p} dt, \tag{5.90}$$

$$G_{2n,s}(0, \ldots, 0, z_n = z, 0, \ldots, 0, z_{2n} = 1) = \tag{5.91}$$

$$= 2^{-\frac{s}{2n}} \Gamma\left(\frac{s-2}{n}\right) \exp\left(\frac{z^2}{8}\right) D_{s/n}\left(\frac{z}{\sqrt{2}}\right) = \int_0^\infty t^{s-1} \exp(-zt^n - t^{2n}) dt.$$

Derivatives of functions $G_{p,s}(z_1, \ldots, z_p)$ with respect to z_i can be easily calculated from their definition, and we obtain

$$\frac{\partial G_{p,s}}{\partial z_i} = -G_{p,s+i}, \quad i = 1, \ldots, p. \tag{5.92}$$

Integrating $G_{p,s}$ by parts we obtain the recurrent functional equations

$$(s-1)G_{p,s} = \sum_{i=1}^p z_i G_{p,s+i}, \quad \Re s > 0, \quad p = 1, 2, \ldots, \tag{5.93}$$

and, in particular, for $s = 1$ the equation

$$\sum_{i=1}^p z_i G_{p,i+1} = 0. \tag{5.94}$$

We remark that in many physical cases the generalized partition function Z can be transformed to the the form

$$Z = C(G_{p,s})^r, \quad r = 1, 2, \ldots, k, \tag{5.95}$$

where C is a known function of the given parameters. E.g.,

a) $k = 1$, $r = 1$, $C = 1$, $s = 1$: this occurs in the theory of polymers [61],

b) $k = 1$, $r = 1$, $C = \frac{(2\pi m)^{3N/2} V^N}{\Gamma(3N/2)}$, $s = 3N/2$: this occurs in the ideal gas of pth order (with observables H^i, $i = 1, \ldots, p$, where H is the kinetic energy) of N particles of mass m in volume V,

c) $k = 1$, $r = 1$, $C = \left(\frac{2\pi}{\omega}\right)^N \frac{1}{(N-1)!}$, $s = N$: this occurs for the "ideal solid state" of N 1-dimensional harmonic oscillators with frequency ω, also with observables H^i, $i = 1, \ldots, p$.

For the evaluation of generalized gamma functions, we reduce them to simpler functions (they have been used primarily by Czajkowski, only with his $k = s - 1$ and different sign of z_i)

$$g_{p,s}(z_1, \ldots, z_{n-1}) = \int_0^\infty t^{s-1} \exp\left(\sum_{i=1}^{p-1} z_i t^i - t^p\right) dt, \tag{5.96}$$

such that, since $\Re z_p > 0$, we can write

$$G_{p,s}(z_1, \ldots, z_p) = (z_p)^{-\frac{s}{p}} g_{p,s}\left(-z_1 (z_p)^{-\frac{1}{p}}, -z_2 (z_p)^{-\frac{2}{p}}, \ldots, -z_{p-1}(z_p)^{-\frac{p-1}{p}}\right). \tag{5.97}$$

Now we introduce the following abbreviations:

$$\mathbf{x} = (x_1, \ldots, x_{p-1}), \quad \mathbf{x} \wedge \mathbf{y} = (x_1 y_1, \ldots, x_{p-1} y_{p-1}), \tag{5.98}$$

$$\frac{\mathbf{x}}{\mathbf{y}} = \left(\frac{x_1}{y_1}, \ldots, \frac{x_{p-1}}{x_{p-1}}\right), \quad \mathbf{x}^{\mathbf{y}} = x_1^{y_1} \cdots x_{p-1}^{y_{p-1}}, \quad \mathbf{x}! = x_1! \cdots x_{p-1}!, \tag{5.99}$$

$$\mathbf{xy} = x_1 y_1 + \ldots + x_{p-1} y_{p-1},$$
$$\mathbf{x} \le \mathbf{y} \quad \text{if} \quad x_i \le y_i \quad \text{for} \quad i = 1, \ldots, p-1, \tag{5.100}$$

$$\mathbf{1} = (1, 1, \ldots, 1), \quad \mathbf{0} = (0, 0, \ldots, 0), \quad \mathbf{N} = (1, 2, \ldots, p-1). \tag{5.101}$$

Substituting in (5.96) $y = t^n$ and expanding $\exp\left(\sum_{i=1}^{p-1} z_i y^{i/p}\right)$ into Taylor series we obtain

$$g_{p,s}(z) = \frac{1}{p} \sum_{k=0}^\infty \frac{z^k}{k!} \Gamma\left(\frac{s + N k}{p}\right). \tag{5.102}$$

Let k_i and m_i be arbitrary positive integers such that $k_i/m_i = i/p$, $i = 1, 2, \ldots, p = 1$ (k_i and m_i may be prime to each other or not). Then (5.102) can be written as follows

$$g_{p,s}(z) = \frac{1}{p} \sum_{i=0}^{p-1} \frac{z^i}{i!} \Gamma\left(\frac{s}{p} + \left(\frac{k}{m}\right)i\right) K_i(z), \qquad (5.103)$$

$$K_i(z) = \sum_{j=0}^{\infty} z^{m \wedge j} \frac{\left(\frac{s}{p} + \left(\frac{k}{m}\right)i\right)_{kj}}{(i+1)_{m \wedge j}}, \qquad (5.104)$$

where the symbol $(\alpha)_k$ denotes $\Gamma(\alpha + k)/\Gamma(\alpha)$, and $(\alpha)_k$ abbreviates $(\alpha_1)_{k_1}, (\alpha_2)_{k_2}, \cdots (\alpha_{p-1})_{k_{p-1}}$.

Now we can remark that in the multidimensional series (5.104) which has the form $\sum_j a_j x_j$ the coefficients a_j have the property that the ratios

$$\frac{a_{j_1, j_2, \ldots, j_{p-1}}}{a_{j_1, \ldots, j_{k-1}, j_k+1, j_{k+1}, \ldots, j_{p-1}}} \qquad (5.105)$$

are for any $k = 1, \ldots, p - 1$ rational functions of $j_1, \ldots, j_{n-1}$. This means that each $K_i(z)$ is a generalized hypergeometrical function of $p-1$ variables, cf. [25], [30], [180].

In [166], there is a discussion of the one-variable case, namely, when we put $z_1 = \ldots = z_{n-1} = 0$, $z_n = z$, $z_{n+1} = \ldots = z_{p-1} = 0$, $n = 1, 2, \ldots, p - 1$. Then denoting the one-dimensional generalized hypergeometrical function by

$$_pF_q \left[\begin{array}{c} \alpha_1, \alpha_2, \ldots, \alpha_p \\ \beta_1, \beta_2, \ldots, \beta_q \end{array} z \right] = \sum_{j=0}^{\infty} \frac{(\alpha_1)_j (\alpha_2)_j \ldots (\alpha_p)_j}{(\beta_1)_j (\beta_2)_j \ldots (\beta_q)_j} \frac{z_j}{j!}, \qquad (5.106)$$

cf. [30], one obtains ($i = i_n$)

$$K_i(z) =_k F_{m-1} \left[\begin{array}{c} \frac{s}{kp} + \frac{i}{m}, \frac{s}{kp} + \frac{i}{m} + \frac{1}{k}, \ldots, \frac{s}{kn} + \frac{i}{m} + \frac{k-1}{k} \quad k^k \left(\frac{z}{m}\right)^m \\ \frac{i+1}{m}, \frac{i+2}{m}, \ldots, \frac{m-1}{m}, \frac{m+1}{m}, \ldots, \frac{m+i}{m} \end{array} \right] \qquad (5.107)$$

where $i = 1, 2, \ldots, m - 1$, and k and m are arbitrary positive integers such that

$$\frac{k}{m} = \frac{n}{p}. \qquad (5.108)$$

E.g., in the particular case $p = m = 3$, $n = k = 1$, $s = 1, 2, \ldots$, we obtain

$$g_{3,1}(z) = \frac{2}{3} \sqrt{\frac{-z}{3}} S_{0,1/3} \left(2 \left(\frac{-z}{3}\right)^{3/2}\right),$$

$$g_{3,s}(z) = \left(\frac{d}{dz}\right)^{s-1} g_{3,1}(z), \qquad (5.109)$$

where $S_{\mu,\nu}(z)$ is the Lommel function, cf. [30].

It is interesting to see how the quantum case differs from the classical one. This can be seen on the examples considered by Antoniewicz in [23]. For the quantum electromagnetic field coherent with respect to energy in pth order he obtained for the quantum partition function

$$Z(\beta_1, \ldots, \beta_p) = \sum_{n=0}^{\infty} \exp(-\beta_1 n - \ldots - \beta_p n^p) \qquad (5.110)$$

the expression

$$Z(\beta_1, \ldots, \beta_p) = 1 + \frac{1}{2\pi i} \int_{\sigma_1 - i\infty}^{\sigma_1 + i\infty} G_{p,s}(\beta_1, \ldots, \beta_p)\zeta(s)ds, \qquad (5.111)$$

where $\zeta(s)$ is the Riemann zeta function, cf. [318]. By this he used the fact that function $G_{p,s}$ (5.84) can be considered as the *Mellin transformation* of the corresponding exponential function:

$$g(s) = \mathcal{M}\{f(t); s\} = \int_0^{\infty} f(t)t^{s-1}dt \qquad (5.112)$$

with the inverse transformation

$$f(t) = \mathcal{M}^{-1}\{g(s); t\} = \frac{1}{2\pi i} \int_{-i\infty}^{i\infty} g(s)t^{-s}\, ds, \qquad (5.113)$$

where the integral is understood in the sense of the Cauchy main value, cf. [232]. Antoniewicz carried out the calculation for the following two cases: $p = 2$, and $p = 3$ with $\beta_2 = 0$. The latter case is rather complicated, so we quote here only the results of the former case putting for simplicity of notation $\alpha = \beta_2$, $\beta = \beta_1$ (this corresponds to the second and the first power of energy, respectively):

$$Z(\beta, \alpha) = \frac{1}{\sqrt{\alpha}} \exp\left(\frac{\beta_2}{4\alpha}\right) \text{Erfc}\left(\frac{\beta}{2\sqrt{\alpha}}\right) - \frac{1}{\beta} + \frac{1}{1 - e^{-\beta}} + \sum_{k=3}^{\infty} \frac{B_{k+1}}{k+1}\omega_k(\alpha.\beta),$$
$$(5.114)$$

where $\text{Erfc}\, x = \int_x^{\infty} e^{-t^2}dt$ and B_k are Bernoulli's numbers $B_1 = -1/2$, $B_{2n} = (-1)^{n-1}\pi^{2n}\int_0^{\infty}(t^{2n}/\sinh^2 t)dt$, $B_{2n+1} = 0$, $n = 1, 2, \ldots$, and

$$\omega_k(\alpha, \beta) = \sum_{m=1}^{(1/2)(k-1)} (-1)^m \frac{\alpha^m}{m!} \frac{\beta^{k-2m}}{(k-2m)!}, \qquad (5.115)$$

so for the first odd k

$$\omega_3 = -\alpha\beta, \; \omega_5 = -\frac{\alpha\beta^3}{1!3!} + \frac{\alpha^2}{2!1!}, \; \omega_7 = -\frac{\alpha\beta^5}{1!5!} + \frac{\alpha^2\beta^3}{2!3!} - \frac{\alpha^3\beta}{3!1!}, \; \ldots. \quad (5.116)$$

Antoniewicz estimated the remainder in the series in (5.114) and from this he concluded that this series converges for $\alpha < \pi/2\sqrt{2}$. Reminding that $\alpha > 0$ we can calculate the limit for of the first term in (5.114) for α going to 0 from the positive side. Using the de l'Hôspital rule we obtain the value $1/\beta$. Hence

$$\lim_{\alpha \to 0+} Z(\beta, \alpha) = \frac{1}{1 - e^{-\beta}} = \sum_{n=0}^{\infty} e^{-\beta n} = Z(\beta) \qquad (5.117)$$

which is exactly equal to the sum of the quantum Planck's states.

5.9. Higher-order states and the thermodynamical limit

One of the most important problems of any new theory is its correspondence with the older, accepted theory in this domain. Such was the case with the Copernican astronomy and the Galileo-Newtonian physics with respect to the Ptolemeian astronomy and the Aristotelian physics, respectively. Such was also the case of the relativistic physics with respect to the Galilean physics, and of quantum mechanics with respect to classical mechanics. In our case, we have the problem how the higher-order thermodynamics can be put in agreement with the hitherto accepted Gibbsian statistical mechanics. This problem was first studied in Toruń by Wojciech Jaworski in a series of papers [167], [168], [169], [170], [171], [172]. We will quote the conclusions of the last paper:

> The main result of this paper is that, from a purely thermodynamical point of view, the information corresponding to the higher-order moments of extensive physical quantities is not essential and can be neglected in the maximum entropy procedure. This is, of course, not an unexpected result. Our work, however, is not a proof of triviality. It can be viewed rather as a test of the maximum entropy formalism and, in fact, of the formula $S_\rho = -\langle \ln \rho \rangle_\rho$. We showed that the maximum entropy inference has a certain 'stability' property with respect to information corresponding to higher-order moments of extensive quantities. This result is reasonable. It can serve as an argument in favour of the maximum entropy method in statistical physics. It also enables us to understand better why these methods are successful.

Is it, however, proved that all the higher-order thermodynamics reduce in the thermodynamical limit to the usual one, the first-order thermodynamics? By no means. Let us note that Jaworski used the term "the higher-order moments of *extensive* physical quantities", i.e., the quantities whose densities are homogeneous in space, the quantities being proportional to the volume. Such is the (mean) energy of an infinite ideal gas or an ideal solid state, but not the energy of a finite bounded body with interactions, e.g., of the Sun or the Earth, or of such complex systems as a biological cell or of mesoscopic systems, or else of the self-organizing nonlinear systems, not speaking about the pumped systems. Jaworski spoke actually about "a purely thermodynamic point of view" meaning by this the point of view of the conventional thermodynamics of extensive quantities and the usual thermodynamic limit. He has shown in [168] that for the second-order energetic temperature ($p = 2$) of an ideal gas there is an inequality limiting the number of particles (we shall show it below in a similar context), but in [169] he has proved that for $p \geq 3$ "macroscopic fluctuations of the energy are possible and the energy distribution has several sharply peaked maxima". So thermodynamics can be generalized to a new field of applications, not yet sufficiently studied. Nevertheless, it is very essential to show a possible correspondence of the higher-order thermodynamics to the conventional thermodynamics, and this is the first condition to consider it as a reasonable theory.

The full method and the proof of Jaworski are rather complicated, both in its rigorous special version (for ideal gases) [169], as well as in its general non-rigorous, "formal" version presented in [172]. Therefore, we shall avoid to give this method here in full, limiting ourselves only to the main idea. Jaworski carries out his reasonings parallelly for the classical and quantum cases (the difference being that in the classical case we have integrals of continuous quantities and in the quantum cases traces of infinite matrices). He considers the case of a system of N identical particles with Hamiltonian H_N and in a volume V_N, and then he goes over to the thermodynamic (Van Hove) limit

$$N \to \infty, \qquad V_N \to \infty, \qquad \frac{V_N}{N} = v = \text{constant independent of } N. \tag{5.118}$$

The macrostate considered is

$$\Phi_N = \left\{ \rho \in \mathfrak{S}_N : \langle H_N^j \rangle_\rho = U_{j,N}, \quad j = 1, 2, \ldots, p \right\}. \tag{5.119}$$

Jaworski assumes the extensivity in the form

$$U_{j,N} = N^j u_j + o(N^j), \qquad j = 1, 2, \ldots, p, \tag{5.120}$$

where the symbol $y_N = o(x_N)$ means that $\lim_{N \to \infty} (y_N/x_N) = 0$, and u_j are constants independent of N:

$$u_j = \lim_{N \to \infty} \frac{U_{j,N}}{N^j}. \tag{5.121}$$

Since in the general case it is practically impossible to calculate the dependence of the generalized temperatures $\beta_{j,N}$ on N, Jaworski uses the following argument: "Phenomenological thermodynamics suggests that the entropy S_N of the state ρ_N^* satisfies

$$S_N = Ns + o(N) \qquad (5.122)$$

where s is a constant

$$s = \lim_{N \to \infty} \frac{S_N}{N} \qquad (5.123)$$

(extesitivity)." Then using relations

$$S_N = \ln Z_N + \sum_{j=1}^{p} \beta_{j,N} U_{j,N} \qquad (5.124)$$

he derives the equations

$$\beta_{j,N} = \frac{\partial S_N}{\partial U_{j,N}} = \lim_{\Delta U_{j,N} \to 0} \frac{\Delta S_N}{\Delta U_{j,N}} = \frac{1}{N^{j-1}} \lim_{\Delta U_{j,N} \to 0} \frac{\Delta S_N/N}{\Delta U_{j,N}/N^j}. \qquad (5.125)$$

Taking into account (5.120) and (5.122) one obtains

$$\beta_{j,N} = \frac{1}{N^{j-1}} (\alpha_j + o(1)), \qquad j = 1, \dots, p, \qquad (5.126)$$

where α_j are constants independent of N. We see that for $N \to \infty$ all higher order temperatures β_j ($j = 2, \dots, p$) disappear. Of course, this is a result of our assumptions about the extensitivity of $U_{j,n}$ and S_N, eqs. (5.120) and (5.122), which are valid only in the realm of traditional thermodynamics and which cannot be proved for the general case of higher order thermodynamics. But this shows the possibility of correspondence with the usual thermodynamics, and also that the general systems in higher-order states are heterogeneous, non-uniform. We have in the same time an explanation of the seeming "paradox" of information: that more information can change radically our macroscopic view. Sometimes the latter is possible, but not in the most cases since our senses are adjusted through biological evolution to the most frequent physical situations, while our phenomenological theories are based on the data of our senses, without making special experiments. So the senses give such "measurements" which usually are biologically safe, but they cannot guarantee security in all situations.

Mathematically, the situation is clear: in probability theory there are the famous "theorems of large numbers". They give an essential limitation for the higher-order temperatures. Indeed, there is a theorem about the limit distribution of sums of independent random variables, cf. [107], in which it is proved, under some conditions, that this limit distribution has the form of a Gaussian. The ideal gas in the thermodynamic limit (5.118) corresponds to these conditions and we get then Gibbs distribution with one temperature. But interactions between particles and a finite number of particles (the *mesoscopic case*, as we call it) can change this result and then the higher-order distribution can be possible. On the other hand, we have to take into account inequalities which can exist between the higher-order moments, e.g., $\sigma^2 = U_2 - U_1^2 > 0$ or stronger, cf. [63], which can limit the solvability of equations for temperature coefficients. Another type of inequalities, depending on N, have been already mentioned in connection with [168], and they will be discussed below. All these limitations, however, cannot exclude the possibility of higher-order temperatures and essentially non-Gaussian distributions in rather special, but actually very important and numerous cases (we usually disregard them just because of the lacking theory and lacking "senses" for them). In Section 5.11 we shall give an example of an empirical case of such a situation taken from a concrete linguistic statistics [164].

5.10. Simple examples and generalized thermodynamic limits

Now we present simplified mathematical and physical models in which all calculations can be carried out explicitly and rigorously to the end enabling the complete physical discussion, cf. [162], [163]. The simplification consists in considering only one higher-order temperature. All the examples will be classical.

To begin with, we discuss the maximization of entropy in a one-dimensional case, where the domain of integration is the whole real line or the right half-line. In the firstcase the entropy is (we put as usually $k_B = 1$ and the constant under the logarithm sign also equal to 1)

$$S[\rho(x)] = -\int_{-\infty}^{\infty} \rho(x) \ln \rho(x) \, dx, \qquad (5.127)$$

and the constraints of our macrostate Φ_{2n} are ($p = 2n, \, n = 1, 2, \ldots, \sigma > 0$, n and σ given)

$$\rho(x) \geq 0, \qquad \int_{-\infty}^{\infty} \rho(x) \, dx = 1, \qquad \langle x^{2n} \rangle = \int_{-\infty}^{\infty} x^{2n} \rho(x) \, dx = \sigma^{2n}.$$

$$(5.128)$$

From the principle of maximum entropy we easily find the representative state

$$\rho^*(x) = \frac{1}{2(2n)^{1/2n}\sigma\Gamma(1+1/2n)} \exp\left(-\frac{x^{2n}}{2n\sigma^{2n}}\right) = f_n(x), \quad -\infty < x < \infty,$$
(5.129)

where

$$\beta_n = \frac{1}{2n\sigma^{2n}}, \quad Z(\beta_n) = \frac{2\Gamma(1+1/2n)}{\beta_n^{1/2n}}.$$
(5.130)

We can check that

$$-\frac{\partial \ln Z(\beta_n)}{\partial \beta_n} = \sigma^{2n} = \langle x^{2n} \rangle,$$
(5.131)

as it should be. We see that for $n = 1$ the function $f_1(x)$ in (5.129) is the Gaussian distribution, and in physics $\beta_1 = \beta$. For arbitrary positive integer n, $f_n(x)$ may be called, as in Chap.2 Sec. 2, a *generalized Gaussian distribution of order n*. For growing n and constant σ, $f_n(x)$ slowly becomes more and more similar to the "rectangular" step function

$$f_\infty(x) = \begin{cases} 1/2\sigma & \text{if } |x| \leq \sigma \\ 0 & \text{if } |x| > \sigma, \end{cases}$$
(5.132)

cf. Fig. 5.1.

The entropy of a representative state has the form

$$S[\rho^*(x)] = \ln[2(2n)^{1/2n}\sigma\Gamma(1+1/2n)] + \frac{1}{2n} = \ln Z(\beta_n) + \frac{1}{2n}.$$
(5.133)

(Attention: in [162], there are two small misprints in the middle expression of Eq. (3.4). They should be corrected according to (5.133).) We may also write for $n \to \infty$

$$S_\infty = \ln(2\sigma), \qquad Z_\infty = 2\sigma, \qquad \beta_\infty = 0$$
(5.134)

which is a singular solution since ρ^* is not a continuous, but a step function. Now, when we have the representative state ρ^*, we can easily calculate all the other positive power moments of x, namely:

$$\langle x^{2k-1} \rangle = 0, \quad \langle x^{2k} \rangle = \frac{(2n)^{k/n}\sigma^{2k}\Gamma(1+(2k+1)/2n)}{(2k+1)\Gamma(1+1/2n)},$$
(5.135)

$k = 1, 2, ...$, which for $k = n$ reduces to σ^{2n} in agreement with (5.131).

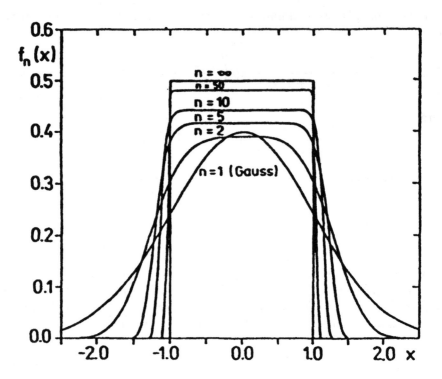

Figure 5.1. Generalized Gauss distribution of order n ($\sigma = 1$).

In the case of non-negative x all the integrations are to be carried out between 0 and ∞ and we can take any positive-integer power moment of x, odd or even, as an auxiliary condition:

$$\langle x^n \rangle = \int_0^\infty x^n \rho(x)\, dx = \sigma^n, \quad \sigma > 0, \quad n = 1, 2, \ldots. \tag{5.136}$$

Then we obtain the representative state

$$\rho^*(x) = \frac{1}{n^{1/n}\sigma\Gamma(1 + 1/n)} \exp\left(-\frac{x^n}{n\sigma^n}\right) = f_n(x), \qquad 0 \le x < \infty, \tag{5.137}$$

as a *generalized exponential distribution (semi-Gaussain or Laplace)*, cf. Fig. 5.2, and

$$\beta_n = \frac{1}{n\sigma^n}, \quad Z(\beta_n) = \frac{\Gamma(1 + 1/n)}{\beta_n^{1/n}}. \tag{5.138}$$

We check

$$-\frac{\partial \ln Z(\beta_n)}{\partial \beta_n} = \sigma^n = \langle x^n \rangle, \tag{5.139}$$

and calculate

$$S[\rho^*(x)] = \ln Z(\beta_n) + \frac{1}{n}, \quad \langle x^k \rangle = \frac{n^{k/n}\sigma^k\Gamma(1+(k+1)/n)}{(k+1)\Gamma(1+1/n)}, \quad (5.140)$$

$k = 1, 2, \ldots$ (By mistake in [162] in the numerator of the expression (3.10) for $\langle x^k \rangle$ the factor σ^k is missing.) Finally, for $n \to \infty$ we have

$$f_\infty(x) = \begin{cases} 1/\sigma & \text{if } 0 \le x \le \sigma, \\ 0 & \text{if } x > \sigma, \end{cases} \quad (5.141)$$

and

$$S_\infty = \ln \sigma, \quad Z_\infty = \sigma, \quad \beta_\infty = 0. \quad (5.142)$$

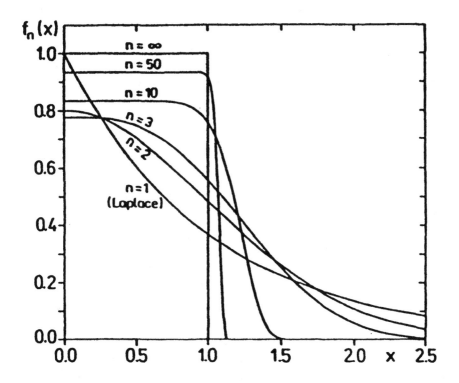

Figure 5.2. Generalized exponential (Laplace) distribution of order n ($\sigma = 1$).

Going over to more physical models let us discuss a classical ideal gas (but with indistinguished particles to avoid the Gibbs paradox) of $N = 1, 2, \ldots$ particles with mass $m > 0$ in a 3-dimensional volume $V > 0$ with one n-order ($n = 1, 2, \ldots$) energetic temparature (total energy = kinetic energy since we neglect the interactions). Because of these assumptions the border surface energy and the shape energy are negligible. (Concerning the shape information and the pattern recognition problem, cf. the papers [216] and [161].) Denoting momenta of the particles by $p = \{p_i, \ i = 1, 2, \dots, 3N\} \in \mathbb{R}^{3N}$ we have

$$H(p) = \frac{1}{2m} \sum_{i=1}^{3N} p_i^2, \qquad \rho(p) \geq 0, \qquad \frac{V^N}{N!} \int_{-\infty}^{\infty} \rho(p) \, d^{3N}p = 1, \quad (5.143)$$

and according to our assumption

$$\langle H^n \rangle = \frac{V^N}{N!} \int_{-\infty}^{\infty} H^n(p) \rho(p) \, d^{3N}p = U^n, \qquad U > 0. \quad (5.144)$$

For the entropy we introduce physical units with the Boltzmann constant $k_B > 1$

$$S[\rho(p)] = -k_B \frac{V^N}{N!} \int_{-\infty}^{\infty} \rho(p) \ln f(p) \, d^{3N}p \quad (5.145)$$

and we obtain from the principle of maximum entropy

$$\rho^*(p) = [Z(\beta_n)]^{-1} \exp[-\beta_n H^n(p)], \quad (5.146)$$

where we denote

$$\beta_n = \frac{1}{(k_B T_n)^n} > 0, \qquad T_n > 0, \quad (5.147)$$

calling T_n an *absolute temperature of order* n. By means of a simple calculation we obtain

$$Z(\beta_n) = Z(\beta_n; V, N) = \frac{V^N (2\pi m)^{3N/2} \Gamma(3N/2n)}{N! n \beta_n^{3N/2n} \Gamma(3N/2)},$$

$$U^n = -\frac{\partial Z}{\partial \beta_n} = \frac{3N}{2n\beta_n} \quad (5.148)$$

or

$$\beta_n = \frac{3N}{2nU^n}. \quad (5.149)$$

Expressing the estimated entropy as a function of $n, U, V,$ and N we get

$$S[\rho^*(p)] = S(U, V, N) = k_B \left(\ln Z + \frac{3N}{2n} \right) =$$

$$k_B \ln \left\{ \frac{V^N}{N!} \left[2\pi m U \left(\frac{2n}{3N} \right)^{1/n} \right]^{3N/2} \frac{\Gamma(1 + (3N)/2n)}{\Gamma(1 + (3N)/2)} \right\} + k_B \frac{3N}{2n}. \quad (5.150)$$

Now we can calculate the main thermodynamic quantities and relations of order n:

a) the temperature of order n, T_n,

$$\frac{1}{T_n} = k_B^{1-1/n} \left(\frac{\partial S(U, V, N,)}{\partial(U^n)}\right)^{1/n} = k_B \frac{1}{U} \left(\frac{3N}{2n}\right) \tag{5.151}$$

(attention: some misprints in the middle expression in eq. (4.9) in [162]) or the total energy of order n, $U = (U_n)^{1/n} > 0$,

$$U = \left(\frac{3N}{2n}\right)^{1/n} k_B T_n. \tag{5.152}$$

(Checking: for $n = 1$ we get the well-known formula of ideal gases $U = (3/2)Nk_BT$.)

b) the pressure of order n, P_n,

$$P_n = \frac{1}{\beta_n} \frac{\partial Z(\beta_n, V, N)}{\partial V} = \frac{N}{\beta_n V} = \frac{2nU^n}{3V} \tag{5.153}$$

which gives the equation of state of order n (equation of n-hyperstate)

$$P_n V = N k_B^n T_n^n \quad\text{or}\quad P_n V = \frac{2n}{3} U^n. \tag{5.154}$$

(Checking: for $n = 1$ we get the well-known formula of ideal gases $PV = (2/3)Nk_BT$.)

c) specific heats of order n by constant volume

$$c_V^{(n)} = \frac{\partial U}{\partial T_n}\bigg|_V = k_B \left(\frac{3N}{2n}\right)^{1/n}, \tag{5.155}$$

by constant pressure

$$c_{P_n}^{(n)} = \frac{\partial(U + P_n V)}{\partial T_n}\bigg|_{P_n} = k_B \left[\left(\frac{3N}{2n}\right)^{1/n} + nNk_B^{n-1}T_n^{n-1}\right] \tag{5.156}$$

which gives the difference

$$c_{P_n}^{(n)} - c_V^{(n)} = nNk_B^n T_n^{n-1}. \tag{5.157}$$

d) the thermal expansion coefficient of order n

$$\alpha_n = \frac{1}{V} \frac{\partial V}{\partial T_n}\bigg|_{P_n} = \frac{nNk_B^n T_n^{n-1}}{P_n V} = \frac{n}{T_n} > 0 \tag{5.158}$$

which gives

$$T_n = \frac{n}{\alpha_n} \quad \text{or} \quad n = \alpha_n T_n \tag{5.159}$$

enabling to calculate T_n provided n and α_n are given (the principle of a generalized gas thermometer), or to calculate n provided α_n and T_n are measured.

We do not claim that this model is realistic. We discussed it only because of its mathematical simplicity and rigorous solvability. Since the Hamiltonian depends only on momenta (no interactions of particles), the distribution in space is homogenous, and mean total energy is additive. If we assume, according to Jaworski, that

$$U = (U_n)^{1/n} = N u_n, \tag{5.160}$$

where u_n is a constant independent of N, we obtain

$$\beta_n = \frac{3}{2 n u_n^n N^{n-1}} \tag{5.161}$$

which only for $n = 1$ is reasonable (non-zero) also in the thermodynamic limit $N \to \infty$. For $n > 1$ it cannot be true in the usual thermodynamic limit since then the result would be contradictory with the mentioned Gaussian limit distribution. But for finite N and V the results are mathematically and physically reasonable, although rather improbable because of the absence of temperatures of lower order than n.

If there is a weak interaction between particles (as gravitational one), a physically reasonable assumption for limiting distributions is not a homogeneous one, but a distribution vanishing in spatial infinity (e,g., the distribution of matter of the Sun or the Earth). This shows a possibility of a generalization of the concept of thermodynamic limit, cf. [162], Sec. 5.

Now we discuss a more elaborated model of an ideal gas in a state with the first and second energetic temperatures, cf. [162], Section 6, and [163]. Since the potential energy is physically determined only up to an arbitrary real constant, it is necessary to consider the central moments of total energy independent of this constant. In our case we have a generalized Gibbs state

$$\rho^* = Z^{-1} \exp[-\beta H - \beta'(H - U)^2], \quad \beta' > 0, \tag{5.162}$$

where

$$Z = Z(\beta, \beta', U) = \operatorname{tr} \exp[-\beta H - \beta'(H - U)^2], \tag{5.163}$$

and the additional conditions giving the statistical moments of energy

$$U = \operatorname{tr}(\rho H), \quad D^2 = \operatorname{tr}[\rho(H - U)^2] > 0, \quad D > 0. \tag{5.164}$$

To elucidate the physical meaning of β' we write the following consequence of (5.162) and the first equation in (5.164)

$$\text{tr}\{(H - U)\exp[-\beta(H - U) - \beta'(H - U)^2]\} = 0. \qquad (5.165)$$

Taking U as a function of β and β' we differentiate both sides of (5.165) with respect to β and obtain

$$-\frac{\partial U}{\partial \beta} - D^2 + 2\beta' D^2 \frac{\partial U}{\partial \beta} = 0. \qquad (5.166)$$

Using the definition of the heat capacity c_V (of the first order)

$$\frac{c_V}{k_B} = \frac{\partial U}{\partial(1/\beta)} = -\beta^2 \frac{\partial U}{\partial \beta} \qquad (5.167)$$

we obtain the relation

$$D^2 = \frac{c_V/k_B}{2\beta' c_V/k_B + \beta^2} \qquad (5.168)$$

or

$$\beta' = \frac{1}{2D^2} - \frac{\beta^2}{2c_V/k_B}. \qquad (5.169)$$

If the heat capacity $c_V > 0$, then from (5.168) and (5.162) we obtain the following inequalities

$$\beta^2 D^2 = \frac{\beta^2 c_V/kB}{2\beta' c_V/k_B + \beta^2} \leq c_V/k_B, \qquad (5.170)$$

$$0 < \beta' D^2 = \frac{\beta' c_V/k_B}{2\beta' c_V/k_B + \beta^2} \leq \frac{1}{2}. \qquad (5.171)$$

Eq. (5.169) suggests how to measure the second-order temperature.

Going over to the partition function of the ideal gas we finally obtain after some calculation, cf. [163] Section 3,

$$Z(\beta, \beta') = \frac{V^N}{N!} \exp\left(-\frac{9N^2(3N\beta' - \beta^2)^2}{32\beta^4}\right) \times$$

$$\times \left(\frac{3\pi m}{\sqrt{2\beta'}}\right)^{3N/2} U_{(3N-1)/2}\left(\frac{\beta^2 - 3N\beta'}{\beta\sqrt{2\beta'}}\right), \qquad (5.172)$$

where $U_a(x) = U(a, x)$ is a parabolic cylinder function (Weber's function) [2], Chapter 19 (in [163], p. 247, it is erroneously called hyperbolic cylinder function). For positive parameter $a = (3N)/2 - 1/2$, as is the case, and large $x^2 + 4a$, which we assume, the Weber function can be presented by the leading term of Darwin's expansion [2], Section 19.10, p. 284,

$$U_a(x) = \frac{(2\pi)^{1/4}}{\sqrt{\Gamma(a+1/2)}} \exp\left(-\frac{1}{4}x\sqrt{x^2+4a} - a\ln\left(\frac{x+\sqrt{x^2+4a}}{2\sqrt{a}}\right)\right).$$

$$(5.173)$$

In this approximation, we obtain the solution, cf. [163],

$$\beta = \frac{3N}{2U}, \qquad \beta' = \frac{1}{D^2} - \frac{3N}{4U^2} > 0. \tag{5.174}$$

From the second expression in (5.174) we see that the physical condition for the existence of this solution (positivity of β') is the inequality

$$N < \frac{2}{3}\left(\frac{U}{D}\right)^2 \tag{5.175}$$

which is a reasonable limitation for N (since for infinite N only the Gibbs distribution with the first-order temperature is possible according to the limit theorem of large numbers). The inequality (5.175) (we can add there the equality sign for the first-order temperature case, $\beta' = 0$) has been obtained for the first time by W. Jaworski in [168], eq. (5.23) p. 655, and it has been proved again in [162].

It is an interesting question how the inequality (5.175) can be influenced by interactions between particles in a finite system. In [163], two simple integrable models have been investigated:

(1) a semiharmonic oscillatory chain of the Toda type [319],

(2) a 3-dimensional harmonic solid.

The results can be summarized by adding a numerical coefficient, α_i, with $i = 0$ for the ideal gas (with finite N and V), $i = 1$ for the model (1), and $i = 2$ for the model (2), to the right-hand side of the inequality (5.175):

$$N \le \alpha_i \left(\frac{U}{D}\right) \tag{5.176}$$

with

$$\alpha_0 = \frac{2}{3}, \qquad \alpha_1 = 1, \qquad \alpha_2 = \frac{1}{3}. \tag{5.177}$$

In the following table we give the value of maximal N, N_{max}, for different ratios D/U expressed by percent (%) and for the three different models: We see that the influence of integrable interactions is rather weak. Probably the non-integrable (chaotic) interactions, as well a complex (e.g., hierarchical) structure and an outside pumping may have a stronger influence on N_{max}. But the orders of magnitude of N_{max} in the table seem to

MODEL	D/U		
	10%	1%	0.1%
ideal gas	66	6666	6.66×10^5
semiharmonic chain	100	10000	10^6
3d harmonic solid	33	3333	3.33×10^5

TABLE 5.2.

be reasonable, e.g., for biological applications. Indeed, the typical number of atoms in a biological molecule (e.g., a protein molecule), the typical number of protein molecules in an organelle, or the typical number of organelles in a biological cell may be of such orders of magnitude. Therefore we can hope that the thermodynamic description by means of the two first energetic temperatures may be reasonable for such "small" systems ("mesoscopic systems"). The N_{max} is as if a "lower Avogadro number", being big enough to fulfil the condition of big X^2 for the Darwin expansion in our approximation. Such a description is in a sense complementary to the so-called extended irreversible thermodynamics, cf. [240], which is a phenomenological macroscopic approach to irreversible phenomena using the concept of local temperatures. Although our temperatures are so far nonlocal and global, they can also slowly change in time and in space (it would be information thermodynamics in nonrelativistic quantum field theory which is not yet developed). Since we add to the usual two conditions by maximization of entropy (normalization + mean energy) some additional conditions, entropy is smaller than the usually considered in equilibrium, therefore this extension may be cosidered as belonging to irreversible thermodynamics in the usual sense. We are fully conscious, however, of the difficulties of going over to the real physical applications in our approach to mesoscopic thermodynamics.

Finally, we would like to mention that T. Nakagomi in the paper [238] proposed another approach to higher order temperatures by means of his mesoscopic version of thermodynamic equilibrium condition. "The idea to combine the microscopic description of the system and the macroscopic one of the reservoir was used as one of the bases of 'mesoscopic thermodynamics' proposed by" him in [237].

Another final remark is that the multi-temperature method is essentially a different mathematical problem than the one-temperature method. That means that, e.g., the first-order energetic temperature is no better approximated by the former than by the latter: in general, it is a completely new quantity for each order, it may even change sign. The main temperature is always that of the highest order and this is always positive to guarantee the normalizability of the distribution (especially in the case of the infinite interval of integration). We may speak about approximation of temperatures only when the number of statistical moments goes to infinity.

In this book we do not use methods of fenomenological thermodynamics for higher order temperatures, for interested readers cf. [143].

5.11. An empirical linguistic example

Since at present there are difficulties, or impossibilities, in the analysis of physical or biological systems with interactions, more realistic than the simplified models treated above, we have been looking for other statistical models having a sure empirical basis. To such cases many statistical investigations of linguistics belong. The case of languages is interesting for us because   elements (words, phoneme   morphemes, etc.) have, in general, strong statistical correlations. This is the most important problem for us: whether statistical correlations can essentially change the probabilistic distribution even in the limiting case of large numbers. General statistical distributions can be treated by the cut-off problem of moments, therefore by means of our method of higher-order temperatures. As we already said, we can surely use the word "temperature" in this application since just the classical linguistics gives the proof of the proper sense of this word: "temperature"(derivated from the Latin language) basically means moderation, mixing, relaxing, and etymologically has no direct relation to energy and thermodynamics. Thus, breaking with the recent tradition, we do not hesitate to apply this concept to the statistics of languages, hoping even that it may help a deeper understanding of this statistics giving new parameters describing global properties of the investigated language.

In the paper [164] we investigated by computer methods the statistics of words and letters (in phonemic writing systems) of indo-european and non-indo-european languages. For letter statistics we took English, German and Russian and for word statistics French, Russian and Japanese which have been available to the authors (R. S. Ingarden and J. Meller). Since probability distributions are given by the published tables of relative frequencies, our problem is now opposite to that investigated above: we do not measure the mean values to find the distribution, but starting from the distribution we seek the mean values (or rather temperatures connected with them). The main problem is whether the given distribution is fixed by one or by many temperatures. The check of the calculations is to evaluate the "true" and "approximated" entropies, namely, the entropy calculated directly from the distribution, and the maximum entropy calculated from the partition function, moments and temperatures. If both values coincide (to some accuracy), it is the proof that the calculation is correct, and the maximum entropy principle can be applied to this case (the languages underlie our "generalized thermodynamics" or "information dynamics").

Our numerical calculations have shown that the answer to our question is positive: the language statistics shows a multi-temperature distribution up to the 5th order at least (this limitation is rather mathematical: only up to the 5th order there exist algebraic solutions of algebraic equations and computer cannot easily calculate higher orders than the 5th, cf. e.g. [289]). Our numerical results are as follows:

a) for letter statistics in Table 5.3 and Table 5.4,

Language	β_0	β_1	β_2	$\beta_3.10^2$	$\beta_4.10^4$	$\beta_5.10^6$
English	1.740	0.1583	0.00705	-0.1317	0.5274	-0.5446
German	0.928	0.8427	-0.15132	1.2732	-4.6791	6.2684
Russian	1.592	0.4018	-0.04254	0.2708	-0.8263	0.9913

TABLE 5.3.

Language	true entropy	approx. entropy	error
English	4.03 bit	4.27 bit	5.9 %
German	4.04 bit	4.30 bit	6.4 %
Russian	4.35 bit	4.54 bit	4.4 %

TABLE 5.4.

b) for word statistics in Table 5.5 and Table 5.6

We add here an analysis of statistical moments and temperature terms in the entropy for word statistics in Table 5.7 and Table 5.8. From the latter table we see how essential is the contribution to the entropy of the corresponding order terms. Hence, we can conclude that for language statistics the higher order temperatures are very essential parameters which describe in a global sense letter or word correlations. Therefore, their analysis seems to be interesting for descriptive and comparative

Language	β_0	$\beta_1 10^3$	$\beta_2 10^6$	$\beta_3 10^9$	$\beta^4 10^{12}$	$\beta_5 10^{15}$
French	3.802	30.49	-72.51	79.34	-39.35	7.200
Russian	4.334	27.87	-67.95	74.94	-37.15	6.768
Japanese	4.531	25.34	-63.14	72.12	-37.16	7.043

TABLE 5.5.

Language	true entropy	approx. entropy	error
French	9.19 bit	8.55 bit	7.0 %
Russian	7.72 bit	8.79 bit	13.9 %
Japanese	7.92 bit	9.06 bit	14.3 %

TABLE 5.6.

linguistic. For us this example presents a model of a statistical system with correlations (interactions).

Moment	French	Russian	Japanese
M_1	$0.240.10^3$	$0.258.10^3$	$0.290.10^3$
M_2	$0.236.10^6$	$0.274.10^6$	$0.307.10^6$
M_3	$0.322.10^9$	$0.382.10^9$	$0.418.10^9$
M_4	$0.495.10^{12}$	$0.593.10^{12}$	$0.633.10^{12}$
M_5	$0.813.10^{15}$	$0.983.10^{15}$	$1.021.10^{15}$

TABLE 5.7.

Term	French	Russian	Japanese
β_0	3.802	4.334	4.531
$\beta_1 M_1$	7.318	7.190	7.349
$\beta_2 M_2$	-17.086	-18.624	-19.372
$\beta_3 M_3$	25.539	28.586	30.124
$\beta_4 M_4$	-19.483	-22.034	-23.530
$\beta_5 M_5$	5.854	6.655	7.191
S	5.944 nat	6.107 nat	6.293 nat

TABLE 5.8.

Chapter 6

Information Thermodynamics II

6.1. Isoentropic motions

The information-theoretical approach to incomplete description of quantum systems (information thermodynamics) has been presented in the preceding Chapter. Now, these ideas will be further developed, c.f. [196], [195].

Let $\mathcal{H}$ be a complex, separable Hilbert space corresponding to a quantum system and let $\mathfrak{S}(\mathcal{H})$ be the set of all density operators (states) on $\mathcal{H}$. Let H be the Hamiltonian of the quantum system. It is assumed for simplicity that H is time independent. The time evolution of the state ρ is described by the von Neumann equation

$$i\hbar \frac{d\rho(t)}{dt} = [H, \rho(t)], \qquad (6.1)$$

with the initial condition $\rho(0) = \rho$. The solution of (6.1) can be written in the form

$$\rho(t) = U_t^* \rho U_t, \qquad (6.2)$$

where

$$U_t = \exp\left(\frac{i}{\hbar} H t\right). \qquad (6.3)$$

Let $\mathbf{A} = (A_1, \ldots, A_p)$ be an incomplete set of thermodynamically regular observables, cf. (5.30), such that not all $A_1, \ldots, A_p$ are constants of motion. If at time $t = 0$ the macrostate $\Phi_{\mathbf{A}}$ is known, i.e., the expectation values $M_{A_1}(\Phi_{\mathbf{A}}), \ldots, M_{A_p}(\Phi_{\mathbf{A}})$ are known, then the estimated value $M_{A_j}^t(\Phi_{\mathbf{A}})$ at time $t \geq 0$ of the observable $A_j \in \mathbf{A}$ $(j = 1, \ldots, p)$ is given by

$$M_{A_j}^t(\Phi_{\mathbf{A}}) = \operatorname{tr}(\rho^*(t) A_j), \quad j = 1, \ldots, p, \qquad (6.4)$$

where $\rho^*(t) = U_t^* \rho^* U_t$, and ρ^* is the representative state for the macrostate $\Phi_{\mathbf{A}}$, c.f., (5.25), (5.26) and (5.27). The macrostate $\Phi_{\mathbf{A}}^t$ at time $t \geq 0$ is defined as follows

$$\Phi_{\mathbf{A}}^t = \{\rho \in \mathfrak{S}(\mathcal{H}) : \operatorname{tr}(\rho A_j) = M_{A_j}^t(\Phi_{\mathbf{A}}), \quad j = 1, \ldots, p\}. \qquad (6.5)$$

Since $M_{A_j}^0(\Phi_A) = M_{A_j}(\Phi_A)$, $j = 1, \ldots, p$, it follows that $\Phi_A^0 = \Phi_A$. From (6.5) one has the relation

$$M_{A_j}^t(\Phi_A) = M_{A_j}(\Phi_A^t), \quad j = 1, \ldots, p. \tag{6.6}$$

For the macrostate Φ_A^t, $t \geq 0$, there exists the represantative state $\rho^*(\Phi_A^t)$ of the form

$$\rho^*(\Phi_A^t) = Z^{-1}[\lambda(t)] \exp\left(-\sum_{j=1}^p \lambda_j(t) A_j\right), \tag{6.7}$$

where

$$Z[\lambda(t)] = \operatorname{tr} \exp\left(-\sum_{j=1}^p \lambda_j(t) A_j\right) \tag{6.8}$$

and the parameters $\lambda_1(t), \ldots, \lambda_p(t)$ are unique functions of $M_{A_1}^t(\Phi_A), \ldots, M_{A_p}(\Phi_A)$ which are given in terms of the equations

$$M_{A_j}^t(\Phi_A) = -\frac{\partial \ln Z[\lambda(t)]}{\partial \lambda_j(t)}, \quad j = 1, \ldots, p. \tag{6.9}$$

The representative state $\rho^*(\Phi_A^t)$ is the operator-valued function of the estimated values $M_{A_1}^t(\Phi_A), \ldots, M_{A_p}^t(\Phi_A)$. It should be noted that the equality

$$\rho^*(\Phi_A^t) = \rho^*(t) = U_t^* \rho^*(\Phi_A) U_t \tag{6.10}$$

holds only for $t = 0$ since not all of $A_1, \ldots, A_p$ are constants of motion.

THEOREM 6.1. *For every macrostate Φ_A and $t \geq 0$ the inequality*

$$S(\Phi_A^t) \geq S(\Phi_A) \tag{6.11}$$

is satisfied, where the entropy $S(\Phi_A^t)$ is defined as follows

$$S(\Phi_A^t) := \sup\{S(\rho) : \rho \in \Phi_A^t\} = -\operatorname{tr}(\rho^*(\Phi_A^t) \ln \rho^*(\Phi_A^t)). \tag{6.12}$$

Proof. It follows from (6.4) and (6.5) that

$$\Phi_A^t = [\rho^*(t)]_A. \tag{6.13}$$

On the other hand, from (6.13) and (6.10) one concludes that

$$S(\Phi_A^t) \geq S(\rho^*(t)) = S(\rho^*(0)) = S(\Phi_A^t). \quad \Box \tag{6.14}$$

The above theorem may be considered as an analogon of Boltzmann's H-theorem in information thermodynamics. In order to calculate the estimated values $M_{A_j}^t(\Phi_\mathbf{A})$, $j = 1, \ldots, p$, one has to calculate $\rho^*(t)$, which, in non-trivial cases, cannot be practically carried on. An approximation for the estimated values $M_{A_j}^t(\Phi_\mathbf{A})$, $j = 1, \ldots, p$, will be given. Its character will be explained in Sec. 6.3. From (6.1) and (6.4) one obtains the following equations

$$i\hbar \frac{dM_{A_j}^t(\Phi_\mathbf{A})}{dt} = \mathrm{tr}(\rho^*(t)[A_j, H]) \tag{6.15}$$

with the initial conditions

$$M_{A_j}^0(\Phi_\mathbf{A}) = M_{A_j}(\Phi_\mathbf{A}), \quad j = 1, \ldots, p. \tag{6.16}$$

The equations (6.15), in general, do not form the closed systems of equations for $M_{A_j}^t(\Phi_\mathbf{A})$, $j = 1, \ldots, p$. However, it is the case when the relations

$$[A_j, H] = \sum_{k=0}^{p} h_{jk} A_k, \quad j = 1, \ldots, p, \tag{6.17}$$

where h_{jk} are complex numbers and

$$A_0 = I \tag{6.18}$$

hold.

In this case, (6.15) and (6.17) one finds

$$i\hbar \frac{dM_{A_j}^t(\Phi_\mathbf{A})}{dt} = \sum_{k=0}^{p} h_{jk} M_{A_k}^t(\Phi_\mathbf{A}), \tag{6.19}$$

where $M_{A_0}^t(\Phi_\mathbf{A}) = 1$.

THEOREM 6.2. *Let the observables $A_1, \ldots, A_p$ satisfy relation (6.17). Then for every macrostate $\Phi_\mathbf{A}$ and $t \geq 0$ the relation*

$$S(\Phi_\mathbf{A}^t) = S(\Phi_\mathbf{A}) \tag{6.20}$$

holds.

Proof. Since $\rho^*(\Phi_\mathbf{A}^t) \in [\rho^*(t)]_\mathbf{A}$, it follows from (6.17) and the form of ρ^*, c.f., (5.25), that

$$\rho^*(\Phi_\mathbf{A}^t) = \rho^*(t). \tag{6.21}$$

From (6.12) and (6.21) one concludes that (6.20) is satisfied. □

In the case when the relations (6.17) are not satisfied, the equations (6.15) can be written in the form

$$i\hbar \frac{dM_{A_j}^t(\Phi_\mathbf{A})}{dt} - \mathrm{tr}(\rho^*(\Phi_\mathbf{A}^t)[A_j, H]) = \tag{6.22}$$

$$= \mathrm{tr}\{(\rho^*(t) - \rho^*(\Phi_\mathbf{A}^t))[A_j, H]\}, \quad j = 1, \ldots, p,$$

The closed system of equations for $M_{A_1}^t(\Phi_A), \ldots, M_{A_p}^t(\Phi_A)$ is obtained provided the right-hand side in (6.21) is equal to zero.

The approximate estimate values $\overline{M}_{A_j}^t(\Phi_A)$, $j = 1, \ldots, p$, are defined as the solution of the equations

$$i\hbar \frac{d\overline{M}_{A_j}^t(\Phi_A)}{dt} = \operatorname{tr}(\rho^*(\overline{\Phi}_A^t)[A_j, H]), \quad j = 1, \ldots, p, \tag{6.23}$$

with the initial condition

$$\overline{M}_{A_j}^0(\Phi_A) = M_{A_j}(\Phi_A), \quad j = 1, \ldots, p, \tag{6.24}$$

where $\rho^*(\overline{\Phi}_A^t)$ is determined by the equations

$$\rho^*(\Phi_A^t) = Z^{-1}[\tilde{\lambda}(t)] \exp\left(-\sum_{j=1}^{p} \tilde{\lambda}_j(t) A_j\right), \tag{6.25}$$

$$Z[\tilde{\lambda}(t)] = \operatorname{tr} \exp\left(-\sum_{j=1}^{p} \tilde{\lambda}_j(t) A_j\right), \tag{6.26}$$

and

$$\overline{M}_{A_j}^t(\Phi_A) = -\frac{\partial Z[\tilde{\lambda}(t)]}{\partial \tilde{\lambda}_j(t)}, \quad j = 1, \ldots, p. \tag{6.27}$$

Moreover, $\rho^*(\overline{\Phi}_A^t)$ is the representative state for the approximative macrostate $\overline{\Phi}_A^t$, $t \geq 0$, which is defined by the relation

$$\overline{\Phi}_A^t = \{\rho \in \mathfrak{S}(\mathcal{H}) : \operatorname{tr}(\rho A_j) = \overline{M}_{A_j}^t(\Phi_A), \ j = 1, \ldots, p\}. \tag{6.28}$$

The equations (6.23) combined with (6.25), (6.26) and (6.27) form the closed system of equations for $\overline{M}_{A_j}^t(\Phi_A)$, $j = 1, \ldots, p$. It should be noted that the equations (6.23) are exact provided the condition (6.17) is satisfied.

THEOREM 6.3. *For every macrostate* Φ_A *and* $t \geq 0$ *the equality*

$$S(\overline{\Phi}_A^t) = S(\Phi_A) \tag{6.29}$$

holds, where

$$S(\overline{\Phi}_A^t) = \sup\{S(\rho) : \rho \in \overline{\Phi}_A^t\} = -\operatorname{tr}(\rho^*(\overline{\Phi}_A^t) \ln \rho^*(\overline{\Phi}_A^t)). \tag{6.30}$$

Proof. Taking into account (6.25), (6.26) and (6.27) it follows from (6.29) that $S(\overline{\Phi}_{\mathbf{A}}^t)$ can be written in the form

$$S(\overline{\Phi}_{\mathbf{A}}^t) = \ln Z[\tilde{\lambda}(t)] + \sum_{j=1}^p \tilde{\lambda}_j(t)\overline{M}_{A_j}^t(\Phi_{\mathbf{A}}). \qquad (6.31)$$

Differentiating (6.31) with respect to time and using (6.23), one obtains

$$\frac{dS(\overline{\Phi}_{\mathbf{A}}^t)}{dt} = \mathrm{tr}\Big\{ H[\rho^\star(\overline{\Phi}_{\mathbf{A}}^t), \sum_{j=1}^p \tilde{\lambda}_j(t)A_j] \Big\} = 0. \qquad (6.32)$$

The last equality follows from (6.25). Since $S(\overline{\Phi}_{\mathbf{A}}^0) = S(\Phi_{\mathbf{A}})$ one gets (6.29).
□

The time evolution of the estimate values $\overline{M}_{A_j}^t(\Phi_{\mathbf{A}})$, $j = 1, \ldots, p$, given in terms of the equations (6.23) – (6.27), due to the equality (6.29) will be called an **A**-*isoentropic motion*.

6.2. Estimation of observables

Let X be an observable and $\mathbf{A} = (A_1, \ldots, A_p)$ be an incomplete set of thermodynamically regular observables such that a **A**-macrostate $\Phi_{\mathbf{A}}$ is given.
    The operator

$$\tilde{X}[\Phi_{\mathbf{A}}] = \sum_{j=0}^p a_j A_j, \qquad (6.33)$$

where $a_0, a_1, \ldots, a_p \in \mathbb{R}^1$, is called the *estimation of the operator* X on the macrostate $\Phi_{\mathbf{A}}$ provided the relation

$$\mathrm{tr}\Big\{\rho^\star(X - \tilde{X}[\Phi_{\mathbf{A}}])^2\Big\} = \min_{(a_0,a_1,\ldots,a_p)} \mathrm{tr}\Big\{\rho^\star(X - \sum_{j=0}^p a_j A_j)^2\Big\}, \qquad (6.34)$$

where $\rho^\star = \rho^\star(\Phi_{\mathbf{A}})$ is the representative state for $\Phi_{\mathbf{A}}$.
THEOREM 6.4. *If the expectation values*

$$K_{mn} = K_{nm} = \mathrm{tr}\rho^\star(A_m A_n + A_n A_m) \qquad (6.35)$$
$$C_m = \mathrm{tr}\rho^\star(X A_m + A_m X), \quad m, n = 0, 1, \ldots, p, \qquad (6.36)$$

and $\mathrm{tr}(\rho^\star X^2)$ *are finite, then there exists unique* $\tilde{X}[\Phi_{\mathbf{A}}]$ *such that for all* $\rho \in \Phi_{\mathbf{A}}$ *the equality*

$$M_X(\Phi_{\mathbf{A}}) \equiv \mathrm{tr}(\rho^\star X) = \mathrm{tr}(\rho \tilde{X}[\Phi_{\mathbf{A}}]) \qquad (6.37)$$

holds.

Proof. Since $A_0 = I$, $A_1, \ldots, A_p$ are linearly independent, for all a_0, $a_1, \ldots, a_p \in \mathbb{R}^1$ such that $\sum_{j=0}^{p} a_j^2 > 0$ the inequality

$$\sum_{m,n=0}^{p} K_{mn} a_m a_n = 2 \operatorname{tr}\left(\rho^*(\sum_{m=0}^{p} a_m A_m)^2\right) > 0 \qquad (6.38)$$

holds. From (6.38) it follows that the matrix $K = \| K_{mn} \|_{n,m=1}^{p}$ is strictly positive and $\det K > 0$. Let F be defined as follows

$$F = F(x_0, x_1, \ldots, x_p) = \operatorname{tr}\left(\rho^*(X - \sum_{j=0}^{p} x_j A_j)^2\right). \qquad (6.39)$$

The conditions for the minimum of F have the form

$$\frac{\partial F}{\partial x_i} = \sum_{j=0}^{p} K_{ij} x_j - C_j = 0, \qquad (6.40)$$

and

$$\sum_{i,j=0}^{p} \frac{\partial^2 F}{\partial x_i \partial x_j} z_i z_j = \sum_{i,j=0}^{p} K_{ij} z_i z_j > 0 \qquad (6.41)$$

for all $z_0, z_1, \ldots, z_p \in \mathbb{R}^1$. The inequality (6.41) is implied by (6.38). Since the matrix K is nonsingular, the equations (6.40) determine uniquely $x_0, x_1, \ldots, x_p$. Putting $i = 0$ in (6.40) and taking into account (6.35), (6.36), one finds that the equality

$$\operatorname{tr}(\rho^* X) = \operatorname{tr}(\rho^* \tilde{X}[\Phi_A]) \qquad (6.42)$$

holds. The relation (6.37) follows from (6.42) and the definition of Φ_A. $\square$

It follows from (6.40), (6.35), and (6.36) that the $x_0, x_1, \ldots, x_p$ are functions of $M_{A_1}(\Phi_A), \ldots, M_{A_p}(\Phi_A)$. Using (6.35) and (6.36) the equations (6.40) can be rewritten in the form

$$\operatorname{tr}(\rho^*\{X - \tilde{X}[\Phi_A], A_j\}) = 0, \quad j = 0, \ldots, p, \qquad (6.43)$$

where $\{A, B\} = AB + BA$. The notion of the estimation $\tilde{X}[\Phi_A]$ of the operator X is convenient since $\tilde{X}[\Phi_A]$ is parameterized in terms of expectation values $M_{A_1}(\Phi_A), \ldots, M_{A_p}(\Phi_A)$ of the observables $A_1, \ldots, A_p \in \mathbf{A}$.

6.3. The estimation of Hamiltonian and isoentropic motion

Let $\mathbf{A} = (A_1, \ldots, A_p)$ be incomplete set of observables and let $\dim \mathcal{H} < \infty$. The linear span of $A_1, \ldots, A_p$ is denoted by $\mathfrak{A}_\mathbf{A}$. Moreover, it is assumed that $\mathfrak{A}_A$ is a Lie algebra, i.e., the observables $A_1, \ldots, A_p$ satisfy the commutation relations

$$[A_k, A_l] = i \sum_{j=1}^{p} c_{kl}^j A_j, \tag{6.44}$$

where

$$c_{kl}^j = -c_{lk}^j = \bar{c}_{kl}^j \tag{6.45}$$

and the Jacobi identity holds. Each element $a \in \mathfrak{A}_A$ can be written in the form

$$a = (\mathbf{a}, \mathbf{A}) = \sum_{j=1}^{p} a_j A_j, \tag{6.46}$$

where $\mathbf{a} = (a_1, \ldots, a_p) \in \mathbb{Z}^p$. Let $b = b^* \in \mathfrak{A}_A$, then the relation

$$(\mathcal{F}\mathbf{a}, \mathbf{A}) = a + e^b a e^{-b} \tag{6.47}$$

defines the linear map $\mathcal{F}: \mathbb{Z}^p \to \mathbb{Z}^p$.

LEMMA 6.1. *The map $\mathcal{F}$ has the inverse.*

Proof. Suppose the opposite, i.e., that there exists a subspace $P \subset \mathbb{Z}^p$ such that

$$\mathcal{F}\mathbf{a} = 0 \tag{6.48}$$

holds for $\mathbf{a} \in P$. Let us consider the case $\dim P = 1$. If $\mathbf{a} \in P$ it follows (6.47) that

$$a + e^b a e^{-b} = 0 \tag{6.49}$$

and

$$[b, a] + e^b [b, a] e^{-b} = 0. \tag{6.50}$$

Since $\dim P = 1$ one has either

$$[b, a] = 0 \tag{6.51}$$

or

$$[b, a] = \lambda a. \tag{6.52}$$

If $[b, a] = 0$, then it follows from (6.49) that $a = 0$, i.e., $\mathbf{a} = 0$ and $\mathcal{F}$ has the inverse. If (6.52) holds, one finds from (6.52) that

$$e^b a e^{-b} = e^\lambda a \tag{6.53}$$

and (6.49) is satisfied, provided

$$\lambda = \pm i\pi(2k+1), \quad k = 0, 1, \ldots. \tag{6.54}$$

Since dim $\mathcal{H} = n < \infty$, and $b = b^*$, there are orthogonal eigenvectors $\varphi_1, \ldots, \varphi_n \in \mathcal{H}$ such that

$$b\varphi_\alpha = \beta_\alpha \varphi_\alpha, \quad \alpha = 1, \ldots, n, \tag{6.55}$$

and the eigenvalues $\beta_1, \ldots, \beta_n$ are real. From (6.52), (6.54) and (6.55) it follows that

$$ba\varphi_\alpha = (\beta_\alpha \pm i\pi(2k+1))a\varphi_\alpha, \tag{6.56}$$

contrary to (6.55), and consequently the relation

$$a\varphi_\alpha = 0 \tag{6.57}$$

should hold for $\alpha = 1, \ldots, n$, i.e., $a = 0$, ($\mathbf{a} = 0$), i.e., $\mathcal{F}$ has the invarse. If dim $\mathcal{H} > \infty$ the proof goes along the same lines. $\square$

LEMMA 6.2. *Let the equations*

$$\mathrm{tr}(e^b X A_j + A_j X) = 0, \quad j = 1, \ldots, p, \tag{6.58}$$

be satisfied, where $b = b^* \in \mathfrak{A}_\mathbf{A}$, $X \in \mathcal{B}(\mathcal{H})$. *Then one has*

$$\mathrm{tr}(e^b X A_j) = \mathrm{tr}e^b(A_j X) = 0, \quad j = 1, \ldots, p. \tag{6.59}$$

Proof. Using (6.47), the equations (6.58) can be rewritten in the form

$$\mathrm{tr}(e^b X (A_j + e^b A_j e^{-b})) = 0, \quad j = 1, \ldots, p, \tag{6.60}$$

or

$$\mathcal{F}^T z = 0, \tag{6.61}$$

where

$$z_j = \mathrm{tr}(e^b X A_j), \quad j = 1, \ldots, p, \tag{6.62}$$

and ^{T} denotes the transposition. From Lemma 3.1. and (6.61) one obtains (6.59). $\square$

The above result can be also formulated in the following manner: Let $b = b^* \in \mathfrak{A}_\mathbf{A}$, $X \in \mathcal{B}(\mathcal{H})$; then the equality

$$\mathrm{tr}(e^b\{X, a\}) = 0 \tag{6.63}$$

for all $a \in \mathfrak{A}_\mathbf{A}$ implies that

$$\mathrm{tr}(e^b[X, a]) = 0. \tag{6.64}$$

As an immediate consequence of Lemma 6.2 one has

THEOREM 6.5. *Let* $\mathbf{A} = (A_1, \ldots, A_p)$ *be the incomplete set of observables such that* $\mathbf{A}$ *generates the Lie algebra* $\mathfrak{A}_\mathbf{A}$. *Then for every macrostate* $\Phi_\mathbf{A}$ *such that* $\rho^* = \rho^*(\Phi_\mathbf{A}) > 0$ *and any observable* X *the equality*

$$\text{tr}(\rho^*[A_j, X]) = \text{tr}(\rho[A_j, \tilde{X}[\Phi_\mathbf{A}]]), \quad j = 1, \ldots, p, \qquad (6.65)$$

holds for all $\rho \in \Phi_\mathbf{A}$, *where* $\tilde{X}[\Phi_\mathbf{A}]$ *is the estimate for* X *on* $\Phi_\mathbf{A}$.

Proof. Taking into account (6.43), the form of the representative state ρ^*, (6.63) and (6.64) one obtains

$$\begin{aligned} \text{tr}(\rho^*[A_j, X]) &= \text{tr}(\rho^*[A_j, \tilde{X}[\Phi_\mathbf{A}]) = \\ &= \text{tr}(\rho[A_j, \tilde{X}[\Phi_\mathbf{A}]]), \quad j = 1, \ldots, p. \end{aligned} \qquad (6.66)$$

The last equality follows from (6.33), the definition of $\Phi_\mathbf{A}$ and the fact that $[A_j, \tilde{X}[\Phi_\mathbf{A}]] \in \mathfrak{A}_\mathbf{A}$ for all $j = 1, \ldots, p$. $\square$

Let us consider the **A**-isoentropic motion introduced in Section 6.1. Using (6.33), (6.43), (6.65), the equations (6.15) can be rewritten in the form

$$i\hbar \frac{dM^t_{A_j}(\Phi_\mathbf{A})}{dt} - \text{tr}(\rho^*(\Phi^t_\mathbf{A})[A_j, \tilde{H}[\Phi^t_\mathbf{A}]]) = \qquad (6.67)$$

$$= \text{tr}\left\{\rho^*(t)[A_j, H - \tilde{H}[\Phi^t_\mathbf{A}]]\right\}, \quad j = 1, \ldots, p,$$

with the initial condition (6.16). One can observe that the equations (6.23) for **A**-isoentropic motion are obtained in zero-order perturbation with respect to the operator

$$V[\Phi^t_\mathbf{A}] = H - \tilde{H}[\Phi^t_\mathbf{A}], \qquad (6.68)$$

and they can be written as follows

$$i\hbar \frac{dM^t_{A_j}(\Phi_\mathbf{A})}{dt} = \text{tr}\left(\rho^*(\Phi^t_\mathbf{A})[A_j, \tilde{H}[\Phi^t_\mathbf{A}]]\right), \quad j = 1, \ldots, p, \qquad (6.69)$$

with the initial condition (6.16).

6.4. Mean field method and A-isoentropic motions

The estimated value $M_X(\Phi_\mathbf{A})$ of the observable X in the macrostate $\Phi_\mathbf{A}$ is given by the formula

$$M_X(\Phi_\mathbf{A}) = \text{tr}(\rho^*(\Phi_\mathbf{A})X), \qquad (6.70)$$

where $\rho^*(\Phi_{\mathbf{A}})$ is the representative state for $\Phi_{\mathbf{A}}$, c.f., (5.25), (5.26) and (5.27). It follows from the definition of $\rho^*(\Phi_{\mathbf{A}})$, that $M_X(\Phi_{\mathbf{A}})$ is the function of the mean values $M_{A_1}(\Phi_{\mathbf{A}}), \ldots, M_{A_p}(\Phi_{\mathbf{A}})$. A macrostate $\Phi_{\mathbf{A}}$ belonging to the family

$$\mathcal{F}_{\mathbf{A}}(X) = \left\{ \Phi_{\mathbf{A}} : \frac{\partial M_X(\Phi_{\mathbf{A}})}{\partial M_{A_j}(\Phi_{\mathbf{A}})} = 0, \quad j = 1, \ldots, p \right\} \tag{6.71}$$

is called a *Hartree-Fock type* $\mathbf{A}$-*macrostate* with respect to the observable X.

Let $\dim \mathcal{H} < \infty$ and $\mathbf{A} = (A_1, \ldots, A_p)$ be the incomplete set of observables such that the linear span $\mathfrak{A}_{\mathbf{A}}$ of $\mathbf{A}$ is a Lie algebra. For fixed $b = b^* \in \mathfrak{A}_{\mathbf{A}}$ and arbitrary $x \in \mathbb{R}^1$ the family of linear maps $\mathcal{F}(x) \colon \mathbb{Z}^p \to \mathbb{Z}^p$ is defined as follows

$$(\mathcal{F}(x)\mathbf{z}, \mathbf{A}) = e^{xb}(\mathbf{z}, \mathbf{A})e^{-xb}. \tag{6.72}$$

LEMMA 6.3. *The map* $\Gamma \colon \mathbb{Z}^p \to \mathbb{Z}^p$, *where*

$$\Gamma = \int_0^1 dx \, \mathcal{F}(x), \tag{6.73}$$

is invertible.

Proof is similar to that of Lemma 6.1.

Under the above assumptions concerning $\mathbf{A}$ one has

THEOREM 6.6. *Let* $\Phi_{\mathbf{A}} \in \mathcal{F}_{\mathbf{A}}(X)$ *be the* $\mathbf{A}$-*macrostate such that* $\rho^*(\Phi_{\mathbf{A}}) > 0$. *Then the relations*

$$\mathrm{tr}(\rho^*(\Phi_{\mathbf{A}})A_j X) = \mathrm{tr}(\rho^*(\Phi_{\mathbf{A}})X)\mathrm{tr}(\rho^*(\Phi_{\mathbf{A}})A_j) \tag{6.74}$$

hold for $j = 1, \ldots, p$.

Proof. Since $\rho^*(\Phi_{\mathbf{A}})$ has the form (5.25) and there is one-to-one correspondence between the Lagrange multipliers $\beta_1, \ldots, \beta_p$ and mean values $M_{A_1}(\Phi_{\mathbf{A}}), \ldots, M_{A_p}(\Phi_{\mathbf{A}})$, it follows that (6.71) can be defined in the following equivalent form

$$\mathcal{F}_{\mathbf{A}}(X) = \left\{ \Phi_{\mathbf{A}} : \frac{\partial M_X(\Phi_{\mathbf{A}})}{\partial \beta_j} = 0, \quad j = 1, \ldots, p \right\}. \tag{6.75}$$

On the other hand, using the formula, c.f. [343],

$$\frac{\partial}{\partial \lambda} \exp(Z) = \exp(Z) \int_0^1 dx \, e^{-xZ} \frac{\partial Z}{\partial \lambda} e^{xZ}, \tag{6.76}$$

and (5.25), (5.26), (5.27), (6.73), the equations

$$\frac{\partial M_X(\Phi_A)}{\partial \beta_j} = 0, \quad j = 1, \ldots, p, \tag{6.77}$$

can be rewritten in the form

$$\Gamma z = 0, \tag{6.78}$$

where

$$z_j = \text{tr}(\rho^*(\Phi_A) A_j X) - \text{tr}(\rho^*(\Phi_A) X) \text{tr}(\rho^*(\Phi_A) A_j), \quad j = 1, \ldots, p. \tag{6.79}$$

Using Lemma 6.3 one obtains (6.74). $\square$

It follows from Theorem 6.6 that if $\Phi_A \in \mathcal{F}_A(X)$ and $\rho^*(\Phi_A) > 0$ the relations

$$\text{tr}(\rho^*(\Phi_A)[A_j, X]) = 0, \quad j = 1, \ldots, p, \tag{6.80}$$

hold.

In the case when $X = H$ — the Hamiltonian of the system, it follows from (6.23) and (6.80) that every $\Phi_A \in \mathcal{F}_A(H)$ such that $\rho^*(\Phi_A) > 0$ is the stationary A-macrostate with respect to the A-isoentropic motion.

6.5. Application to the system of N fermions

As an illustration of the formalism developed in the preceding sections, the system of N identical fermions with spin S ($S = 1/2, 3/2, \ldots$) will be considered. The Hilbert space $\mathcal{H}_N^\wedge \equiv \mathcal{H}_N^\wedge(V, S)$ corresponding to the system in question consists of skew-symmetric functions of N-variables $x_1, \ldots, x_N$, where $x = (\vec{r}, \alpha)$, $\vec{r} \in V \subseteq \mathbb{R}^3$, $\alpha = \mathbb{Z}_S = \{-S, \ldots, S\}$ and the scalar product $(\varphi, \psi)_N$ of elements $\varphi, \psi \in \mathcal{H}_N^\wedge$ has the form

$$(\varphi, \psi)_N = \int d^N x \overline{\varphi}(x_1, \ldots, x_N) \psi(x_1, \ldots, x_N), \tag{6.81}$$

where

$$\int d^N x(\cdots) = \sum_{\alpha_1 = -S}^{S} \int_V d^3 r_1 \cdots \sum_{\alpha_N = -S}^{S} \int d^3 r_N(\cdots). \tag{6.82}$$

The set of all density operators on $\mathcal{H}_N^\wedge$ will be denoted by $\mathfrak{S}_N$

Each element $\varphi \in \mathcal{H}_N^\wedge$ is mapped by the operator $\rho_N \in \mathfrak{S}_N$ into the element $\rho_N \varphi \in \mathcal{H}_N^\wedge$ given by the expression

$$(\rho_N \varphi)(x_1, \ldots, x_N) = \int d^N y \rho_N(x_1, \ldots, x_N | y_1, \ldots, y_N) \varphi(y_1, \ldots, y_N). \tag{6.83}$$

Since the operator $\rho_N \in \mathfrak{S}_N$ is a positive definite nuclear operator, it can be represented in the form

$$\rho_N(x_1, \ldots, x_N | y_1, \ldots, y_N) = \sum_{i=1}^{\infty} \lambda_i \varphi_i(x_1, \ldots, x_N) \overline{\varphi}_i(y_1, \ldots, y_N), \quad (6.84)$$

where $\lambda_i \geq 0$, $(i = 1, \ldots)$, $\sum_{i=1}^{\alpha} \lambda_i = 1$, and the functions $\varphi_i(x_1, \ldots, x_N)$, $i = 1, 2, \ldots$, form an orthonormal and complete base in $\mathcal{H}_N^{\wedge}$.

Let $\mathcal{H}_p \equiv \mathcal{H}_p^{\wedge}(V, S)$ be the corresponding Hilbert space of p fermions $(1 \leq p \leq N)$, and $\mathfrak{S}_p$ be the set of all p-fermions density operators. To every density operator $\rho_N \in \mathfrak{S}_N$ one can associate the density operator $\rho_N^p \in \mathfrak{S}_p$ by means of the formula

$$\rho_N^p(x_1, \ldots, x_p | x_1', \ldots, x_p') =$$
$$\int d^{(N-p)} y \rho_N(x_1, \ldots, x_p, y_{p+1}, \ldots, y_N | x_1', \ldots, x_p', y_{p+1}, \ldots, y_N). \quad (6.85)$$

The relation (6.85) defines the map $T_N^p : \mathfrak{S}_N \to \mathfrak{S}_p$. The operator ρ_N^p is called the *reduced p-fermion density operator*, cf. [57].

The problem of characterization of the set $T_N^p \mathfrak{S}_N \subset \mathfrak{S}_p$, called the *N-representability problem* of p-fermion density operators, cf. [57], [207], is solved completely for $p = 1$. 1-fermion density operator $\rho_1 \in \mathfrak{S}_1$ with the spectral decomposition

$$\rho_1(x | x') = \sum_{i=1}^{\infty} \mu_i \varphi_i(x) \overline{\varphi}_i(x') \quad (6.86)$$

is an element of $T_N^1 \mathfrak{S}_N$ iff the eigenvalues $\mu_1, \mu_2, \ldots$ obey the inequalities

$$0 \leq \mu_i \leq \frac{1}{N}, \quad i = 1, 2, \ldots \quad (6.87)$$

and the normalization condition

$$\sum_{i=1}^{\infty} \mu_i = 1. \quad (6.88)$$

In particular, it follows from (6.87) and (6.88) that the operator $\rho_1 \in T_N^1 \mathfrak{S}_N$ with the minimal number of non-zero eigenvalues has the decomposition

$$\rho(x | x') = \frac{1}{N} \sum_{i=1}^{N} \varphi_i(x) \overline{\varphi}_i(x'). \quad (6.89)$$

A linear operator $L : \mathcal{H}_N^{\wedge} \to \mathcal{H}_N^{\wedge}$ is defined in terms of the corresponding kernel $L(x_1, \ldots, x_N | x_1', \ldots, x_N')$. In most quantum mechanical problems, the essential role is played by 1- and 2-particles observables. The most general form of *one-particle observable* is the following

$$O_1(x_1,\ldots,x_N|x'_1,\ldots,x'_N) = \tag{6.90}$$

$$\frac{1}{N!}\sum_{i=1}^{N}\begin{vmatrix} \delta(x_1 - x'_1), & \cdots & ,\omega(x_1|x'_i), & \cdots & ,\delta(x_1 - x'_N) \\ \vdots & & \vdots & & \vdots \\ \delta(x_N - x'_1), & \cdots & ,\omega(x_N|x'_i), & \cdots & ,\delta(x_N - x'_N) \end{vmatrix},$$

where

$$\delta(x - x') = \delta(\vec{r} - \vec{r}')\delta_{\alpha\alpha'}, \qquad (x = (\vec{r},\alpha), \quad x' = (\vec{r}',\alpha')) \tag{6.91}$$

and

$$\omega(x|x') = \overline{\omega}(x'|x) \tag{6.92}$$

is an observable in $\mathcal{H}_1$.

In particular, if $\omega(x|x') = \delta(x - x')\omega(x)$ the observable O_1 has the form

$$O_1(x_1,\ldots,x_N|x'_1,\ldots,x'_N) = [\omega(x_1)+\ldots+\omega(x_N)]I_N(x_1,\ldots,x_N|x'_1,\ldots,x'_N),$$
$$\tag{6.93}$$

where I_N is the identity operator on $\mathcal{H}_N^{\wedge}$:

$$I_N(x_1,\ldots,x_N|x'_1,\ldots,x'_N) = \frac{1}{N!}\det\|\,\delta(x_i - x'_j)\,\|_{i,j=1}^{N}\,. \tag{6.94}$$

The name of 1-particle operator is justified by the fact that its expectation value $\mathrm{tr}(\rho_N O_1)$ is determined in terms of the 1-fermion reduced density operator ρ_N^1, i.e.,

$$\mathrm{tr}(\rho_N O_1) = N\mathrm{tr}(\omega\rho_N^1) = \int dx\,dx'\rho_N^1(x|x')\omega(x'|x)\,. \tag{6.95}$$

LEMMA 6.4. *Let the operator* $\omega\colon \mathcal{H}_1 \to \mathcal{H}_1$ *have the spectral decomposition*

$$\omega(x|x') = \sum_{i=1}^{\infty}\omega_i\varphi_i(x)\overline{\varphi}_i(x')\,, \tag{6.96}$$

where $\omega_i = \overline{\omega}_i$, *and the functions* $\varphi_i(x)$, $i = 1,2,\ldots$, *form an orthonormal and complete base (CONS) in* $\mathcal{H}_N^{\wedge}$. *Then the 1-particle operator* O_1, *defined by (6.90), can be written in the form*

$$O_1(x_1,\ldots,x_N|x'_1,\ldots,x'_N) = \tag{6.97}$$

$$\sum_{1\leq i_1<\ldots<i_N<\infty}(\omega_{i_1}+\ldots+\omega_{i_N})\varphi_{[i_1}(x_1)\ldots\varphi_{i_N]}(x_N)\overline{\varphi}_{[i_1}(x'_1)\ldots\overline{\varphi}_{i_N]}(x'_N)\,,$$

where

$$\varphi_{[i_1}(x_1)\dots\varphi_{i_N]}(x_N) = \frac{1}{\sqrt{N!}}\begin{vmatrix} \varphi_{i_1}(x_1), & \dots & ,\varphi_{i_N}(x_1) \\ \vdots & & \vdots \\ \varphi_{i_1}(x_N), & \dots & ,\varphi_{i_N}(x_N) \end{vmatrix} \qquad (6.98)$$

is the Hartree-Fock state in $\mathcal{H}_N^{\wedge}$.

Proof. Taking into account (6.90) and the relation

$$\delta(x - x') = \sum_{i=1}^{\infty} \varphi_i(x)\overline{\varphi}_i(x'), \qquad (6.99)$$

one obtains (6.97). □

An example of 2-particle observable is the operator O_2:

$$O_2(x_1,\dots,x_N|x_1',\dots,x_N') = I_N(x_1,\dots,x_N|x_1',\dots,x_N') \sum_{1\le i<j\le N} V(x_i,x_j),$$
$$(6.100)$$

where $V(x,x')$ is a real function of $x, x' \in V \odot \mathbb{Z}_S$, such that

$$V(x,x') = V(x',x). \qquad (6.101)$$

For 2-particle observable O_2 the relation

$$\mathrm{tr}(\rho_N O_2) = \frac{N(N-1)}{2}\int dx\,dx'\,\rho_N^2(x,x'|x,x')V(x,x') \qquad (6.102)$$

holds.

In order to illustrate the general considerations given in the preceding sections, the incomplete set $\mathbf{A}$ of observables will be chosen in the form

$$\mathbf{A} = \{(x|A_1|x') : x, x' \in V \odot \mathbb{Z}_S\}, \qquad (6.103)$$

where the family of 1-particle operators $(x|A_1|x')$ is given in the form

$$(x|A_1|x')(x_1,\dots,x_N|x_1',\dots,x_N') = \qquad (6.104)$$

$$\frac{1}{N}\frac{1}{N!}\sum_{j=1}^{N}\begin{vmatrix} \delta(x_1 - x_1'), & \dots & ,\delta(x_1 - x')\delta(x_j' - x), & \dots & ,\delta(x_1 - x_N') \\ \vdots & & \vdots & & \vdots \\ \delta(x_N - x_1'), & \dots & ,\delta(x_N - x')\delta(x_j' - x), & \dots & ,\delta(x_N - x_N') \end{vmatrix}.$$

The operators $(x|A_1|x')$ have the following properties

$$(x|A_1^*|x') = (x'|A_1|x), \tag{6.105}$$

$$\int dx (x|A_1|x) = I_N, \tag{6.106}$$

$$[(x|A_1|x'),(y|A_1|y')] = \frac{1}{N}\{\delta(x-y)(y|A_1|x') + \\ - \delta(y-x')(x|A_1|y')\}, \tag{6.107}$$

i.e., they generate a Lie algebra, and

$$\mathrm{tr}(\rho_N(x|A_1|x')) = \rho_N^1(x|x'). \tag{6.108}$$

It means that the incomplete description of the system of N-fermions will be given in terms of 1-fermion reduced density operators.

The **A**-macrostates with respect to the set **A** will be denoted by $\Phi, \Psi, \ldots$ A **A**-macrostate containing state $\rho_N \in \mathfrak{S}_N$ will also be denoted by $[\rho_N]$. N-representable 1-fermion density operator $\rho_1 \in T_N^1 \mathfrak{S}_N$ is said to be *regular* if there are non-negative numbers $\kappa_1, \kappa_2, \ldots$, such that

$$\sum_{i_1,\ldots,i_N=1} \kappa_{i_1} \ldots \kappa_{i_N} \delta_{i_1,\ldots,i_N}^{i_1,\ldots,i_N} = 1, \tag{6.109}$$

where

$$\delta_{j_1,\ldots,j_N}^{i_1,\ldots,i_N} = \frac{1}{N!} \begin{vmatrix} \delta_{j_1}^{i_1}, & \cdots & , \delta_{j_N}^{i_1} \\ \vdots & & \vdots \\ \delta_{j_1}^{i_N}, & \cdots & , \delta_{j_N}^{i_N} \end{vmatrix}, \tag{6.110}$$

and the eigenvalues $\mu_1, \mu_2, \ldots$ of the density operator ρ_1 can be written in the form

$$\mu_i = \kappa_i \sum_{i_2,\ldots,i_N=1}^{\infty} \kappa_{i_2} \ldots \kappa_{i_N} \delta_{i,i_2,\ldots,i_N}^{i,i_2,\ldots,i_N}. \tag{6.111}$$

In particular, if one chooses

$$\kappa_1 = \ldots = \kappa_M = \left(\frac{M}{N}\right)^{-\frac{1}{N}},$$

$$\kappa_{M+k} = 0, \quad k = 1,2,\ldots, \quad M \geq N, \tag{6.112}$$

then the relation (6.109) is satisfied and from (6.110) it follows that one has

$$\mu_1 = \ldots = \mu_M = \frac{1}{M}, \quad \mu_{M+k} = 0, \quad k = 1,2,\ldots. \tag{6.113}$$

In what follows, the notion

$$M_{\mathbf{A}}(\Phi) \;=\; \rho_1 \in T_N^1 \mathfrak{S}_N \qquad\qquad (6.114)$$

will be used instead of

$$M_{(x|A_1|x')}(\Phi) \;=\; \rho_1(x|x') . \qquad\qquad (6.115)$$

A macrostate Φ is said to be *regular* provided $M_{\mathbf{A}}(\Phi)$ is a regular 1-fermion N-representable density operator.

THEOREM 6.7. *Let Φ be a regular $\mathbf{A}$-macrostate, i.e.,*

$$\Phi \;=\; \{\rho_N \in \mathfrak{S}_N : \mathrm{tr}(\rho_N(x|A_1|x'))= \rho_1(x|x') \in T_N^1 \mathfrak{S}_N\}, \qquad (6.116)$$

where ρ_1 has the spectral decomposition

$$\rho_1(x|x') \;=\; \sum_{j=1}^{\infty} \mu_j \varphi_j(x) \overline{\varphi}_j(x') , \qquad\qquad (6.117)$$

and its eigenvalues can be written in the form (6.111). Then there exists the representative state $\rho_N^*(\Phi)$ for the macrostate Φ which has the form

$$\rho_N^*(\Phi)(x_1,\ldots,x_N|x'_1,\ldots,x'_N) \;=\; \frac{1}{N!}\det\Big(\,\|\,D(x_i|x'_j)\,\|_{i,j=1}^{N}\,\Big), \quad (6.118)$$

where

$$D(x|x') \;=\; \sum_{j=1}^{\infty} \kappa_j \varphi_j(x) \overline{\varphi}_j(x') \qquad\qquad (6.119)$$

and $\kappa_1,\kappa_2,\ldots$ are non-negative numbers satisfying the conditions (6.109), (6.111).

Proof. Using the method of Lagrange multipliers for finding the maximum of the entropy $S(\rho_N)$ on the macrostate Φ, one obtains that the representative state should have the form

$$\rho_N^*(\Phi) \;=\; Z^{-1} \exp\Big[-\int dx\,dx'\lambda(x|x')(x|A_1|x')\Big], \qquad (6.120)$$

where

$$Z \;=\; Z[\lambda] \;=\; \mathrm{tr}\exp\Big[-\int dx\,dx'\lambda(x|x')(x|A_1|x')\Big]. \qquad (6.121)$$

The Lagrange multiplier $\lambda(x'|x)$ being the self-adjoint operator on $\mathcal{H}_1$, i.e.,

$$\lambda(x|x') \;=\; \overline{\lambda(x'|x)}, \qquad\qquad (6.122)$$

is determined by the equation

$$-\frac{\delta \ln Z[\lambda]}{\delta \lambda(x'|x)} = \rho_1(x|x'), \qquad (6.123)$$

and the self-adjoint operator

$$\Lambda = \int dx\,dx'\lambda(x'|x)(x|A_1|x') \qquad (6.124)$$

is a 1-particle observable on $\mathcal{H}_N^\wedge$. Taking into account (6.103) and (6.104) one obtains

$$\Lambda(x_1,\ldots,x_N|x_1',\ldots,x_N') = \qquad (6.125)$$

$$\frac{1}{N}\frac{1}{N!}\sum_{j=1}^{N}
\begin{vmatrix}
\delta(x_1 - x_1'), & \cdots & ,\lambda(x_1|x_j'), & \cdots & ,\delta(x_1 - x_N') \\
\vdots & & \vdots & & \vdots \\
\delta(x_N - x_1'), & \cdots & ,\delta(x_N|x_j'), & \cdots & ,\delta(x_N - x_N')
\end{vmatrix}.$$

The existence of $Z[\lambda]$ implies that the operator Λ should have the point spectrum, i.e., the operator $\lambda(x|x')$ should have the following spectral decomposition

$$\lambda(x|x') = \sum_{j=1}^{\infty} \lambda_j \psi_j(x)\overline{\psi}_j(x'), \qquad (6.126)$$

where functions $\psi_i(x)$ form an orthonormal and complete base in $\mathcal{H}_1$. From Lemma 6.4 it follows that Λ can be written in the form

$$\Lambda(x_1,\ldots,x_N|x_1',\ldots,x_N') = \qquad (6.127)$$

$$\frac{1}{N}\sum_{1\le i_1 < \ldots < i_N < \infty}(\lambda_{i_1}+\ldots+\lambda_{i_N})\psi_{[i_1}(x_1)\ldots\psi_{i_N]}(x_N)\overline{\psi}_{[i_1}(x_1')\ldots\overline{\psi}_{i_N]}(x_N').$$

Taking into account (6.127) and (6.121), one finds

$$Z[\lambda] = \sum_{i_1,\ldots,i_N=1}^{\infty} \delta_{i_1,\ldots,i_N}^{i_1,\ldots,i_N} \exp\left[-\frac{1}{N}(\lambda_{i_1}+\ldots+\lambda_{i_N})\right]. \qquad (6.128)$$

Using the relation

$$\frac{\delta\lambda_i}{\delta\lambda(x'|x)} = \psi_i(x)\overline{\psi}_i(x') \qquad (6.129)$$

which follows from (6.126), the equation (6.123) takes the form

$$Z^{-1}\sum_{i_1,\ldots,i_N=1}\delta_{i_1,\ldots,i_N}^{i_1,\ldots,i_N}\exp[-\frac{1}{N}(\lambda_{i_1}+\ldots+\lambda_{i_N})]\psi_{i_1}(x)\overline{\psi}_{i_1}(x') = \rho_1(x|x').$$

$$(6.130)$$

Introducing the parameters $\kappa_1, \kappa_2, \ldots$ by the relation

$$\kappa_i = \left[Z^{-N} e^{-\lambda_i}\right]^{\frac{1}{N}}, \quad i = 1, 2, \ldots, \tag{6.131}$$

the equation (6.130) can be rewritten in the form

$$\sum_{i_1,\ldots,i_N=1}^{\infty} \delta_{i_1,\ldots,i_N}^{i_1,\ldots,i_N} \kappa_{i_1} \ldots \kappa_{i_N} \psi_{i_1}(x) \overline{\psi}_{i_N}(x') = \rho_1(x|x') = \sum_{i=1}^{\infty} \mu_i \varphi_i(x) \overline{\varphi}_i(x'). \tag{6.132}$$

It follows from (6.132) that $\psi_i(x) = \varphi_i(x)$ and μ_i are related to κ_i by means of the formula (6.111). Using (6.120), (6.121), (6.124), (6.127), (6.128), (6.131) and (6.111) the representative state $\rho_N^{\star}(\Phi)$ takes the form

$$\rho_N^{\star}(\Phi)(x_1,\ldots,x_N|x_1',\ldots,x_N') = \tag{6.133}$$
$$\sum_{1 \leq i_1 < \ldots < i_N < \infty} \kappa_{i_1} \ldots \kappa_{i_N} \varphi_{[i_1}(x_1) \ldots \varphi_{i_N]}(x_N) \overline{\varphi}_{[i_1}(x_1') \ldots \overline{\varphi}_{i_N]}(x_N')$$

which implies (6.118).$\Box$

From (6.133) it follows that reduced p-fermion density operator $\rho_N^{\star p}(\Phi)$ has the form

$$\rho_N^{\star p}(\Phi)(x_1,\ldots,x_p|x_1',\ldots,x_p') = \tag{6.134}$$
$$\sum_{i_1,\ldots,i_N=1}^{\infty} \delta_{i_1,\ldots,i_N}^{i_1,\ldots,i_N} \kappa_{i_1} \ldots \kappa_{i_N} \varphi_{[i_1}(x_1) \ldots \varphi_{i_p]}(x_p) \overline{\varphi}_{[i_1}(x_1') \ldots \overline{\varphi}_{i_p]}(x_p') \quad 1 \leq p \leq N.$$

The representative state $\rho_N^{\star}(\Phi)$ defined by (6.118) and (6.119) depends implicitly on the density operator ρ_1 since the parameters $\kappa_1, \kappa_2, \ldots$ are functions of the eigenvalues $\mu_1, \mu_2, \ldots$ of ρ_1 given by (6.111).

Let Φ_M $(M \geq N)$ be the macrostate of the form

$$\Phi_M = \left\{ \rho_N \in \mathfrak{S}_N : \mathrm{tr}(\rho_N(x|\Lambda_1|x')) = \overline{\rho}_1(x|x') \right.$$
$$= \frac{1}{M} \sum_{i=1}^{M} \varphi_i(x) \overline{\varphi}_i(x'), \ M \geq N \bigg\}, \tag{6.135}$$

then taking into account (6.112), (6.113), (6.118) and (6.119) the representative state $\rho_N^{\star}(\Phi_M)$ can be written in the form

$$\rho_N^{\star}(\Phi_M)(x_1,\ldots,x_N|x_1',\ldots,x_N') = \frac{M^N}{N!\binom{M}{N}} \det(\| \overline{\rho}_1(x_i|x_j') \|_{i,j=1}^{N}). \tag{6.136}$$

One can easily check that if $M = N$ then $\rho_N^\star(\Phi_N)$ is the projector on the Hartree-Fock vector $\varphi_{[1}(x_1)\ldots\varphi_{N]}(x_N)$, while in the case $M > N$, $\binom{M}{N}\rho_N^\star(\Phi_M)$ is the projector on $\binom{M}{N}$-dimensional subspace of $\mathcal{H}_N^\wedge$ spanned by the Hartree-Fock vectors $\varphi_{[i_1}(x_1)\ldots\varphi_{i_N]}(x_N)$, $1 \leq i_1 < \ldots < i_N \leq M$. Moreover, it follows from (6.136) that the reduced p-fermion density operators $\rho_N^p(\Phi_M)$, $1 \leq p \leq N$, are given by the formula

$$\rho_N^p(\Phi_M)(x_1,\ldots,x_p|x_1',\ldots,x_p') = \frac{M^p}{p!\binom{M}{p}} \det(\| \bar{\rho}_1(x_i|x_j') \|_{i,j=1}^p). \quad (6.137)$$

It should be noted that the relation

$$\rho_N^1(\Phi_M)(x|x') = \bar{\rho}(x|x') \quad (6.138)$$

holds.

In order to apply the concepts developed in Sections 6.1, 6.2 and 6.3, the Hamiltonian of N-fermion system is assumed in the form

$$H_N(x_1,\ldots,x_N|x_1',\ldots,x_N') = \quad (6.139)$$
$$I_N(x_1,\ldots,x_N|x_1',\ldots,x_N')\Big[\sum_{1\leq i\leq N} H(x_i) + \sum_{1\leq i<j\leq N} V(x_i,x_j)\Big].$$

The results will be presented without proofs which are elementary.

THEOREM 6.8. *Let $\{\Phi_t; t \geq 0\}$ be the family of regular A-macrostates defined by the relation*

$$\Phi_t = \{\rho_n \in \mathfrak{S}_N : \operatorname{tr}(\rho_N(x|A|x'))= \rho_1(x|x';t) \in T_N^1\mathfrak{S}_N\}, \quad (6.140)$$

where $\rho_1(x|x';t)$ has the spectral decomposition (6.117) with time-dependent eigenvalues and eigenfunctions. Then the equations of A-isoentropic motion have the form

$$i\hbar\frac{\partial}{\partial t}\rho_1(x|x';t) = [H(x) - H(x')]\rho_1(x|x';t)$$
$$+ (N - 1) \int dy[V(x,y) - V(x',y)]\rho_N^{\star 2}(\Phi_t)(x,y|x',y), \quad (6.141)$$

where $\rho_N^{\star 2}(\Phi_t)$ is the reduced 2-fermion density operator obtained from the representative state $\rho_N^\star(\Phi_t)$ which is, by the construction, the function of $\rho_1(x|x';t) \equiv \rho_N^{\star 1}(\Phi_t)(x,x')$.
Moreover, the entropy $S(\Phi_t)$, given by the formula

$$S(\Phi_t) = -\operatorname{tr}\rho_N^\star(\Phi_t) \ln \rho_N^\star(\Phi_t) \quad (6.142)$$

is time independent.

It is worthwhile to note that the equations (6.141) are *Vlasov-type equations*, cf. [209].

THEOREM 6.9. *Let Φ_M ($M \geq N \geq 2$) be the A-macrostate and H_N be the Hamiltonian of N-fermion system given by (6.135) and (6.139), respectively. Then the estimated Hamiltonian $\tilde{H}_N[\Phi_M]$ is given by the formula*

$$\tilde{H}_N[\Phi_M] = I_N(x_1,\ldots,x_N|x_1',\ldots,x_N') \sum_{1\leq i\leq N} H(x_i) \qquad (6.143)$$

$$+ \frac{1}{N!} \sum_{1\leq i\leq N} \begin{vmatrix} \delta(x_1 - x_1'), & \cdots & ,W_M(x_1|x_i'), & \cdots & ,\delta(x_1 - x_N') \\ \vdots & & \vdots & & \vdots \\ \delta(x_N - x_1'), & \cdots & ,W_M(x_N|x_i'), & \cdots & ,\delta(x_N - x_N') \end{vmatrix},$$

where

$$W_N(x|x') = -N(N-1)(N-\tfrac{1}{2})\rho_N^{*1}(\Phi_N)(x|x')\int dydz\rho_N^{*2}(\Phi_N)(y,z|y,z)V(y,z$$

$$+ N(N-1)\int dy\rho_N^{*2}(\Phi_N)(x,y|x',y)[V(x,y)+V(x',y)]$$

$$+ N(N-1)(N-2)\int dydz\rho_N^{*3}(\Phi_N)(x,y,z|x',y,z)V(y,z) \qquad (6.144$$

for $M = N \geq 2$, and

$$W_N(x|x') = -\frac{1}{2}M^2(N_1)\rho_N^{*1}(\Phi_M)(x|x')\int dydz\rho_N^{*2}(\Phi_M)(y,z|y,z)V(y,z)$$

$$+ M(N-1)\int dy\rho_N^{*2}(\Phi_M)(x,y|x',y)[V(x,y)+V(x',y)]$$

$$+ \frac{1}{2}M(M-3)(N-1)\int dydz\rho_N^{*3}(\Phi_M)(x,y,z|x',y,z)V(y,z) \quad (6.145)$$

for $M > N \geq 2$.

It follows from (6.95), (6.102), (6.136), (6.137), (6.143), (6.144) and (6.145) that the relation

$$\text{tr}(\rho_N^*(\Phi_M)H_N) = \text{tr}(\rho_N^*(\Phi_M)\tilde{H}_N[\Phi_M]) \qquad (6.146)$$

holds, according to the general result (6.42).

Let π_M be the set of all 1-fermion N-representable density operators of the form

$$\rho_1(x|x') = \sum_{i_1,\ldots,i_N=1}^{M} \delta_{i_1,\ldots,i_N}^{i_1,\ldots,i_N} \kappa_{i_1} \ldots \kappa_{i_N} \varphi_{i_1}(x)\overline{\varphi}_{i_1}(x'), \quad M \geq N, \quad (6.147)$$

where $\kappa_1,\ldots,\kappa_M$ are arbitrary non-negative numbers satisfying (6.109), and $\varphi_1(x),\ldots,\varphi_M(x)$ are arbitrary orthonormal functions belonging to $\mathcal{H}_1$.

For every macrostate $\Phi_M^* = [\rho_1]$, $\rho_1 \in \pi_M$, the representative state $\rho_N^*(\Phi_M^*)$ has the form (6.133), and the estimated value

$$M_{H_N}(\Phi_M^*) = \mathrm{tr}(\rho_N^*(\Phi_M^*)H_N) \tag{6.148}$$

of the Hamiltonian H_N, c.f. (6.139) is the functional of $\kappa_1, \ldots, \kappa_M$ and $\varphi_1(x), \ldots, \varphi_M(x) \in \mathcal{H}_1$ subject to the condition (6.109) and

$$\int \varphi_i(x)\overline{\varphi}_j(x)dx = \delta_{ij}, \quad i, j = 1, \ldots, M \geq N. \tag{6.149}$$

It should be noted that in the case $M = N$, $\kappa_1 = \ldots = \kappa_N = 1$.

According to (6.71) a macrostate $\Phi \in [\rho_1]$, $\rho_1 \in \pi_M$, is a Hartree-Fock type if the following conditions are satisfied:

$$\frac{\partial M_{H_N}(\Phi)}{\partial \kappa_m} + \lambda N \sum_{i_2,\ldots,i_N=1}^{M} \delta_{m,i_2,\ldots,i_N}^{m,i_2,\ldots,i_N} \kappa_{i_2} \ldots \kappa_N = 0, \tag{6.150}$$

$$\frac{\delta M_{H_N}(\Phi)}{\delta \varphi_m(z)} + \sum_{i=1}^{M} \overline{\varphi}_i(z)\lambda_{im} = 0, \tag{6.151}$$

$$\frac{\delta M_{H_N}(\Phi)}{\delta \overline{\varphi}_m(z)} + \sum_{i=1}^{M} \varphi_i(z)\lambda_{im} = 0, \tag{6.152}$$

$$m = 1, \ldots, M,$$

where λ and $\lambda_{ij} = \overline{\lambda}_{ji}$ are the Lagrange multipliers.

THEOREM 6.10. A macrostate $\Phi = [\rho_1]$ is a Hartree-Fock type provided $\kappa_1, \ldots, \kappa_M$ and $\varphi_1(x), \ldots, \varphi_M(x) \in \mathcal{H}_1$ satisfy the equations

$$\sum_{i_1,\ldots,i_N=1}^{M} \delta_{i_1,\ldots,i_N}^{i_1,\ldots,i_N} \frac{\partial}{\partial \kappa_m}(\kappa_{i_1} \ldots \kappa_{i_N}) \left[N \int \overline{\varphi}_{i_1}(x)H(x)\varphi_{i_1}(x)dx + \lambda \right.$$

$$+ N(N-1)\int dx\,dy\,V(x,y)|\varphi_{[i_1}(x)\varphi_{i_2]}(y)|^2 = 0 \tag{6.153}$$

$$NH(z) \sum_{i_2,\ldots,i_N=1}^{M} \delta_{m,i_2,\ldots,i_N}^{m,i_2,\ldots,i_N} \kappa_m \kappa_{i_2} \cdots \kappa_{i_N} \varphi_m(z)$$

$$+ N(N-1) \sum_{i_1,\ldots,i_N=1}^{M} \delta_{i_1,\ldots,i_N}^{i_1,\ldots,i_N} \kappa_{i_1} \cdots \kappa_{i_N} \int dx\,\varphi_{[i_1}(z)\varphi_{i_2]}(x)\delta_{m[i_1}\overline{\varphi}_{i_2]}(x)V(z,x)$$

$$+ \sum_{i=1}^{M} \varphi_i(x)\lambda_{im} = 0, \tag{6.154}$$

$$NH(z') \sum_{i_2,\ldots,i_N=1}^{M} \delta_{m,i_2,\ldots,i_N}^{m,i_2,\ldots,i_N} \kappa_m \kappa_{i_2} \cdots \kappa_{i_N} \overline{\varphi}_m(z')$$

$$+ N(N-1) \sum_{i_1,\ldots,i_N=1}^{M} \delta_{i_1,\ldots,i_N}^{i_1,\ldots,i_N} \kappa_{i_1} \cdots \kappa_{i_N} \int dx V(z',x) \varphi_{[i_1}(x) \delta_{i_2]m} \overline{\varphi}_{[i_1}(x) \overline{\varphi}_{i_2]}(z')$$

$$+ \sum_{i=1}^{M} \lambda_{mi} \overline{\varphi}_i(x) = 0, \tag{6.155}$$

$m = 1, \ldots, M$.

It is worthwhile to point out that in the case $M = N$, the equations (6.153) become irrelevant, while (6.154) and (6.155) are Hartree-Fock equations, cf. [128], [98]. On the other hand, if $\rho_1 = \overline{\rho_1}$ c.f. (6.135), i.e., $M > N$, (6.154), (6.155) represent a many-state extension of the Hartree-Fock method considered in [287]. Eliminating the Lagrange multipliers λ_{ij} from (6.154) and (6.155) one obtains the relation

$$[H(z) - H(z')]\rho_N^{*1}(\Phi)(z|z') + \tag{6.156}$$
$$(N-1) \int dx[V(z,x) - V(z',x)]\rho_N^{*2}(\Phi)(z,x|z',x) = 0$$

i.e., each Hartree-Fock type macrostate is stationary with respect to the A-isoentropic motion, cf. (6.114). One can check that the relation between the estimated Hamiltonian $\tilde{H}_N[\Phi_t]$ and A-isoentropic motion is such as described in Sec. 6.3.

6.6. Fermi field

The second example that will be considered is the fermion system with indefinite number of particles. Let $\mathcal{H}_1$ be the 1-particle Hilbert space defined in Sec. 6.5 and $\mathcal{H}_F$ be the Fock space, cf. [299]. The algebra of observables is generated by the annihilation and creation operators $a(f)$, $a^*(f)$ on $\mathcal{H}_F$, $f \in \mathcal{H}_1$ which satisfy the following anticommutation relations

$$\{a(f), a(g)\} = \{a^*(f), a^*(g)\} = 0,$$
$$\{a(f), a^*(g)\} = \mathbb{1}_F(g,f), \quad g, f \in \mathcal{H}_1, \tag{6.157}$$

where $\{A, B\} = AB + BA$, $\mathbb{1}_F$ is the identity operator on $\mathcal{H}_F$, and $(\cdot, \cdot)$ is the scalar product in $\mathcal{H}_1$. Moreover, it is assumed that there exists the unique element $\Omega \in \mathcal{H}_F$ (vacuum state) such that

$$a(f)\Omega = 0 \tag{6.158}$$

for all $f \in \mathcal{H}_1$. The transformation of the field operators

$$\tilde{a}(x) = \int dz[A(x,z)a(z) + B(x,z)a^*(z)]$$

$$\tilde{a}^*(x) = \int dz[\overline{B}(x,z)a(z) + \overline{A}(x,z)a^*(z)] \qquad (6.159)$$

is called the *canonical transformation* provided (6.159) is invertible and the operators $\tilde{a}(f)$, $\tilde{a}^*(f)$ satisfy the same commutation relations as $a(f)$, $a^*(f)$.

Transformation (6.159) is a canonical one iff the relations

$$\int dz[A(x,z)B(y,z) + B(x,z)A(y,z)] = 0,$$

$$\int dz[\overline{A}(x,z)\overline{B}(y,z) + \overline{B}(x,z)\overline{A}(y,z)] = 0,$$

$$\int dz[A(x,z)\overline{A}(y,z) + B(x,z)\overline{B}(y,z)] = \delta(x-y). \qquad (6.160)$$

hold.

A canonical transformation (6.159) is called *unitary implemented* if there exists the unitary operator U_{AB} on $\mathcal{H}_F$ such that

$$\tilde{a}(f) = U_{AB}a(f)U_{AB}^*,$$

$$\tilde{a}^*(f) = U_{AB}a^*(f)U_{AB}^*. \qquad (6.161)$$

It is known [37] that the canonical transformation is unitary implemented if B is a Hilbert-Schmidt operator, i.e.,

$$\int dx dy |B(x,y)|^2 < \infty. \qquad (6.162)$$

In what follows, two incomplete sets of operators will be considered

$$\mathbf{A}_0 = \{a^*(x)a(y): x,y \in V \otimes \mathbb{Z}_S\}, \qquad (6.163)$$

and

$$\mathbf{A} = \{a^*(x)a(y), a(x)a(y), a^*(x)a^*(y): x,y \in V \otimes \mathbb{Z}_S\}. \qquad (6.164)$$

It follows from (6.157) that the commutation relations

$$[a^*(x)a(y), a^*(x')a(y')] = \delta(x'-y)a^*(x)a(y')$$
$$- \delta(x-y')a^*(x')a(y) \qquad (6.165)$$

and

$$[a(x)a(y), a(x')a(y')] = [a^*(x)a^*(y), a^*(x')a^*(y')] = 0$$

$$[a^*(x)a(y), a(x')a(y')] = \delta(x - x')a(y')a(y)$$
$$- \delta(x - y')a(x')a(y),$$

$$[a^*(x)a(y), a^*(x')a^*(y')] = \delta(y - x')a^*(x)a^*(y')$$
$$- \delta(y - y')a^*(x')a^*(x'),$$

$$[a(x)a(y), a^*(x')a^*(y')] = 1\!\!1_F(\delta(x - y')\delta(y - x') - \delta(x - x')\delta(y - y'))$$
$$+ \delta(x - x')a^*(y')a(y) + \delta(y - y')a^*(x')a(x)$$
$$- \delta(x - y')a^*(x')a(y) - \delta(y - x')a^*(y')a(x) \qquad (6.166)$$

are satisfied, i.e., the sets $\mathbf{A}_0$ and $\mathbf{A}$ generate the Lie algebras.

Let Φ_0 be $\mathbf{A}_0$-macrostate, i.e.,

$$\Phi_0 = \{\rho \in \mathfrak{S} : \operatorname{tr}(\rho a^*(x)a(y)) = G_0(x, y)\}, \qquad (6.167)$$

where G_0 is a positive definite operator on $\mathcal{H}_1$. A macrostate Φ_0 is said to be *regular* if G_0 satisfies the following conditions

$$0 \le (f, G_0 f) \le (f, f) \qquad (6.168)$$

for all $f \in \mathcal{H}_1$, and

$$\operatorname{tr} G_0 = \int dx G_0(x, x) < \infty. \qquad (6.169)$$

From (6.168) and (6.169) it follows that G_0 has the following spectral decomposition

$$G_0(x, y) = \sum_{k=1}^{\infty} n_k \bar{g}_k(x) g_k(y), \qquad (6.170)$$

where $g_1, g_2, \dots$ form an orthonormal and complete base (CONS) in $\mathcal{H}_1$ and

$$0 \le n_k \le 1, \quad k = 1, 2, \dots \qquad (6.171)$$

$$\sum_{k=1}^{\infty} n_k < \infty. \qquad (6.172)$$

Let $\mathcal{H}_1^0$ and $\mathcal{H}_1^1$ be the subspaces of $\mathcal{H}_1$ being the eigensubspaces of G_0 corresponding to eigenvalues 0 and 1, respectively. Due to (6.172) the subspace $\mathcal{H}_1^1$ is finitely dimensional. In the subspace $(\mathcal{H}_1^0 \oplus \mathcal{H}_1^1)^\perp$ the operator G_0 is strictly positive definite and the operator $\ln(G_0^{-1} - 1)$ exists. The operator G_0 in $(\mathcal{H}_1^0 \oplus \mathcal{H}_1^1)^\perp$ will be denoted by $G_0^{(+)}$.

Let Φ be a macrostate with respect to the set $\mathbf{A}$, i.e.,

$$\Phi = \{\rho \in \mathfrak{S} : \text{tr}(\rho a^*(x)a(y)) = G(x,y), \tag{6.173}$$
$$\text{tr}(\rho a(x)a(y)) = \Delta(x,y), \text{tr}(\rho a^*(x)a^*(y)) = -\overline{\Delta}(x,y)\},$$

where $G = G^* \geq 0$, $\Delta(x,y) = -\Delta(y,x)$. A macrostate Φ is said to be *regular* if it can be written in the form

$$\Phi = \{U_{AB}^* \rho U_{AB} : \rho \in \Phi_0\}, \tag{6.174}$$

where U_{AB} generates the canonical transformation (6.159), (6.160), i.e.,

$$\begin{aligned}
G(x,y) &= \text{tr}(\rho \tilde{a}^*(x)\tilde{a}(y)), \\
\Delta(x,y) &= \text{tr}(\rho \tilde{a}(x)\tilde{a}(y)),
\end{aligned} \tag{6.175}$$

for all $\rho \in \Phi_0$. By virtue of (6.159), G and Δ can be expressed in terms of G_0, A and B.

Let $\mathcal{L}$ be a subspace of $\mathcal{H}_1$ and $f_1, \ldots, f_d$, $d = \dim \mathcal{L}$ be an orthonormal and complete base in $\mathcal{L}$. The projection operators $\pi_0(\mathcal{L})$ and $\pi_1(\mathcal{L})$ on $\mathcal{H}_F$ are defined as

$$\pi_0(\mathcal{L}) = \prod_{k=1}^{d}(1 - a^*(f_k)a(f_k)) \tag{6.176}$$

and

$$\pi_1(\mathcal{L}) = \prod_{k=1}^{d} a^*(f_k)a(f_k), \tag{6.177}$$

respectively.

It should be noted that in the case $\mathcal{L} = \mathcal{H}_1$, $\pi_1(\mathcal{H}_1)$ is the projection operator on the vacuum state $\Omega \in \mathcal{H}_F$.

LEMMA 6.5. *Let $\mathcal{L}$ be a subspace of $\mathcal{H}_1$. Then the relations*

$$\lim_{\epsilon_1,\ldots,\epsilon_d \to +0} \prod_{k=1}^{d}(1 - \epsilon_k)\exp[-a^*(f_k)a(f_k)\ln(\frac{1}{\epsilon_k} - 1)] = \pi_0(\mathcal{L}) \tag{6.178}$$

and

$$\lim_{\epsilon_1,\ldots,\epsilon_d \to 1} \prod_{k=1}^{d}(1 - \epsilon_k)\exp[-a^*(f_k)a(f_k)\ln(\frac{1}{\epsilon_k} - 1)] = \pi_1(\mathcal{L}) \tag{6.179}$$

hold.

Proof. Let $f \in \mathcal{H}_1$ and $(f, f) = 1$. Since $a^*(f)a(f)$ is a projection on $\mathcal{H}_F$ one finds that the equality

$$(1 - \epsilon)\exp[-a^*(f)a(f)\ln(\frac{1}{\epsilon} - 1)] = \quad (6.180)$$
$$= (1 - \epsilon)(\mathbb{1}_F - a^*(f)a(f)) + \epsilon a^*(f)a(f)$$

holds. Equality (6.180) implies (6.178) and (6.179). $\square$

THEOREM 6.11. *Let Φ_0 be a regular $\mathbf{A}_0$-macrostate. Then there exists the representative state $\rho^*(\Phi_0)$ of the form*

$$\rho^*(\Phi_0) = Z_0^{-1}\exp\left[-\int dx\,dy\,a^*(x)\Lambda_0(x, y)a(y)\right]\pi_0(\mathcal{H}_1^0)\pi_1(\mathcal{H}_1^1), \quad (6.181)$$

where

$$Z_0 = \exp[-\mathrm{tr}(1 - G_0^{(+)})], \quad (6.182)$$
$$\Lambda_0 = \ln[(G_0^{(+)})^{-1} - 1], \quad (6.183)$$

$\mathcal{H}_1^0$ *and* $\mathcal{H}_1^1$ *are the eigenspaces of G_0 corresponding to the eigenvalues 0 and 1, respectively, while $G_0^{(+)}$ is the operator G_0 on $(\mathcal{H}_1^0 \oplus \mathcal{H}_1^1)^\perp$.*

Proof. Using the Lagrange multipliers method to maximize the entropy $S(\rho) = -\mathrm{tr}\rho\ln\rho$ on Φ_0, one finds that the representative state should have the form

$$\rho^*(\Phi_0) = Z^{-1}\exp\left[-\int dx\,dy\,a^*(x)\Lambda(x, y)a(y)\right], \quad (6.184)$$

where

$$Z = \mathrm{tr}\exp\left[-\int dx\,dy\,a^*(x)\Lambda(x, y)a(y)\right]. \quad (6.185)$$

The Lagrange multiplier $\Lambda(x, y)$ is determined by the relation

$$G_0(x, y) = \frac{\delta\ln Z}{\delta\Lambda(x, y)}. \quad (6.186)$$

The partition function Z exists iff Λ has only purely point spectrum, i.e., Λ should have the form

$$\Lambda(x, y) = \sum_{k=1}^{\infty} \lambda_k \overline{f}_k(x)f_k(y), \quad (6.187)$$

where $f_1, f_2, \ldots$ form an orthonormal and complete base in $\mathcal{H}_1$. Inserting (6.187) into (6.185) one obtains

$$Z = \prod_{k=1}(1 + e^{-\lambda_k}) \tag{6.188}$$

or

$$\ln Z = \operatorname{tr} \ln(1 + e^{-\Lambda}). \tag{6.189}$$

Taking into account (6.186), it follows that

$$G_0 = (e^{\Lambda} + 1)^{-1} \tag{6.190}$$

and

$$\Lambda = \ln(G_0^{-1} - 1) \tag{6.191}$$

provided G_0 has no eigenvalues 0 and 1. From (6.170), (6.187) and (6.191) it turns out that $f_k = g_k$ and $\lambda_k = \ln(1/n_k - 1)$, i.e., the representative state can be written in the form

$$\rho^*(\Phi_0) = \prod_{k=1}^{\infty}(1 - n_k) \exp\left[- a^*(g_k)a(g_k) \ln\left(\frac{1}{n_k} - 1\right)\right]. \tag{6.192}$$

Using Lemma 6.5 and (6.176), (6.177) one concludes that (6.181), (6.182), (6.183) hold. $\square$

One can observe that if $G_0 = 0$, i.e., $\mathcal{H}_1^0 = \mathcal{H}_1$, it follows that

$$\rho^*(\Phi_0) = \pi_0(\mathcal{H}_1), \tag{6.193}$$

i.e., $\rho^*(\Phi_0)$ is the projection on the vacuum state $\Omega \in \mathcal{H}_F$.

Moreover, it follows from (6.181), (6.182) and (6.183) that the entropy $S(\Phi_0)$ is given by the formula

$$S(\Phi_0) = -\operatorname{tr}[(1 - G_0^{(+)}) \ln(1 - G_0^{(+)}) + G_0^{(+)} \ln G_0^{(+)}]. \tag{6.194}$$

THEOREM 6.12. *Let Φ be a regular A-macrostate. The representative state $\rho^*(\Phi)$ has the form*

$$\rho^*(\Phi) = U_{AB}^* \rho^*(\Phi_0)U_{AB}. \tag{6.195}$$

Proof. The form of $\rho^*(\Phi)$ follows immediately from (6.174). $\square$

It should be noted that the representative states $\rho^*(\Phi_0)$ and $\rho^*(\Phi)$ are just quasi-free states, cf. [83], and the following relations are satisfied

$$\mathrm{tr}(\rho^*(\Phi_0)a^*(x_1)\cdots a^*(x_m)a(y_n)\cdots a(y_1)) =$$
$$= \delta_{mn}\det(\|\,G_0(x_i,y_j)\,\|_{ij=1}^m\,) \qquad (6.196)$$

while all remaining expectations vanish,

$$\mathrm{tr}(\rho^*(\Phi)a^*(x_1)\cdots a^*(x_m)a(y_n)\cdots a(y_1)) = 0 \qquad (6.197)$$

for $m + n = 2k + 1$, and, in particular,

$$\mathrm{tr}(\rho^*(\Phi)a^*(x_1)a^*(x_2)a(y_2)a(y_1)) = \qquad (6.198)$$
$$= G(x,y_1)G(x_2,y_2) - G(x_1,y_2)G(x_2,y_1) + \overline{\Delta}(x_1,x_2)\Delta(y_1,y_2) \qquad (6.199)$$

Let us assume that the Hamiltonian of the fermi system has the form

$$H = H_1 + H_2, \qquad (6.200)$$

where

$$H_1 = \int dx dy\ a^*(x)h(x,y)a(y) \qquad (6.201)$$

and

$$H_2 = \int dx_1 dx_2 dy_1 dy_2\ a^*(x_1)a^*(x_2)V(x_1,x_2|y_1,y_2)a(y_2)a(y_1). \qquad (6.202)$$

THEOREM 6.13. *Let $\Phi^t, t \geq 0$ be the family of regular $\mathbf{A}$-macrostates*

$$\Phi^t = \{\rho \in \mathfrak{S}:\ \mathrm{tr}(\rho a^*(x)a(y)) = G(x,y;t)\,, \qquad (6.203)$$
$$\mathrm{tr}(\rho a(x)a(y)) = \Delta(x,y;t)\,,\mathrm{tr}(\rho a^*(x)a^*(y)) = -\overline{\Delta}(x,y;t)\}.$$

The equations of $\mathbf{A}$-isoentropic motion have the form

$$i\hbar\frac{\partial}{\partial t}G(x,y;t) = \mathrm{tr}\Big(\rho^*(\Phi^t)[a^*(x)a(y), H]\Big), \qquad (6.204)$$

$$i\hbar\frac{\partial}{\partial t}\Delta(x,y;t) = \mathrm{tr}\Big(\rho^*(\Phi^t)[a(x)a(y), H]\Big). \qquad (6.205)$$

The right-hand sides of (6.204) and (6.205) can be expressed in terms of G and Δ due to the form of $\rho^*(\Phi^t)$. For the special case of $\mathbf{A}_0$-isoentropic motion, one obtains

$$i\hbar\frac{\partial}{\partial t}G_0(x,y;t) = \int dz[h(y,z)G_0(x,z;t) - h(z,x)G_0(z,y;t)]$$
$$+ 4\int dx_1 dy_1 dy_2 V(y,x_1|y_1,y_2)G_0(x,y_1;t)G_0(x_1,y_2;t)$$
$$- 4\int dx_1 dy_1 dy_2 V(y_1,y_2|x_1,x)G_0(y_2,y;t)G_0(y_1,x_1;t). \qquad (6.206)$$

THEOREM 6.14. *Let* Φ *be a regular* **A**-*macrostate. The estimated Hamiltonian* $\tilde{H}[\Phi]$ *for the Hamiltonian II given by (6.200) has the form*

$$\tilde{H}[\Phi] = H_1 + \int dx\, dy\, a^*(x) U(x,y) a(y) \tag{6.207}$$
$$+ \int dx_1 dx_2 [B(x_1, x_2) a(x_1) a(x_2) - \overline{B}(x_1, x_2) a^*(x_1) a^*(x_2)] + h \mathbb{1}_F,$$

where

$$U(x,y) = 4 \int dx'dy' V(x, x'|y, y') G(x', y'), \tag{6.208}$$

$$B(x_1, x_2) = \int dy_1 dy_2 \overline{\Delta}(y_1, y_2) V(y_1, y_2 | x_1, x_2) \tag{6.209}$$

and

$$h = -\int dx_1 dx_2 dy_1 dy_2 V(x_1, x_2 | y_1, y_2)[2\, G(x_1, y_1) G(x_2, y_2)$$
$$+ \overline{\Delta}(x_1, x_2) \Delta(y_1, y_2). \tag{6.210}$$

Finally, the estimated value of H on Φ has the form

$$M_H(\Phi) = \text{tr}(\rho^*(\Phi) H) = \int dx\, dy\, h(x,y) G(x,y)$$
$$+ \int dx_1 dx_2 dy_1 dy_2 V(x_1, x_2 | y_1, y_2)[G(x_1, y_1) G(x_2, y_2)$$
$$- G(x_1, y_2) G(x_2, y_1) + \overline{\Delta}(x_1, x_2) \Delta(x_1, x_2)]. \tag{6.211}$$

Chapter 7

Open Systems

7.1. Positive dynamical semigroups

Let $\mathcal{H}$ be a separable complex Hilbert space corresponding to a physical system and let $\mathcal{T}(\mathcal{H})$ be the Banach space of trace class operators on $\mathcal{H}$. The *positive cone*, i.e., the set of all positive operators from $\mathcal{T}(\mathcal{H})$, will be denoted by $\mathcal{V}^+(\mathcal{H})$. The set of all states (density operators) can be embedded in $\mathcal{T}(\mathcal{H})$. More precisely,

$$\mathfrak{S}(\mathcal{H}) = \{\rho \in \mathcal{V}^+(\mathcal{H}) : \operatorname{tr}\rho = 1\}, \tag{7.1}$$

cf. [197], [198].

A *positive dynamical semigroup* (in the Shrödinger picture) is a family of linear maps [1] $\{\Lambda_t : \mathcal{T}(\mathcal{H}) \to \mathcal{T}(\mathcal{H}); t \geq 0\}$ such that

(i) $\Lambda_t : \mathcal{V}^+(\mathcal{H}) \to \mathcal{V}^+(\mathcal{H})$, $t \geq 0$,
(ii) $\operatorname{tr}(\Lambda_t\rho) = \operatorname{tr}\rho$, for all $\rho \in \mathcal{T}(\mathcal{H})$, $t \geq 0$,
(iii) $\Lambda_t\Lambda_s = \Lambda_{t+s}$, $t, s \geq 0$,
(iv) $\lim_{t \downarrow 0} \| \Lambda_t\rho - \rho \|_1 = 0$, for all $\rho \in \mathcal{T}(\mathcal{H})$ (*strong continuity*), where $\| \cdot \|_1$ is the trace norm.

Since $\Lambda_t \colon \mathfrak{S}(\mathcal{H}) \to \mathfrak{S}(\mathcal{H})$, $t \geq 0$, it may be considered as describing the time evolution of an *open system*, while the time evolution of an *isolated system* is given in terms of the Hamiltonian dynamics

$$\Lambda_t^0\rho = e^{-\frac{i}{\hbar}Ht}\rho e^{\frac{i}{\hbar}Ht}, \quad -\infty < t < \infty, \tag{7.2}$$

where H is the Hamiltonian of the system, and Λ_t^0 is a one-parameter group.

In order to translate (i) – (iv) to the Heisenberg picture, one defines the *dual time evolution* Λ_t^*, $t \geq 0$, by means of the relation

$$\operatorname{tr}[a(\Lambda_t\rho)] = \operatorname{tr}[(\Lambda_t^*a)\rho] \tag{7.3}$$

---

[1] In Chap. 7 the Schrödinger picture semigroups is denoted by Λ and the Heisenberg picture semigroup by Λ^* while in Chap. 4 vice versa.

for all $\rho \in T(\mathcal{H})$ and $a \in B(\mathcal{H}) = T^*(\mathcal{H})$, and one obtains:

A *positive dynamical semigroup* (in the Heisenberg picture) is a family of linear maps $\{\Lambda_t^* : B(\mathcal{H}) \to B(\mathcal{H}); \ t \geq 0\}$ such that

  (i) Λ_t^* is *positive*, i.e., it maps positive elements of $B(\mathcal{H})$ into positive ones,

  (ii) $\Lambda_t^* I \doteq I$, $t \geq 0$,

  (iii) $\lim_{t \downarrow 0} \text{tr}[\rho(\Lambda_t^* a - a)] = 0$ for all $\rho \in T(\mathcal{H})$ and $a \in B(\mathcal{H})$,

  (iv) $\Lambda_t^* \Lambda_s^* = \Lambda_{t+s}^*$, $t, s \geq 0$,

  (v) Λ_t^* is *normal*, i.e., if the sequence $(n = 1, 2, \ldots)$ $\text{tr}(\rho a_n) \to \text{tr}(\rho a)$, then also the sequence $\text{tr}[\rho(\Lambda_t^*)a_n)] \to \text{tr}[\rho(\Lambda_t^* a)]$, (*ultraweak continuity of* Λ_t^*).

One notes that Λ_t and Λ_t^* are *contractions*, i.e.,

$$\| \Lambda_t \rho \|_1 \leq \| \rho \|_1, \tag{7.4}$$

and

$$\| \Lambda_t^* a \| \leq \| a \|, \tag{7.5}$$

where $\| \cdot \|$ is the norm in $B(\mathcal{H})$.

From the definitions of positive dynamical semigroups Λ_t and Λ_t^*, $t \geq 0$, it follows that there exist densely defined operators L and L^* on $T(\mathcal{H})$ and $B(\mathcal{H})$, such that

$$\frac{d}{dt} \Lambda_t \rho = L \Lambda_t \rho = \Lambda_t L \rho, \quad \rho \in D(L), \tag{7.6}$$

and

$$\frac{d}{dt} \Lambda_t^* a = L^* \Lambda_t^* a = \Lambda_t^* L^* a, \quad a \in D(L^*), \tag{7.7}$$

respectively ($D(L)$ is the domain of L).

The equations (7.6) and (7.7) are called *quantum Markovian master equations* in the Schrödinger and Heisenberg pictures, respectively.

There is a problem to find the general form of the generator L (L^*) of a positive dynamical semigroup. In the case when L^* (L) is a bounded generator, i.e., the corresponding dynamical semigroup is norm-continuous, in [198] it has been proved the following

THEOREM 7.1. *Let* $P = \{P_1, P_2, \ldots\}$ *be a family of projection operators on finite-dimensional and mutually orthogonal subspaces of* $\mathcal{H}$, *such that* $\sum_{i=1}^{\infty} P_i = I$. *A bounded linear operator* L *on* $T(\mathcal{H})$ *generates a positive dynamical semigroup iff for every* P *the relations*

$$a_{ij}(P) \geq 0, \quad i \neq j = 1, 2, \ldots, \tag{7.8}$$

and

$$\sum_{i=1}^{\infty} a_{ij}(P) = 0, \quad j = 1, 2, \ldots, \tag{7.9}$$

are satisfied, where

$$a_{ij}(P) = \text{tr}[P_i(LP_j)]. \tag{7.10}$$

The conditions (7.8) and (7.9) can be considered as quantum analogues of the *Kolmogorov conditions* for discrete Markov processes, cf. [184].

The only explicit characterization of a bounded generator of a positive dynamical semigroup in the Heisenberg picture is given by the following

THEOREM 7.2 (Lindblad [220]). *Let* Φ^* *be a normal positive bounded map on* $\mathcal{B}(\mathcal{H})$ *and* H *be a bounded selfadjoint operator on* $\mathcal{H}$. *Then the bounded operator*

$$L^* a = \Phi^*(a) - \frac{1}{2} \left(\Phi^*(I)a + a\Phi^*(I) \right) + i[H, a] \tag{7.11}$$

is the generator of a positive dynamical semigroup.

Proof. Taking into account Theorem 7.1 and the relation

$$\text{tr}[P_i(LP_j)] = \text{tr}[(L^*P_i)P_j] = a_{ij}(P), \tag{7.12}$$

one finds

$$\sum_{i=1}^{\infty} a_{ij}(P) = 0, \quad j = 1, 2, \ldots, \tag{7.13}$$

since $L^*I = 0$, and

$$a_{ij}(P) = \text{tr}[\Phi^*(P_i)P_j] \geq 0, \quad i \neq j = 1, 2, \ldots, \tag{7.14}$$

due to positivity of Φ^*. $\square$

Unfortunately, the general form of positive maps on $\mathcal{B}(\mathcal{H})$ is unknown. However, there is a class of positive maps, called *decomposable maps*, cf. [347], [314], which have the form

$$\Phi^*(a) = \Lambda_1^* a + \Lambda_2^* a^T, \tag{7.15}$$

where Λ_1^*, Λ_2^* are *completely positive maps*, cf. Chapter 3, on $\mathcal{B}(\mathcal{H})$ and T denotes the transposition (with respect to some CONS, a complete orthonormal system or base). In the case $M_2(\mathbb{C})$ (the C^*-algebra of all complex matrices 2×2), Φ^* given by (7.15) is the general form of a positive map, cf. [54]. Let $s_\alpha = \frac{1}{2}\sigma_\alpha$, $\alpha = 1, 2, 3$, where σ_1, σ_2, σ_3 are the Pauli matrices, and let the bold-faced letters denote symbolic 3-vectors (with numerical and matrix components) $\hat{a} = sa$, $a \in \mathbb{R}^3$. Then in [199] it has been shown

THEOREM 7.3. *A generator L of a positive dynamical semigroup on $M_2(\mathbb{C})$ in the Schrödinger picture on has the form*

$$L\rho = -i[\hat{h}, \rho] + \frac{1}{2} \sum_{m,n=1}^{3} (e_m, De_n + iAe_n) \{[e_m\rho, e_n] + [e_m, \rho e_n]\}, \quad (7.16)$$

where $\{e_n\}$ is a CONS in $\mathbb{R}^3$, $\hat{e}_n = se_n$, $\hat{h} = hs$, $h \in \mathbb{R}^3$. D and A are linear operators on $\mathbb{R}^3$ of the form

$$D = I\,\mathrm{tr}F - 2F, \qquad (7.17)$$

$$Az = (Fm_0 + m_0 \times h) \times z, \qquad (7.18)$$

$$Fx = \sum_{n=1}^{3} \gamma_n(e_n x)e_n, \quad \gamma_1, \gamma_2, \gamma_3 \geq 0, \qquad (7.19)$$

and $m_0 \in S \subseteq \mathbb{R}^3$, where

$$S = \begin{cases} \{0\} & for \quad \gamma_1\gamma_2\gamma_3 = 0, \\ \{z \in \mathbb{R}^3 : \inf_{\|x\|=1} [(Fx, x - z) + x(h \times z)] \geq 0\} & for \quad \gamma_1\gamma_2\gamma_3 > 0. \end{cases}$$
$$(7.20)$$

Putting $\rho(t) = \Lambda_t \rho$, and $m(t) = \mathrm{tr}(\rho(t)s)$, where ρ is a density matrix, and using the equations (7.6) and (7.16), one obtains the following equation for $m(t)$:

$$\frac{d}{dt}m(t) = h \times (m(t) - m_0) - F(m(t) - m_0) \qquad (7.21)$$

which has the form of the Bloch equation used in the paramagnetic resonance theory, cf. [1].

7.2. Completely positive dynamical semigroups

Let $\{\Lambda_t^* : B(\mathcal{H}) \to B(\mathcal{H}); \ t \geq 0\}$ be a dynamical semigroup in the Heisenberg picture. We imagine that there is another system associated with the n-dimensional Hilbert space $\mathcal{H}_n$ which is regarded as dynamically uncoupled to the system in question. The dynamics of the joint system is described by the dynamical semigroup $\{\Lambda_t^* \otimes \mathbb{1}_n : B(\mathcal{H}) \otimes M_n \to B(\mathcal{H}) \otimes M_n; \ t \geq 0\}$, where $M_n = B(\mathcal{H}_n)$ is the C^*-algebra of all complex $n \times n$ matrices and $\mathbb{1}_n$ is the identity map of M_n. The map Λ_t^* is called then n-positive if $\Lambda_t^* \otimes \mathbb{1}_n$ is positive, while Λ_t^* is said to be *completely positive* if it is n-positive for any (positive integer) n, cf. [28], [311], and also Chapter 3. The importance of complete positivity has been pointed out by Kraus [202], Lindblad [220], and Accardi [4]. The problem of the form of the generator of a completely positive dynamical semigroup has been studied in [220], [108].

THEOREM 7.4 (Lindblad [220]). *A bounded operator L on $T(\mathcal{H})$ is the generator of norm-continuous completely positive dynamical semigroup* $\{\Lambda_t \colon T(\mathcal{H}) \to T(\mathcal{H}), \ t \geq 0\}$ *iff it has the form*

$$
\begin{aligned}
L\rho &= -i[H, \rho] + \frac{1}{2} \sum_{j \in A} \left\{ [V_j, \rho V_j^*] + [V_j \rho, V_j^*] \right\} \\
&= -i[H, \rho] + \sum_{j \in A} V_j \rho V_j^* - \frac{1}{2} \Big[\sum_{j \in A} V_j^* V_j, \rho \Big]_+, \quad \rho \in T(\mathcal{H}), (7.22)
\end{aligned}
$$

where H is a bounded selfadjoint operator, $[A, B]_+ = AB + BA$, $\{V_j\}_{j \in A}$ is a sequence of bounded operators, $\sum_{j \in A} V_j^* V_j$ converges ultraweakly, and the right hand side converges in the trace norm.

The generator L^* of the dual semigroup $\{\Lambda_t^*; \ t \geq 0\}$ in the Heisenberg picture is given by

$$
\begin{aligned}
L^* a &= i[H, a] + \frac{1}{2} \sum_{j \in A} \left\{ [V_j^*, a] V_j + V_j^* [a, V_j] \right\} \\
&= i[H, a] + \Phi(a) - \frac{1}{2} [\Phi(I), a]_+, \quad a \in \mathcal{B}(\mathcal{H}), \qquad (7.23)
\end{aligned}
$$

where the convergence is ultraweak, and

$$
\Phi(a) = \sum_{j \in A} V_j^* a V_j \qquad (7.24)
$$

is the general form of a completely positive ultraweakly continuous map of $\mathcal{B}(\mathcal{H})$ into itself, cf. [202].

THEOREM 7.5 (cf. [108]). *In the case* $\dim \mathcal{H} = n$, *the relations (7.22) and (7.23) can be given in the form*

$$
L\rho = -i[H, \rho] + \frac{1}{2} \sum_{i,j=1}^{n^2-1} c_{ij} \left\{ [F_i, \rho F_j^*] + [F_i \rho, F_j^*] \right\} \qquad (7.25)
$$

and

$$
L^* a = i[H, a] + \frac{1}{2} \sum_{i,j=1}^{n^2-1} c_{ij} \left\{ [F_j^*, a] F_i + F_j^* [a, F_i] \right\}, \qquad (7.26)
$$

where $H = H^*$, $\operatorname{tr} H = 0$, $\operatorname{tr} F_j = 0$, $\operatorname{tr}(F_i F_j^*) = \delta_{ij}$, and (c_{ij}) is a positive definite complex matrix. Moreover, the decomposition into the Hamiltonian part (the first terms) and the dissipative part (the second terms) is unique.

An elementary proof of the above theorem can be based on the following lemmas:

LEMMA 7.1. *Dynamical semigroup* $\Lambda_t^*: M_n \to M_n$ *is a completely positive one iff* $\Lambda_t^* \otimes \mathbb{1}_n : M_n \otimes M_n \to M_n \otimes M_n$ *is positive.*

Proof. It follows from Theorem 5 of [53] that a linear map $\Gamma: M_n \to M_n$ is completely positive iff $\Gamma \otimes \mathbb{1}_n$ is positive. Moreover, from the equality $(\Gamma \otimes \mathbb{1}_n)(I \otimes I) = (\Gamma(I) \otimes I)$, where I stands for the identity of M_n, it follows that $\Gamma \otimes \mathbb{1}_n$ preserves the identity iff $\Gamma(I) = 1$. $\square$

Since $B(\mathcal{H}_n) = T(\mathcal{H}_n)$, dynamical semigroup Λ_t in the Schrödinger picture is completely positive iff $\Lambda_t \otimes \mathbb{1}_n$ is positive.

LEMMA 7.2. *Let* Γ *be a linear map* $M_n \to M_n$ *and let* $\{F_\alpha\}_{\alpha=1}^{n^2}$ *be a CONS in* M_n, *i.e.,* $(F_\alpha, F_\beta) = \mathrm{tr}(F_\alpha^* F_\beta) = \delta_{\alpha\beta}$. *Then* Γ *can be uniquely written in the form*

$$\Gamma : a \to \Gamma a = \sum_{\alpha,\beta=1}^{n^2} c_{\alpha\beta} F_\alpha a F_\beta^*, \quad a \in M_n, \qquad (7.27)$$

and if $\Gamma a^* = (\Gamma a)^*$, *then* $c_{\alpha\beta} = \bar{c}_{\beta\alpha}$.

Proof. First note that

$$\sum_{\alpha=1}^{n^2} F_\alpha^* a F_\alpha = I \,\mathrm{tr}\, a, \quad a \in M_n. \qquad (7.28)$$

Since the left hand side of (7.28) is invariant under the transformation of CONS $F_\alpha \to G_\alpha$, and choosing $\{F_\alpha\}_{\alpha=1}^{n^2} = \{e_{ij}\}_{i,j=1}^n$, where $e_{ij}x = e_i(e_j, x)$ and $\{e_i\}_{i=1}^n$ is a CONS in $\mathcal{H}$, one obtains

$$\sum_{i,j=1}^{n} e_{ij}^* a e_{ij} = \sum_{i,j=1}^{n} e_{ji} a e_{ij} = \left(\sum_{j=1}^{n} e_{jj}\right)\left(\sum_{i=1}^{n} a_{ii}\right) = I\,\mathrm{tr}\, a. \qquad (7.29)$$

Let $\mathcal{L}(M_n, M_n)$ denote the vector space of linear operators $M_n \to M_n$, and let $\{G_\alpha\}$ be a CONS in M_n. $\mathcal{L}(M_n, M_n)$ becomes a unitary space with the inner product

$$\langle \Gamma, \Phi \rangle = \sum_{\alpha=1}^{n^2} (\Gamma G_\alpha, \Phi G_\alpha) = \sum_{\alpha=1}^{n^2} \mathrm{tr}\left[(\Gamma G_\alpha)^*(\Phi G_\alpha)\right]. \qquad (7.30)$$

Define

$$\Gamma_{\alpha\beta} : a \to \Gamma_{\alpha\beta} a = F_\alpha a F_\beta^*, \quad \alpha, \beta = 1, 2, \ldots, n^2. \qquad (7.31)$$

Then $\{\Gamma_{\alpha\beta}\}$ is a CONS in $\mathcal{L}(M_n, M_n)$. Using (7.28) one finds

$$\langle \Gamma_{\alpha\beta}, \Gamma_{\mu\nu} \rangle = \sum_{\lambda=1}^{n^2} \text{tr}\left[(\Gamma_{\alpha\beta}G_\lambda)^*(\Gamma_{\mu\nu}G_\lambda)\right] = \sum_{\lambda=1}^{n^2} \text{tr}\left[(F_\alpha G_\lambda F_\beta^*)^*(F_\mu G_\lambda F_\nu^*)\right]$$

$$= \text{tr}\left[F_\beta(\sum_{\lambda=1}^{n^2} G_\lambda^* F_\alpha^* F_\mu G_\lambda)F_\nu^*\right] = \text{tr}(F_\alpha^* F_\mu)\text{tr}(F_\beta F_\nu^*) = \delta_{\alpha\mu}\delta_{\beta\nu}. \quad (7.32)$$

The last assertion of the lemma can be easily verified. $\square$

LEMMA 7.3. Let $\{F_\alpha\}$, $\alpha = 1, 2, \ldots, n^2$ be a CONS in M_n such that $F_{n^2} = \frac{1}{\sqrt{n}}\mathbb{1}$ and let L be a linear operator $M_n \to M_n$ such that $L\rho^* = (L\rho)^*$ and $\text{tr}(L\rho) = 0$ for all $\rho \in M_n$. Then L can be uniquely written in the form

$$L\rho = -i[H, \rho] + \frac{1}{2}\sum_{i,j=1}^{n^2-1} c_{ij}\left\{[F_i, \rho F_j^*] + [F_i\rho, F_j^*]\right\}, \quad (7.33)$$

where $H = H^*$, $\text{tr}H = 0$, and $c_{ij} = \bar{c}_{ji}$.

Proof. From (7.27) one has

$$L\rho = \frac{1}{n}c_{n^2 n^2}\rho + \frac{1}{\sqrt{n}}\sum_{i=1}^{n^2-1}(c_{in^2}F_i\rho + c_{n^2 i}\rho F_i^*) + \sum_{i,j=1}^{n^2-1} c_{ij}F_i\rho F_j^*$$

$$= -i[H, \rho] + [G, \rho]_+ + \sum_{i,j=1}^{n^2-1} c_{ij}F_i\rho F_j^*, \quad (7.34)$$

where $H = H^* = \frac{1}{2i}(F^* - F)$ and $G = G^* = \frac{1}{2n}c_{n^2 n^2}\mathbb{1} + \frac{1}{2}(F + F^*)$, with $F = \frac{1}{\sqrt{n}}\sum_{i=1}^{n^2-1} c_{in^2}F_i$. Now

$$0 = \text{tr}(L\rho) = \text{tr}\left[(2G + \sum_{i,j=1}^{n^2-1} c_{ij}F_j^* F_i)\right] \quad (7.35)$$

for all $\rho \in M_n$ implies

$$G = -\frac{1}{2}\sum_{i,j=1}^{n^2-1} c_{ij}F_i^* F_j,$$

whence (7.33) follows. The uniqueness follows from the dimensionality considerations, since $\text{tr}(L\rho) = 0$ for all $\rho \in M_n$ implies n^2 conditions on L.

$\square$

LEMMA 7.4. *Let* $\{F_\alpha\}$, $\alpha = 1, 2, \ldots, n^2$, *be a CONS in* M_n. *Then*

$$\hat{P}^{(\alpha)} = \sum_{i,j=1}^{n} P_{ij}^{(\alpha)} \otimes e_{ij}, \tag{7.36}$$

where

$$P_{ij}^{(\alpha)} = F_\alpha e_{ij} F_\alpha^*, \quad \alpha = 1, 2, \ldots, n^2, \tag{7.37}$$

is a complete family of mutually orthogonal projections in $M_n \otimes M_n$.

Proof. An element $\hat{P} = \sum_{i,j=1}^{n} P_{ij} \otimes e_{ij}$ of $M_n \otimes M_n$ is a projection iff

$$P_{ij}^* = P_{ij}, \quad \sum_{l=1}^{n} P_{il} P_{lj} = P_{ij}, \quad i, j = 1, 2, \ldots, n. \tag{7.38}$$

Two such projections $\hat{P}$ and $\hat{Q}$ are orthogonal iff

$$\sum_{l=1}^{n} P_{il} Q_{lj} = 0, \quad (i, j = 1, 2, \ldots, n). \tag{7.39}$$

One has

$$P_{ij}^{*(\alpha)} = (F_\alpha e_{ij} F_\alpha^*)^* = F_\alpha e_{ij}^* F_\alpha^* = P_{ji}^{(\alpha)} \tag{7.40}$$

and

$$\begin{aligned}
\sum_{l=1}^{n} P_{il}^{(\alpha)} P_{lj}^{(\beta)} &= \sum_{l=1}^{n} F_\alpha e_{il} F_\alpha^* F_\beta e_{lj} F_\beta^* \\
&= F_\alpha e_{ij} F_\beta^* \mathrm{tr}(F_\alpha^* F_\beta) = \delta_{\alpha\beta} P_{ij}^{(\alpha)}. \quad \square
\end{aligned} \tag{7.41}$$

Proof of Theorem 7.5 The "if" part: If Λ_t is the semigroup generated by (7.25), the generator of the semigroup $\Lambda_t \otimes \mathbb{1}_n$ is $L \otimes \mathbb{1}_n$. By Lemma 7.1 one has to show that $\{c_{\alpha\beta}\} \geq 0$ implies that $L \otimes \mathbb{1}_n$ satisfies the conditions of Theorem 7.1. Since $L \otimes \mathbb{1}_n$ preserves the trace on $M_n \otimes M_n$ one has to check that

$$\mathrm{tr}\left\{ \hat{P}^{(1)}[(L \otimes \mathbb{1}_n) \hat{P}^{(2)}] \right\} \geq 0 \tag{7.42}$$

for all pairs $\hat{P}^{(1)}$, $\hat{P}^{(2)}$ of mutually orthogonal projections in $M_n \otimes M_n$. Using (7.38) and (7.39) one obtains

$$\mathrm{tr}\left\{ \hat{P}^{(1)}[(L \otimes \mathbb{1}_n) \hat{P}^{(2)}] \right\} = \sum_{i,j=1}^{n} \mathrm{tr}[P_{ij}^{(1)}(L P_{ij}^{(2)})]$$

$$= -i \sum_{i,j=1}^{n} \text{tr}\left(P_{ij}^{(1)}[H, P_{ij}^{(2)}]\right) + \sum_{\alpha,\beta=1}^{n^2-1} c_{\alpha\beta} \sum_{i,j=1}^{n} \left[\text{tr}(P_{ij}^{(1)} F_\alpha P_{ji}^{(2)} F_\beta^*)\right.$$

$$\left. - \frac{1}{2}\text{tr}(P_{ij}^{(1)} F_\beta^* F_\alpha P_{ji}^{(2)} + P_{ij}^{(1)} P_{ji}^{(2)} F_\beta^* F_\alpha)\right]$$

$$= \sum_{\alpha,\beta=1}^{n^2-1} c_{\alpha\beta} \sum_{i,j=1}^{n} \text{tr}(P_{ij}^{(1)} F_\alpha P_{ji}^{(2)} F_\beta^*)$$

$$= \sum_{k,l=1}^{n} \sum_{\alpha,\beta=1}^{n^2-1} c_{\alpha\beta}\text{tr}\left[\left(\sum_{j=1}^{n} P_{kj} F_\alpha P_{jl}\right)\left(\sum_{j=1}^{n} P_{kj} F_\beta P_{jl}\right)^*\right] \geq 0 \quad (7.43)$$

since $c_{\alpha\beta} \geq 0$.

The "only if" part: If a linear operator $L: M_n \rightarrow M_n$ generates a completely positive dynamical semigroup on M_n one should have $\text{tr}(L\rho) = 0$ and $L\rho^* = (L\rho)^*$ for all $\rho \in M_n$. Hence, by Lemma 7.3 it can be written in the form (7.33). Since the matrix $(c_{\alpha\beta})$ is selfadjoint, one can choose another set of orthogonal traceless matrices $\{G_1, G_2, \ldots, G_{n^2-1}\}$ such that

$$L\rho = -i[H, \rho] + \frac{1}{2}\sum_{\alpha=1}^{n^2-1} \lambda_\alpha \{[G_\alpha, \rho G_\alpha^*] + [G_\alpha\rho, G_\alpha^*]\}. \quad (7.44)$$

Define

$$\widehat{P}^{(\alpha)} = \sum_{i,j=1}^{n} (G_\alpha e_{ij} G_\alpha^*) \otimes e_{ij}, \quad \alpha = 1, 2, \ldots, n^2 - 1 \quad (7.45)$$

and

$$\widehat{P} = \frac{1}{n}\sum_{i,j=1}^{n} e_{ij} \otimes e_{ij}. \quad (7.46)$$

Then, by Lemma 7.4, Theorem 7.1 and (7.29) one has

$$0 \leq n\text{tr}\left\{\widehat{P}^{(\alpha)}(L \otimes \mathbb{1}_n)|^P\right\} = \sum_{\beta=1}^{n^2-1} \lambda_\beta \sum_{i,j=1}^{n} \text{tr}\left(G_\alpha e_{ij} G_\alpha^* G_\beta e_{ji} G_\beta^*\right)$$

$$= \sum_{\beta=1}^{n^2-1} \lambda_\beta \text{tr}(G_\alpha^* G_\beta)\text{tr}(G_\alpha G_\beta^*) = \lambda_\alpha, \quad \alpha = 1, 2, \ldots, n^2 - 1. \; \square \quad (7.47)$$

In the case dim $\mathcal{H} = n$, the conditions $||c_{ij}||_{i,j=1}^{n^2-1} \geq 0$ implicitly express the inequalities satisfied among the physical parameters characterizing the dynamical evolution (such as relaxation times and equilibrium states) which would be weaker or even nonexisting if just positivity were required. In the case $n = 2$, the generator of a positive dynamical semigroup is given by (7.16). In this case, the condition of complete positivity, expressed by the positive definiteness of the matrix $D + iA$, imposes further restrictions on the inverse relaxation times $\gamma_1, \gamma_2, \gamma_3$ and the range of the variation of m_0. In particular, it implies the inequalities

$$\gamma_1 + \gamma_2 \geq \gamma_3, \quad \gamma_2 + \gamma_3 \geq \gamma_1, \quad \gamma_3 + \gamma_1 \geq \gamma_2. \tag{7.48}$$

For the sake of simplicity, in what follows, the dynamical semigroup is understood as a completely positive dynamical semigroup.

The ergodic properties of dynamical semigroups have been investigated in [65], [306], [86], [99], [101].

A state $\rho \in \mathfrak{S}$ is said to be *invariant (stationary)* with respect to dynamical semigroup $\Lambda_t : \mathcal{T}(\mathcal{H}) \to \mathcal{T}(\mathcal{H})$ if

$$\Lambda_t \rho = \rho, \quad t \geq 0. \tag{7.49}$$

and *relaxing* if there exists an invariant state $\rho_0 \in \mathfrak{S}$ such that

$$\lim_{t \to \infty} \operatorname{tr}(a\Lambda_t \rho) = \operatorname{tr}(a\rho_0) \tag{7.50}$$

for every $a \in \mathcal{B}(\mathcal{H})$ and $\rho \in \mathfrak{S}$. The state ρ_0 is called an *equilibrium state*.

Sufficient conditions for the existence of equilibrium states are known:

THEOREM 7.6 (Spohn [306]). *Let $e^{tL} : \mathcal{B}(\mathcal{H}) \to \mathcal{B}(\mathcal{H})$, $\dim \mathcal{H} = n$, be a dynamical semigroup in the Schrödinger picture, where L is given by (7.25), and let the positive definite matrix $\|c_{ij}\|_{i,j=1}^{n^2-1}$ has a p-fold degenerate eigenvalue zero. If $p < \frac{n}{2}$, then the dynamical semigroup e^{tL} is relaxing and has a unique equilibrium state.*

THEOREM 7.7 (Frigerio [101]). *Let $e^{tL} : \mathcal{T}(\mathcal{H}) \to \mathcal{T}(\mathcal{H})$ be a dynamical semigroup with a bounded generator given by (7.22). If there exists an invariant state ρ_0 with $\overline{\operatorname{Range} \rho_0} = \mathcal{H}$ (i.e., ρ_0 has no zero eigenvalue), the following two properties are equivalent:*

(i) ρ_0 is the unique invariant state,
(ii) $M(L) = \{H, V_j, V_j^*; j \in A\}' = \{\mathbb{C} \cdot \mathbb{1}\}$,
where $\{\}'$ denotes the commutant of the set $\{\}$.

7.3. Diffusion processes on groups

There exists a connection between certain classes of dynamical semigroups and Markov processes on groups.

Let G be a separable locally compact group and $\mathcal{M}(G)$ be the family of all finite measures on G. A measure μ is a *probability measure* if $\mu(G) = 1$.

For any two measures $\mu, \nu \in \mathcal{M}(G)$, the *convolution* $\mu * \nu$ is defined as the set function

$$(\mu * \nu)(A) = \int_G \mu(Ag^{-1})\nu(dg) = \int_G \nu(g^{-1}A)\mu(dg), \qquad (7.51)$$

where $A \in \mathcal{B}_G$, $\mathcal{B}_G$ being the *Borel σ-algebra*, i.e., the smallest σ-algebra of subsets of G which contains all open subsets of G. For each element $g \in G$, the measure degenerated at g is denoted by π_g. For any $g \in G$ and $f \in C(G)$ (the algebra of continuous function on G) one has

$$\int_G f d\pi_g = f(g) \qquad (7.52)$$

and

$$(\mu * \pi_g)(A) = \mu(Ag^{-1}). \qquad (7.53)$$

For the measure π_e, e being the unity of G, one has

$$\pi_e * \mu = \mu * \pi_e = \mu. \qquad (7.54)$$

It is well-known [272] that $\mathcal{M}(G)$ is a topological semigroup with unity π_e under the operation $(\mu, \nu) \rightarrow \mu * \nu$. A *one-parameter convolution semigroup* of a Markov process in G is a family $\{\mu_t; t \geq 0\}$ of probability measures from $\mathcal{M}(G)$ such that

$$\mu_t * \mu_s = \mu_{t+s}, \quad t, s \geq 0, \qquad (7.55)$$

and μ_t converges weakly to π_e when $t \rightarrow 0$. Let G be a locally compact separable real p-dimensional Lie group and $x_1(g), x_2(g), \ldots, x_p(g)$ be canonical coordinates in a given neighbourhood of the unity $e \in G$.

A *diffusion process* or a *Brownian motion* in a locally compact separable real p-dimensional Lie group G is a one-parameter convolution semigroup $\{\mu_t; t \geq 0\}$ of probability measures from $\mathcal{M}(G)$ satisfying the conditions

(i) $\lim\limits_{t \downarrow 0} t^{-1}\mu_t(G - E) = 0$,

(ii) $\lim\limits_{t \downarrow 0} t^{-1}\int_E x_i(g)\mu_t(dg) = a_i, \quad i = 1, \ldots, p$,

(iii) $\lim\limits_{t \downarrow 0} t^{-1}\int_E x_i(g)x_j(g)\mu_t(dg) = b_{ij} = b_{ji}, \quad i, j = 1, \ldots, p$

for any neighbourhood E of the unity $e \in G$, where $a_1, a_2, \ldots, a_p$ are real numbers, and the symmetric matrix $||b_{ij}||_{i,j=1}^p$ is semipositive definite, cf. [113].

Suppose that $(V, \mathcal{H})$ is the representation of G into a group of uniformly bounded operators on a separable Hilbert space $\mathcal{H}$. A mapping

$$\Lambda^* : g \rightarrow \Lambda^*(g) : \mathcal{B}(\mathcal{H}) \rightarrow \mathcal{B}(\mathcal{H}), \qquad (7.56)$$

where

$$\Lambda^*(g)a = V^*(g)aV(g), \quad a \in \mathcal{B}(\mathcal{H}) \tag{7.57}$$

is a representation of G in $\mathcal{B}(\mathcal{H})$. Let $\{\mu_t;\ t \geq 0\}$ be a diffusion process in a locally compact separable Lie group. Then the family $\{\Lambda_t^*:\mathcal{B}(\mathcal{H}) \to \mathcal{B}(\mathcal{H}),\ t \geq 0\}$, where

$$\Lambda_t^* = \int_G V^*(g)aV(g)\mu_t(dg) \tag{7.58}$$

is a dynamical semigroup provided the condition

$$\Lambda_t^* \mathbb{1} = \int_G V^*(g)V(g)\mu_t(dg) = \mathbb{1} \tag{7.59}$$

is satisfied for $t \geq 0$.

Defining the generators of the representations $(V, \mathcal{H})$ and $(V^*, \mathcal{H})$ by

$$X_j = \frac{\partial V(g)}{\partial x_j(g)}\Big|_{g=\epsilon} \tag{7.60}$$

and

$$X_j^* = \frac{\partial V^*(g)}{\partial x_j(g)}\Big|_{g=e}, \tag{7.61}$$

and making the formal calculations (neglecting the problems of domains of X_j, X_j^*) one obtains that the generator L^* of the semigroup (7.58) can be formally written in the form

$$L^* = i[H, a] + \frac{1}{2}\sum_{m,n=1}^{p}\{X_m^*[a, X_n] + [X_m^*, a]X_n\} + \frac{1}{2}[Q, a]_+, \tag{7.62}$$

where

$$H = \frac{i}{2}\sum_{m=1}^{p} a_m(X_m - X_m^*) + \frac{i}{4}\sum_{m,n=1}^{p}(X_m X_n - X_m^* X_n^*) \tag{7.63}$$

and

$$Q = \sum_{m=1}^{p} a_m(X_m + X_m^*) + \frac{1}{2}\sum_{m,n=1}^{p}\{X_m^*(X_n + X_n^*) + (X_m + X_m^*)X_n\}. \tag{7.64}$$

L^* is formally the generator of a dynamical semigroup if $Q = 0$.

In the case $G = \mathrm{SU}(n)$ (*special unitary group in n dimensions*, i.e., the group on n-dimensional unitary matrices with determinant 1) and dim $\mathcal{H} =$

n, the generator L^* of the dynamical semigroup induced by a diffusion process on $SU(n)$ has the form

$$L^* a = i[H, a] - \frac{1}{2} \sum_{k,l=1}^{n^2-1} b_{kl} \{P_k[a, P_l] + [P_k, a]P_l\}, \qquad (7.65)$$

where

$$H = \sum_{k=1}^{n^2-1} a_k P_k \qquad (7.66)$$

and $P_1, P_2, \ldots, P_n$ are the generators of $SU(n)$. If G is a locally compact separable group, $\{\mu_t, \ t \geq 0\}$ is a convolution semigroup of probability measures on G and $(U, \mathcal{H})$ is a unitary representation of G in $\mathcal{H}$, then

$$\Lambda_t^* a = \int U^*(g) a U(g) \mu_t(dg) \qquad (7.67)$$

is a dynamical semigroup.

Let $x \to U(x)$, $x = (x_1, x_2, \ldots, x_n) \in \mathbb{R}^n$ be a strongly continuous unitary representation of the vector group in $\mathbb{R}^n$ in the Hilbert space $\mathcal{H}$, i.e,.

$$\begin{aligned} U(x + y) &= U(x)U(y), \\ U^{-1}(x) &= U(x^*) = U^*(x), \quad x, y \in \mathbb{R}^n. \end{aligned} \qquad (7.68)$$

Moreover, the infinitesimal generators $P_1, P_2, \ldots, P_n$ of the representation $(U, \mathcal{H})$ are strongly commuting selfadjoint operators on $\mathcal{H}$, cf. [295]. The family $\{\mu_t; \ t \geq 0\}$ of probability measures on $\mathbb{R}^n$ of the form

$$\mu_t(\epsilon) = \int_\epsilon p_t(x) d^n x, \qquad (7.69)$$

where ϵ is a Borel subset of $\mathbb{R}^n$, and

$$p_t(x) = \frac{1}{(2\pi t)^{n/2}} \frac{1}{\sqrt{\det B}} \exp\left\{ -\frac{1}{2t} \sum_{j,k=1}^n c_{jk}(x_j - a_j t)(x_k - a_k t) \right\}, \qquad (7.70)$$

where $a_1, a_2, \ldots, a_n$ are real numbers, and $B = (b_{ij}) = (c_{ij})^{-1}$ is a strictly positive definite real matrix of the n-th order, is a (Gaussian) one-parameter convolution semigroup of probability measures.

The dynamical semigroup $\{\Lambda_t^*; \ t \geq 0\}$ induced by a diffusion process in the vector group $\mathbb{R}^n$ has the form

$$\Lambda_t^* a = \int_{\mathbb{R}^n} U^*(x) \, a \, U(x) p_t(x) d^n x, \qquad (7.71)$$

and its generator

$$L^*a = i\Big[\sum_{j=1}^{n} a_j P_j, a\Big] - \frac{1}{2}\sum_{j,k=1}^{n} b_{jk}[P_j, [P_k, a]]. \tag{7.72}$$

Let G be the group $GL(n, \mathbb{C})$, i.e., the *general linear group in n complex dimensions* or the group of all complex $n \times n$ matrices. Let $F_k = F_k^*$, $\mathrm{tr}(F_k F_l) = \delta_{kl}$, $k, l = 1, 2, \ldots, n^2$, and $F_{n^2} = \frac{1}{\sqrt{n}}$ be a CONS in M_n. The canonical coordinates $x_k(V), y_k(V)$, $k = 1, 2, \ldots, n^2$, in $GL(n, \mathbb{C})$ are introduced by means of the relations

$$V = \exp\Big\{\sum_{k=1}^{n^2} z_k(V) F_k\Big\}, \tag{7.73}$$

where

$$z_k(V) = x_k(V) + iy_k(V) \tag{7.74}$$

for an element $V \in GL(n, \mathbb{C})$ from a neighbourhood of unity $\mathbb{1}$ of $GL(n, \mathbb{C})$. One can easily prove the following

THEOREM 7.8. *Let* $\{\mu_t; \ t \geq 0\}$ *be the following diffusion process on* $GL(n, \mathbb{C})$:

(i) $\lim\limits_{t\downarrow 0} t^{-1}\mu_t(G - E) = 0$,

(ii) $\lim\limits_{t\downarrow 0} t^{-1}\int_E z_k(V)\mu_t(dV) = a_k - ib_k$,

(iii) $\lim\limits_{t\downarrow 0} t^{-1}\int_E \bar{z}_k(V)z_l(V)\mu_t(dV) = d_{kl} = \bar{d}_{lk}$,

(iv) $\lim\limits_{t\downarrow 0} t^{-1}\int_E z_k(V)z_l(V)\mu_t(dV) = 0$,

where

$$a_{n^2} = b_{n^2} = 0, \tag{7.75}$$

$$d_{jn^2} = d_{n^2 j} = 0, \quad j = 1, 2, \ldots, n^2 \tag{7.76}$$

the hermitian matrix $(d_{kl})_{k,l=1}^{n^2-1}$ *is positive definite, and*

$$a_k = -\frac{1}{2}\sum_{r,s=1}^{n^2-1} \mathrm{tr}(F_r F_s F_k). \tag{7.77}$$

Then $\{\mu_t, \ t \geq 0\}$ *induces the dynamical semigroup*

$$\Lambda_t^* a = \int_{GL(n,\mathbb{C})} V^* a V \mu_t(dV), \quad a \in M_n \tag{7.78}$$

whose generator L^* has the form

$$L^* a = i\left[\sum_{k=1}^{n^2-1} b_k F_k, a\right] + \frac{1}{2} \sum_{k,l=1}^{n^2-1} d_{kl} \left\{ F_k[a, F_l] + [F_k, a]F_l \right\}. \tag{7.79}$$

This means that each dynamical semigroup in M_n is induced by a diffusion process in $GL(n, \mathbb{C})$.

7.4. Comments and examples

In order to discuss the dynamical semigroup describing the harmonic oscillator in a heat bath, the concepts developed in the preceding sections have to be generalized. Let G, K be locally compact separable groups, with K abelian. A K-*multiplier* for G, cf. [332], is a Borel map

$$m : (g, g') \mapsto m(g, g') \tag{7.80}$$

of $G \times G$ into K such that

(i) $m(g, u)m(gu, z) = m(g, uz)m(u, z)$ for all $g, u, z \in G$,
(ii) $m(g, e) = m(e, g) = 1$ for all $g \in G$.

K is written multiplicatively and 1 is the identity of K. If $K = \mathbb{C}$ — the multiplicative group of complex numbers, the prefix K- is omitted and one speaks simply of *multipliers*. In what follows the case $K = \mathbb{C}$ will be considered. Suppose that m is a multiplier for G. A mapping

$$V : g \mapsto V(g) \tag{7.81}$$

of G into a group of uniformly bounded operators on a separable Hilbert space $\mathcal{H}$ is said to be m-*representation*, cf. [332], if

(i) $g \mapsto V(g)$ is Borel, i.e., $g \mapsto (x, V(g)y)$ is Borel for all $x, y \in \mathcal{H}$,
(ii) $V(e) = \mathbb{1}$,
(iii) $V(g)V(u) = m(g, u)V(g, u)$, $g, u \in G$.

An m-representation $(V, m, \mathcal{H})$ of G in $\mathcal{H}$ induces an $|m|^2$-representation of G in $T(\mathcal{H})$:

$$\Lambda : g \mapsto \Lambda(g) : T(\mathcal{H}) \to T(\mathcal{H}), \tag{7.82}$$

where

$$\Lambda(g)\rho = V(g)\rho V^*(g), \quad \rho \in T(\mathcal{H}). \tag{7.83}$$

Let $\mathcal{M}_m(G, V)$ be the family of all finite measures on G satisfying the condition

$$\int_G V^*(g)V(g)\mu(dg) = \mathbb{1}. \tag{7.84}$$

It is easy to verify that $\mathcal{M}_m(G,V)$ is a semigroup with the unity π_e under the convolution $\bigcirc_m$ given by the formula

$$(\mu \bigcirc_m \nu)(A) = \int_G \int_G \chi_A(gu)|m(g,u)|^2 \mu(dg)\nu(du), \qquad (7.85)$$

where $\chi_A(g)$ stands for the characteristic function of the Borel set $A \in \mathcal{B}_G$.

In the case $|m(g,u)|^2 = 1$ for all $g, u \in G$, the convolution $\bigcirc_m$ coincides with the convolution $*$ defined by (7.53). A family $\{\mu_t;\ t \geq 0,\ \mu_t \in \mathcal{M}_m(G,V)\}$ is said to be (m,V)-Markov process in G provided

(i) $\mu_t \bigcirc_m \mu_s = \mu_{t+s},\quad t, s \geq 0,$
(ii) μ_t weakly converges to π_e when $t \to 0$.

From the above definition it follows:

THEOREM 7.9. *Let G be a locally compact separable group, $(V,m,\mathcal{H})$ be an m-representation of G and $\{\mu_t;\ t \geq 0\}$ be (m,V)-Markov process in $\mathcal{M}_m(G,V)$. Then the family $\{\Lambda_t: \mathcal{T}(\mathcal{H}) \to \mathcal{T}(\mathcal{H}),\ t \geq 0\}$ where*

$$\Lambda_t \rho = \int_G V(g)\rho V^*(g)\mu_t(dg) \qquad (7.86)$$

is a dynamical semigroup.

It should be pointed out that the assumption concerning uniform boundness of the representation V of G in $\mathcal{H}$ can be relaxed. Theorem 7.9 will remain true provided operators $V(g)$ are defined on a common dense set $D(V)$ in $\mathcal{H}$ and supports of measures $\{\mu_t;\ t \geq 0\}$ are contained in the set $\{g \in G: \|V(g)\| < \infty\}$.

Let $\{\mu_t;\ t \geq 0\}$ be an (m,V)-Markov process in G, and let μ_r be a right invariant measure in G. If $\mu_t,\ t \geq 0$, is absolutely continuous with respect to the measure μ_r, then by Radon-Nikodym theorem, cf. [125], there exists Radon-Nikodym derivative $p_t(g)$ of μ_t with respect to μ_r. In this case formulae (7.84), (7.85) and (7.86) can be rewritten in the form

$$\int_G V^*(g)V(g)p_t(g)\mu_r(dg) = \mathbb{1}, \qquad (7.87)$$

$$(p_t \bigcirc_m p_s)(g) = \int_G |m(gu^{-1},u)|^2 p_t(gu^{-1})p_s(u)\mu_r(du), \qquad (7.88)$$

and

$$\Lambda_t \rho = \int_G V(g)\rho V^*(g)p_t(g)\mu_r(dg), \qquad (7.89)$$

respectively.

The time evolution of the harmonic oscillator in a heat bath is described by the dynamical semigroup $\{\Lambda_t: T(\mathcal{H}) \to T(\mathcal{H}); \ t \geq 0\}$, $L^2(\mathbb{R}^1)$ whose infinitesimal generator L is given in the form, cf. [67], [66], [121], [197], [18],

$$L = L_0 + x_1 L_1 + x_2 L_2, \quad x_1 \geq x_2 \geq 0, \tag{7.90}$$

where

$$L_0 \rho = -i[\omega a^* a, \rho] + \frac{\sigma}{2} ([a^* a, \rho a^* a] + [a^* a \rho, a^* a]), \tag{7.91}$$

$$L_1 \rho = \frac{1}{2} ([a, \rho a^*] + [a \rho, a^*]), \tag{7.92}$$

$$L_2 \rho = \frac{1}{2} ([a^*, \rho a] + [a^* \rho, a]), \tag{7.93}$$

where a, a^* are Bose creation and annihilation operators defined on the Schwartz space $S(\mathbb{R}^1)$. The opretators a, a^* and $a^* a$ satisfy the following commutation relations

$$[a, a^*] = \mathbb{1}, \qquad [a^* a, a] = -a, \qquad [a^* a, a^*] = a^*, \tag{7.94}$$

i.e., the operators $\mathbb{1}, a, a^*$ and $a^* a$ generate a Lie algebra defined on $S(\mathbb{R}^1)$.

The Campbell-Hausdorff formula for the Lie algebra generated by means of the operators $\mathbb{1}, a, a^*$ and $a^* a$ has the form

$$W(\gamma_1, \alpha_1, \bar{\beta}_1) W(\gamma_2, \alpha_2, \bar{\beta}_2) = \tag{7.95}$$

$$= W(\gamma_1 + \gamma_2, e^{\gamma_1} \alpha_2 + \alpha_1, e^{-\gamma_1} \bar{\beta}_2 + \bar{\beta}_2) \exp \left\{ \frac{1}{2}(e^{-\gamma_1} \alpha_1 \bar{\beta}_2 - e^{\gamma_1} \bar{\beta}_1 \alpha_2) \right\},$$

where

$$W(\gamma, \alpha, \bar{\beta}) = \exp \left\{ \gamma \left(a^* + \frac{\bar{\beta}}{e^{-\gamma} - 1} \right) \left(a + \frac{\alpha}{e^{\gamma} - 1} \right) + \frac{\alpha \beta (e^{\gamma} + 1)}{2(e^{\gamma} - 1)} \right\}, \tag{7.96}$$

and $\gamma, \alpha, \bar{\beta}$ are complex numbers. The formula (7.95) has been derived by Wilcox, cf. [343], and is given here in a different form.

Let G be the group of matrices of the form

$$g = \begin{pmatrix} e^{\gamma} & \alpha & 0 & 0 \\ 0 & 1 & 0 & 0 \\ 0 & 0 & e^{-\gamma} & \bar{\beta} \\ 0 & 0 & 0 & 1 \end{pmatrix}, \tag{7.97}$$

where $\gamma, \alpha, \bar{\beta}$ are complex numbers. Taking into account (7.95) one concludes that the mapping $G \ni g \mapsto W(\gamma, \alpha, \bar{\beta})$ is an m-representation of the group G in the Hilbert space $\mathcal{H} = L^2(\mathbb{R}^1)$.

The construction of the semigroup $\{\Lambda_t = e^{tL}, \quad t \geq 0\}$ goes in the following steps.

a) Let G_0 be the subgroup of G consisting of all matrices of the form

$$g = \begin{pmatrix} e^{i\gamma_2} & 0 & 0 & 0 \\ 0 & 1 & 0 & 0 \\ 0 & 0 & e^{-i\gamma_2} & 0 \\ 0 & 0 & 0 & 1 \end{pmatrix}, \tag{7.98}$$

where $\gamma_2 \in \mathbb{R}^1$.

It follows from (7.96) that $W(i\gamma_2, 0, 0)$ is a unitary operator on $L^2(\mathbb{R}^1)$ and the map $G_0 \ni g \mapsto W(i\gamma_2, 0, 0)$ is the representation of G_0 in $L^2(\mathbb{R}^1)$. The right invariant measure on G_0 is simply $\mu_r(dg) = d\gamma_2$ and the diffusion process on G_0 is given by the probability density

$$r_t(\gamma_2) = \frac{1}{\sqrt{2\pi t \sigma}} \exp\left\{ -\frac{(\gamma_2 + \omega t)^2}{2\sigma t} \right\}, \qquad \sigma > 0. \tag{7.99}$$

The corresponding dynamical semigroup $\{\Theta_t = e^{tL_0}; \ t \geq 0\}$

$$\Theta_t \rho = \int_{G_0} W(i\gamma_2, 0, 0) \rho W^*(i\gamma_2, 0, 0) r_t(\gamma_2) d\gamma_2, \tag{7.100}$$

has the generator L_0 in the form (7.91).

b) Let G_1 be the subgroup of G consisting of all matrices of the form

$$g = \begin{pmatrix} 1 & \alpha & 0 & 0 \\ 0 & 1 & 0 & 0 \\ 0 & 0 & 1 & \bar{\alpha} \\ 0 & 0 & 0 & 1 \end{pmatrix} \tag{7.101}$$

where $\alpha \in \mathbb{C}$. It follows from (7.96) that $W(0, \alpha, \bar{\alpha})$ is a unitary operator on $L^2(\mathbb{R}^1)$ (Weyl operator) and the map $G_1 \ni g \mapsto W(0, \alpha, \bar{\alpha})$ is an m-representation of G_1 and $|m|^2 = 1$. The right invariant measure μ_r on G_1 has the form $\mu_r(dg) = d^2\alpha = d\alpha_1 d\alpha_2, \ \alpha = \alpha_1 + i\alpha_2$. A diffusion process on G_1 has the probability density

$$s_t(\alpha) = \frac{1}{\pi \kappa_2 t} \exp\left\{ -\frac{|\alpha|^2}{\kappa_2 t} \right\}, \tag{7.102}$$

and the corresponding semigroup

$$\Sigma_t \rho = \int_{G_1} W(0, \alpha, \bar{\alpha}) \rho W^*(0, \alpha, \bar{\alpha}) s_t(\alpha) d^2\alpha \tag{7.103}$$

has the generator

$$K = \kappa_2(L_1 + L_2) = -\kappa_2[a, [a^*, \cdot]], \tag{7.104}$$

where L_1 and L_2 are given by (7.92) and (7.93), respectively. Moreover, one can check that the relations

$$[L_0, L_1] = [L_0, L_2] = 0 \tag{7.105}$$

and

$$[L_1, L_1 + L_2] = -(L_1 + L_2) \tag{7.106}$$

hold.

It follows from (7.105) that the semigroups Θ_t and Σ_t commute, and $\Theta_t\Sigma_t$ is a semigroup whose generator is $L_0 + K$. One has to note that in both cases the condition (7.87) is satisfied, since $W(i\gamma_2, 0, 0)$ and $W(0, \alpha, \bar{\alpha})$ are unitary operators.

c) Let G_2 be the subgroup of G consisting of all matrices of the form

$$g = \begin{pmatrix} e^{\gamma_1} & 0 & 0 & 0 \\ 0 & 1 & 0 & 0 \\ 0 & 0 & e^{-\gamma_1} & \beta \\ 0 & 0 & 0 & 1 \end{pmatrix}, \tag{7.107}$$

where $\gamma_1 \in \mathbb{R}^1$, $\beta \in \mathbb{C}$. It follows from (7.95) that the mapping $G_2 \ni g \mapsto W(\gamma_1, 0, \beta)$ is a nonunitary representation of G_2, and $m(g, u) = 1$ for $g, u \in G_2$. Since the right invariant measure on G_2 has the form $\mu_r(dg) = d\gamma_1 d^2\beta = d\gamma_1 d\beta_1 d\beta_2$, $\beta = \beta_1 + i\beta_2$. $(1, -W)$-Markov process in G_2 is defined by means of the relations

$$\int_{G_2} q_t(\gamma_1 - \gamma_1', \bar{\beta} - e^{-(\gamma_1 - \gamma_1')}\bar{\beta}')q_s(\gamma_1', \bar{\beta}')d\gamma_1' d^2\beta' = q_{t+s}(\gamma_1, \bar{\beta}), \quad t, s \geq 0 \tag{7.108}$$

and the condition

$$\int_{G_2} W^*(\gamma_1, 0, \bar{\beta})W(\gamma_1, 0, \bar{\beta})q_t(\gamma_1, \bar{\beta})d\gamma_1 d^2\beta = \mathbb{1}. \tag{7.109}$$

From (7.95) and (7.96) one obtains the following relation

$$W^*(\gamma_1, 0, \bar{\beta})W(\gamma_1, 0, \bar{\beta}) = \exp(-e^{\gamma_1}\bar{\beta}a^*)\exp(2\gamma^1 a^*a)\exp(-e^{-\gamma_1}\bar{\beta}a). \tag{7.110}$$

Using coherent states and relation (7.110), one can easily verify that the probability density

$$q_t(\gamma_1, \bar{\beta}) = \frac{1}{\pi(e^{\kappa_1 t} - 1)}\exp\left\{-\frac{|\beta|^2}{e^{\kappa_1 t} - 1}\right\}\delta(\gamma_1 + \kappa_1 t) \tag{7.111}$$

satisfies (7.108) and (7.109). The semigroup

$$\Phi_t \rho = \int_{G_2} W(\gamma_2, 0, \bar{\beta}) \rho W^*(\gamma_2, 0, \bar{\beta}) q_t(\gamma_1, \bar{\beta}) d\gamma_1 d^2\beta \qquad (7.112)$$

has the generator L_1, cf. (7.92). Due to (7.105) semigroups Θ_t and Φ_t commute.

   d) In order to construct dynamical semigroup $\{\Lambda'_t = \exp[t(\kappa_1 L_1 + \kappa_2 L_2)],$ $t \geq 0\}$ its generator is written in the form

$$L' = \kappa_1 L_1 + \kappa_2 L_2 = (\kappa_1 - \kappa_2)L_1 + \kappa_2(L_1 + L_2) = (\kappa_1 - \kappa_2)L_1 + \kappa_2 K. \quad (7.113)$$

it follows from (7.106) that generators L_1 and K generate the Lie algebra which is isomorphic to the Lie algebra generated by a^*a and a with the correspondence $L_1 \to a^*a$, $K \to a$. Using (7.95) and (7.96) one obtains the relation

$$\exp[t(\kappa_1 - \kappa_2)L_1 + \kappa_2 K] = \exp(\lambda_1 L_1)\exp(\lambda_2 K) \qquad (7.114)$$

where

$$\lambda_1 = (\kappa_1 - \kappa_2)t, \qquad (7.115)$$

$$\lambda_2 = \frac{\kappa_2}{\kappa_1 - \kappa_2}\left(e^{(\kappa_1 - \kappa_2)t} - 1\right), \quad \kappa_1 - \kappa_2 \geq 0. \qquad (7.116)$$

Taking into account (7.114), (7.112), (7.93) and (7.95) one obtains

$$\Lambda'\rho = \int_{G_1} W(\gamma_1, \alpha, \bar{\beta}) \rho W(\gamma_1, \alpha, \bar{\beta}) p'_t(\gamma_1, \alpha, \beta) d\gamma_1 d^2\alpha d^2\beta, \qquad (7.117)$$

where (m, W)-Markov process on the group G' with

$$p'_t(\gamma_1, \alpha, \bar{\beta}) = \frac{\kappa_1 - \kappa_2}{\pi^2 \kappa_2} \frac{e^{(\kappa_1 - \kappa_2)t}}{(e^{(\kappa_1 - \kappa_2)t} - 1)^2} \qquad (7.118)$$

$$\times \exp\left\{-\frac{|\alpha - \beta|^2}{e^{(\kappa_1 - \kappa_2)t} - 1} - \frac{(\kappa_1 - \kappa_2)|\alpha|^2}{\kappa_2(e^{(\kappa_1 - \kappa_2)t} - 1)} - |\alpha|^2 + \frac{1}{2}(\alpha\bar{\beta} + \bar{\alpha}\beta)\right\}$$

$$\times \delta\left(\gamma_1 + \frac{\kappa_1 - \kappa_2}{2}t\right),$$

and G' is the subgroup of G consisting of all matrices

$$\begin{pmatrix} e^{\gamma_1} & \alpha & 0 & 0 \\ 0 & 1 & 0 & 0 \\ 0 & 0 & e^{-\gamma_1} & \bar{\beta} \\ 0 & 0 & 0 & 1 \end{pmatrix} \qquad (7.119)$$

with $\gamma_1 \in \mathbf{R}^1$, $\alpha, \bar{\beta} \in \mathbf{C}$.

Since the semigroup θ_t commutes with λ'_t, one finally obtains that the semigroup $\{\Lambda_t = e^{tL},\ t \geq 0\}$ can be written in the form

$$\Lambda_t \rho = \int_G W(\gamma, \alpha, \bar{\beta}) \rho W^*(\gamma, \alpha, \beta) p_t(\gamma, \alpha, \beta) d^2\gamma d^2\alpha d^2\beta, \qquad (7.120)$$

where

$$p_t(\gamma, \alpha, \bar{\beta}) = p'_t(\gamma_2, \alpha, \bar{\beta}) r_t(i\gamma_2), \quad \gamma = \gamma_1 + i\gamma_2, \qquad (7.121)$$

is the Radon-Nikodym derivative of (n, W)-Markov process on G, $\{\mu_t;\ t \geq 0\}$, with respect to the right invariant measure $\mu_r(dg) = d^2\gamma d^2\beta d^2\alpha$ on G which satisfies the relations

$$\int_G p_t(\gamma - \gamma', \alpha - e^{\gamma-\gamma'}\alpha', \beta - e^{-(\gamma-\gamma')}\bar{\beta}') p_s(\gamma', \alpha', \bar{\beta}')$$
$$\times a(\gamma, \gamma', \alpha, \alpha', \beta, \beta') d^2\gamma' d^2\alpha' d\beta' = p_{t+s}(\gamma, \alpha, \bar{\beta}) \quad (7.122)$$

with

$$a(\gamma, \gamma', \alpha, \alpha', \bar{\beta}, \bar{\beta}') = |m(gg'^{-1}, g')|^2 \qquad (7.123)$$
$$= \exp\left\{\frac{1}{2}[(e^{-(\gamma-\gamma')}\alpha - \alpha')\bar{\beta}' - (e^{\gamma-\gamma'}\beta - \beta')\bar{\alpha}' + \text{c.c.}]\right\}.$$

The action of the semigroup $\{\Lambda_t;\ t \geq 0\}$ can also be described in terms of the expectation values of normally ordered products of the creation and annihilation operators.

Let

$$f_{m,s}(t) = \text{tr}\left[(a^*)^m a^{m+s}(\Lambda_t \rho)\right], \qquad (7.124)$$

then one obtains

$$f_{m,s}(t) = \sum_{r=0}^{m} \binom{m}{r}\binom{m+s}{r} r! \left[\frac{\kappa_2}{\kappa_1 - \kappa_2}(1 - e^{-\gamma t})\right]^r e^{-\gamma t(m-r)}$$
$$\times f_{m-r,s}(0) \exp\left[-s(i\omega + \gamma)t - \frac{\sigma}{2}s^2 t^2\right], \qquad (7.125)$$

where

$$\gamma = \kappa_1 - \kappa_2. \qquad (7.126)$$

From (7.125) it follows that

$$\lim_{t\to\infty} f_{m,s}(t) = \left(\frac{\kappa_2}{\kappa_1 - \kappa_2}\right)^m m!\, \delta_{s,0}, \qquad (7.127)$$

i.e., there exists the equilibrium state ρ_e such that

$$\lim_{t \to \infty} \Lambda_t \rho = \rho_e \qquad (7.128)$$

for all $\rho \in \mathfrak{S}$ and

$$\rho_e = \left(1 - e^{-\beta}\right) e^{-\beta a^* a}, \qquad (7.129)$$

where

$$\beta = \ln \left(\frac{\kappa_2}{\kappa_1}\right). \qquad (7.130)$$

Further developments and applications of dynamical semigroups can be found in [72], [19].

7.5. Detailed balance

In this section, which is based on the results from [17] and [201], a quantum version of the detailed balance condition for a dynamical semigroup is discussed. Since quantum dynamical semigroups are the analogues of classical discrete Markov processes, the detailed balance condition for a Markovian master equation suggests itself as a natural generalization of the corresponding definition in the classical case. The Chapman-Kolmogorov equation for a discrete Markov process on the set $\{1, 2, \ldots, N\}$,

$$\frac{dp_i}{dt} = Lp, \qquad (7.131)$$

where $p = (p_1, p_2, \ldots, p_N)$, $p_i \geq 0$, $i = 1, 2, \ldots, N$, $\sum_{i=1}^{N} p_i = 1$, and

$$L_{ik} = a_{ik} - \delta_{ik} \sum_{m=1}^{N} a_{mi} \qquad (7.132)$$

satisfies *detailed balance* with respect to a stationary state $p^0 = (p_1^0, p_2^0, \ldots, p_N^0)$, $p_i^0 \geq 0$, $\sum_{i=1}^{N} p_i^0 = 1$, if

$$a_{ij} p_j^0 = a_{ji} p_i^0. \qquad (7.133)$$

The algebra of observables is the set U of sequences $f = (f_1, f_2, \ldots, f_N)$ with $f_i \in \mathbb{C}$, and the state p^0 defines on U an inner product as

$$\langle f, g \rangle = \sum_{i=1}^{N} p^0 \bar{f}_i g_i. \qquad (7.134)$$

It is straightforward to check that (7.133) is equivalent to

$$\langle f, L^* g \rangle = \langle L^* f, g \rangle, \qquad (7.135)$$

where $L^* = L^T$ is the generator of the dual dynamics.

Consider a strongly continuous dynamical semigroup $\{\Lambda_t \colon \mathcal{T}(\mathcal{H}) \to \mathcal{T}(\mathcal{H}), \ t \geq 0\}$ with a densely defined generator L, admitting a faithful stationary state ρ^0. One can define, in analogy to (7.135), an inner product in $\mathcal{B}(\mathcal{H})$ as

$$\langle a, b \rangle = \mathrm{tr}[\rho^0 a^* b], \tag{7.136}$$

and denote by $L^2(\mathcal{H}, \rho^0)$ the separable Hilbert space which is the completion of $\mathcal{B}(\mathcal{H})$ with respect to (7.136). The elements of $L^2(\mathcal{H}, \rho^0)$ are of the form $a = h(\rho^0)^{-1/2}$, h being a Hilbert-Schmidt operator.

A Heisenberg semigroup Λ_t^* is not, in general, strongly continuous on $\mathcal{B}(\mathcal{H})$, but it can be extended to a strongly continuous contraction semigroup on $L^2(\mathcal{H}, \rho^0)$. Indeed, using Kadison's inequality, cf. [110], and the invariance of ρ^0, one has

$$
\begin{aligned}
\langle \Lambda_t^* a, \Lambda_t^* a \rangle &= \mathrm{tr}\left[\rho^0 (\Lambda_t^*)^*(\Lambda_t a)\right] \\
&\leq \mathrm{tr}\left[\rho^0 \Lambda_t^*(a^* a)\right] = \mathrm{tr}[\rho^0 a^* a] = \langle a, a \rangle. \tag{7.137}
\end{aligned}
$$

The functions $t \mapsto \langle a, \Lambda_t^* b \rangle = \mathrm{tr}[\Lambda_t(\rho^0 a) b]$, $a, b \in \mathcal{B}(\mathcal{H})$ are continuous. Therefore, since $\mathcal{B}(\mathcal{H})$ is dense in $L^2(\mathcal{H}, \rho^0)$, $t \mapsto \Lambda_t^*$ is strongly measurable on $L^2(\mathcal{H}, \rho^0)$, hence strongly measurable and strongly continuous, cf. [179].

Let L^* denote the densely defined generator of the extension of $\{\Lambda_t^*; \ t \geq 0\}$ to $L^2(\mathcal{H}), \rho^0)$. In general, $D(L^*) \cap \mathcal{B}(\mathcal{H})$ is not dense in $\mathcal{B}(\mathcal{H})$ unless $t \mapsto \Lambda_t^*$ is strongly continuous on $\mathcal{B}(\mathcal{H})$.

It would not do to generalize (7.135) by demanding L^* to be selfadjoint with respect to the inner product (7.136) (ρ^0-selfadjoint), since this would rule out the case of Hamiltonian dynamics. Indeed, if H is a selfadjoint operator on $\mathcal{H}$ commuting with ρ^0, then $L_H^* = i[H, \cdot]$ is skewadjoint with respect to (7.136) (ρ^0-skewadjoint). Hence one proposes the following:

A quantum dynamical semigroup $\{\Lambda_t^*, \ t \geq 0\}$ satisfies the *detailed balance condition* with respect to a faithful stationary ρ^0 if L^* can be written as a sum

$$L^* = L_a^* + L_s^*, \tag{7.138}$$

where

(i) $L_a^* = i[H, \cdot]$, H — selfadjoint, $[H, \rho^0] = 0$, then L_a^* is ρ^0-skewadjoint and generates a group of unitaries in $L^2(\mathcal{H}, \rho^0)$;

(ii) L_a^* is ρ^0-selfadjoint.

Since ρ^0 is a faithful normal state in $\mathcal{B}(\mathcal{H})$, the set of functionals $\{\varphi_x, \ x \in L^2(\mathcal{H}, \rho^0)\}$, where

$$\varphi_x(a) = \langle x, a \rangle = \mathrm{tr}\left[(\rho^0 x)^* a\right] \tag{7.139}$$

is dense in the space of normal functionals on $\mathcal{B}(\mathcal{H})$, cf. [350]. Therefore, V_t and Γ_t defined as

$$\text{tr}\left[aV_t(\rho^0 x)\right] = \langle x, e^{tL_a^*}a\rangle, \tag{7.140}$$

$$\text{tr}[a\Gamma_t(\rho^0 x)] = \langle x, e^{tL_s^*}a\rangle, \tag{7.141}$$

where $x \in L^2(\mathcal{H}, \rho^0)$, $a \in \mathcal{B}(\mathcal{H})$, can be extended to dynamical semigroups of $\mathcal{T}(\mathcal{H})$ with generators $L_a = -i[H, \cdot]$ and L_s, respectively, such that

$$\text{tr}\left[L_a(\rho^0 x^*)\right] = -\langle L_a^* x, a\rangle, \tag{7.142}$$

$$\text{tr}\left[L_s(\rho^0 x^*)\right] = \langle L_s^* x, a\rangle, \tag{7.143}$$

where $a \in \mathcal{B}(\mathcal{H})$, $x \in D(L^*)$, and

$$L = L_a + L_s. \tag{7.144}$$

If a decomposition of the formula (7.144) exists, it is clearly unique, and it coincides in the finite-dimensional case with the one given by formula (7.25) with the restriction that the semigroup Γ_t is norm-continuous; it is possible then to give the general form of the generator L of a dynamical semigroup satisfying detailed balance. This is provided by the following

THEOREM 7.10 ([201]). *In order for* $\{\Lambda_t;\ t \geq 0\}$ *to be a dynamical semigroup of* $\mathcal{T}(\mathcal{H})$ *satisfying detailed balance with respect to a faithful stationary state* ρ^0 *and having a norm-continuous dissipative part, it is necessary and sufficient that its Heisenberg generator* L^* *can be written in the form (7.138), with*

$$L_a^* a = i[H, a], \tag{7.145}$$

$$L_s^* a = w^* - \lim_{M\to\infty} \frac{1}{2} \sum_{r,r',s,s'=1}^{M} c_{rr',ss'} \left\{ P_{rr'}[a, P_{s's}] + [P_{rr'}, a]P_{s's} \right\}, \tag{7.146}$$

where H *is selfadjoint and commutes with* ρ^0, $P_{rs} = \varphi_r(\varphi_s, \cdot)$, *where* $\{\varphi_r\}$ *is a CONS of eigenvectors of* ρ^0, *namely,* $\rho^0\varphi_r = \rho_r^0\varphi_r$; $\{c_{rr',ss'}\}$ *is positive in the sense that*

$$\sum_{rr',ss'} \bar{a}_{rr'}c_{rr',ss'}a_{ss'} \geq 0 \tag{7.147}$$

for all $\{a_{rr'}\}$; *and* $\sum_{rr'ss'=1}^{N} c_{rr',ss'}P_{rr'}P_{ss'}$ *converges ultraweakly as* $N \to \infty$;

$$c_{rr',ss'}\rho_s^0 = c_{s's,r'r}\rho_{s'}^0, \tag{7.148}$$

or, equivalently,

$$c_{rr',ss'}\rho_r^0 = c_{s's,r'r}\rho_{r'}^0. \tag{7.149}$$

If $\mathcal{H}$ is N-dimensional, Alicki [17] has proved that L_s^* can be given a simple diagonal form as follows

$$L_s^* a = \sum_{i,j=1}^{N} d_{ij}\{X_{ij}^*[a, X_{ij}] + [X_{ij}^*, a]\},\tag{7.150}$$

where

$$\mathrm{tr}[X_{ij}^* X_{kl}] = \delta_{ik}\delta_{jl},\tag{7.151}$$

$$\rho^0 X_{ij}(\rho^0)^{-1} = \rho_i^0(\rho_j^0)^{-1} X_{ij},\tag{7.152}$$

$$X_{ij}^* = X_{ji},\tag{7.153}$$

$$d_{ij} \geq 0, \quad d_{ji}\rho_j^0 = d_{ji}\rho_i^0.\tag{7.154}$$

7.6. Reduced dynamics

The master equation is a tool for extracting the dynamics of a subsystem of a larger system by the use of projection techniques on Banach spaces, cf. [288], [355], [88], [281], [120], [16], [210], [280], [85], [89], [6], [220]. One is interested in a spatially confined quantum system S, with underlying Hilbert space $\mathcal{H}^S$ and the algebra of observables $\mathcal{B}(\mathcal{H}^S)$. The "reservoir" R is taken to be an infinite quantum system, with algebra of observables U_R and Hilbert space $\mathcal{H}_\omega^R$ determined by the GNS representation π_ω induced by a suitable reference state ω^R on U_R which is assumed to be stationary under the free evolution of R. It is assumed that π_ω is faithful and in the notation used here an element of U_R and its representative $\pi_\omega(A)$ will not be distinguished. The system $S + R$ is considered to be isolated, so that its time evolution is determined by a selfadjoint Hamiltonian H acting on $\mathcal{H}^S \otimes \mathcal{H}_\omega^R$:

$$H = H^S \otimes \mathbb{1}^R + \mathbb{1}^S \otimes H^R + \lambda H^{SR} = H^0 + \lambda H^{SR},\tag{7.155}$$

where H^S is the free Hamiltonian of S, H^R is the free Hamiltonian of R in the representation induced by the stationary state ω^R, and

$$H^{SR} = \sum_j V_j^S \otimes V_j^R\tag{7.156}$$

is the interaction Hamiltonian, with V_j^S selfadjoint on $\mathcal{H}^S$ and V_j^R selfadjoint on $\mathcal{H}_\omega^R$.

The Banach space of trace class operators on $\mathcal{H}^S \otimes \mathcal{H}_\omega^R$ is denoted by $T(\mathcal{H}^S \otimes \mathcal{H}_\omega^R)$ which is homomorphic with the space of normal functionals on $B(\mathcal{H}^S) \otimes \pi_\omega(U^R)''$ according to the map

$$\varphi : T(\mathcal{H}^S \otimes \mathcal{H}_\omega^R) \;\to\; \left[B(\mathcal{H}^S) \otimes \pi_\omega(U^R)'' \right]$$

$$\varphi(\sigma)(a) = \mathrm{tr}_{R+S}(\sigma a), \quad a \in B(\mathcal{H}^S) \otimes \pi_\omega(U^R)''. \tag{7.157}$$

In the following the same notation will be used for σ and $\varphi(\sigma)$.

The capital letters with tilde, and capital Greek letters, will be used to denote operators acting on the spaces $T(\mathcal{H}^S)$, $T(\mathcal{H}_\omega^R)$, $T(\mathcal{H}^S \otimes \mathcal{H}_\omega^R)$. The identity maps acting on these spaces will be denoted by $\tilde{I}^R$, $\tilde{I}^S$ and $\tilde{I}$, respectively. The dynamics on $T(\mathcal{H}^S \otimes \mathcal{H}_\omega^R)$ is induced by

$$\tilde{H} = \tilde{H}^S \otimes \tilde{I}^R + \tilde{I}^S \otimes \tilde{H}^R + \lambda \tilde{H}^{SR} = \tilde{H}^0 + \lambda \tilde{H}^{SR}, \tag{7.158}$$

where $\tilde{H}^S = [H^S, \cdot]$, $\tilde{H}^R = [H^R, \cdot]$, $\tilde{H}^{SR} = [H^{SR}, \cdot]$. One defines two one-parameter groups of automorphisms of $T(\mathcal{H}^S \otimes \mathcal{H}_\omega^R)$:

$$U_t^0 : U_t^0 \sigma = \exp[-i\tilde{H}^0 t]\sigma. \tag{7.159}$$

$$U_t \equiv U_t^{(\lambda)} : U_t \sigma = \mathrm{ext}[-i\tilde{H} t]\sigma, \tag{7.160}$$

where $\sigma \in T(\mathcal{H}^S \otimes \mathcal{H}_\omega^R)$, describing the free and global time evolutions, respectively.

One is interested in the reduced dynamics of S, under the assumption that the initial state $S + R$ is of the form $\sigma(0) = \rho \otimes \omega^R$, where ρ is any normalized positive element of $T(\mathcal{H}^S)$. The operation of partial trace with respect to R is defined as

$$\mathrm{tr}_R : T(\mathcal{H}^S \otimes \mathcal{H}_\omega^R) \to T(\tilde{H}^S),$$

$$\mathrm{tr}_S \left[\mathrm{tr}_R[\sigma]a \right] = \mathrm{tr}_{S+R} \left[\sigma(a \otimes \mathbb{1}_R) \right], \quad a \in B(\mathcal{H}^S). \tag{7.161}$$

The *amplification* [323] or *lifting* [324] is the linear operator

$$\mathcal{A} : T(\mathcal{H}^S) \to T(\mathcal{H}^S \otimes \mathcal{H}_\omega^R), \quad \mathcal{A}\rho = \rho \otimes \omega^R. \tag{7.162}$$

Then

$$\mathcal{E} = \mathcal{A}\mathrm{tr}_R \tag{7.163}$$

is a bounded idempotent (projection) on $T(\mathcal{H}^S \otimes \mathcal{H}_\omega^R)$, which projects onto the subspace $T(\mathcal{H}^S) \otimes \omega^R$, isomorphic to $T(\mathcal{H}^S)$.

The *reduced dynamics* Λ_t of S is defined in the Schrödinger picture as follows:

$$\mathrm{tr}_S \left[(\Lambda_t \rho)a \right] = \mathrm{tr}_{S+R} \left[U_t(\rho \otimes \omega^R)(a \otimes \mathbb{1}_R) \right], \tag{7.164}$$

$\rho \in \mathcal{T}(\mathcal{H}^S)$, $a \in \mathcal{B}(\mathcal{H}^S)$, namely,

$$\Lambda_t \rho = \text{tr}_R[U_t A \rho] \quad \text{or} \quad A \Lambda_t \rho = \mathcal{E} U_t A \rho. \tag{7.165}$$

The dual Heisenberg dynamics Λ_t^* is given by

$$\text{tr}_S[\rho(\Lambda_t^* a)] = \text{tr}_{S+R}[(\rho \times \omega^R) U_t^*(a \otimes \mathbb{1}_R)], \tag{7.166}$$

$\rho \in \mathcal{T}(\mathcal{H}^S)$, $a \in \mathcal{B}(\mathcal{H}^S)$, that is,

$$\Lambda_t^* a = \mathcal{F} U_t^*(a \otimes \mathbb{1}_R), \quad a \in \mathcal{B}(\mathcal{H}^S), \tag{7.167}$$

where $\mathcal{F}$ is defined as

$$\mathcal{F} : \mathcal{B}(\mathcal{H}^S) \otimes \pi_\omega(U^R)'' \to \mathcal{B}(\mathcal{H}^B),$$

$$\text{tr}_{S+R}[(A\rho)a] = \text{tr}_S[\rho(\mathcal{F}a)], \tag{7.168}$$

and it can be shown to be a conditional expectation onto $\mathcal{B}(\mathcal{H}^S)$, cf. [220], [330], [331].

The reduced dynamics in the Heisenberg picture Λ_t^* of a system S coupled to a reservoir R has the remarkable property that Λ_t^* is completely positive for all t. Indeed, from (7.166) it follows that for any sequence $x_1, x_2, \ldots, x_n$ of elements of $\mathcal{H}^S$ one has the inequality

$$\sum_{i,j=1}^n \langle x_i, \Lambda_t^*(a_i^* a_j) x_j \rangle = \text{tr}_{S+R}\left[\sum_{i,j=1}^n (d_{ij} \otimes \omega_R) U_t^*(a_i^* \otimes \mathbb{1}_R) U_t^*) \right] \geq 0 \tag{7.169}$$

since the matrix $d = (d_{ij})_{i,j=1}^n$ with entries

$$d_{ij} = \bar{x}_i \langle x_j, \cdot \rangle \tag{7.170}$$

is semipositive definite.

Starting from the Von Neumann equation for the density operator of the global system $S + R$,

$$\frac{d\sigma(t)}{dt} = -i\tilde{H}\sigma, \tag{7.171}$$

one can formally derive an exact equation for the reduced density operator $\rho(t) = \text{tr}_R \sigma(t)$ of S, called the *master equation*, which under the assumption that $\sigma(0) = \rho \otimes \omega^R$, has the form, cf. [281] and [210],

$$\frac{d}{dt}\rho(t) = -i\tilde{H}_{\text{eff}}\rho(t) + \lambda^2 \int_0^t ds K(s)\rho(t-s), \tag{7.172}$$

where

$$\tilde{H}_{\text{eff}} = [H_{\text{eff}}, \cdot], \quad H_{\text{eff}} = H^S + \lambda \sum_j \omega^R(V_j^R)V_j^S, \tag{7.173}$$

and

$$K(s) = -\text{tr}_R \left[\tilde{H}^{SR}(\tilde{I} - \tilde{P})U_S'(\tilde{I} - \tilde{P})\tilde{H}^{SR}\tilde{A} \right], \tag{7.174}$$

where

$$U_t' = \exp[-i\tilde{H}'t], \quad \tilde{H}' = (\tilde{I} - \tilde{P})\tilde{H}(\tilde{I} - \tilde{P}). \tag{7.175}$$

By setting

$$\tilde{H}^{SR}(t) = U_{-t}^0 \tilde{H}^{SR} U_t^0 = [e^{iH_0 t}H^{SR}e^{-iH_0 t}, \cdot] = [H^{SR}(t), \cdot] \tag{7.176}$$

one observes that the integral kernel $K(s)$ admits a formal power series expansion in the coupling constant λ of the form

$$K(s) = e^{-i\tilde{H}^S s}[K_0(s) + \sum_{0 \geq t_n \geq \ldots \geq t_1 \geq s} K_n(s|t_1, \ldots, t_n)], \tag{7.177}$$

where

$$K_0(s) = -\text{tr}_R \left[\tilde{H}^{SR}(s)(\tilde{I} - \tilde{P})\tilde{H}^{SR}\tilde{A} \right], \tag{7.178}$$

$$K_n(s|t_1, \ldots, t_n) = -\text{tr}_R[\tilde{H}^{SR}(s)(\tilde{I} - \tilde{P})\tilde{H}^{SR}(t_1)(\tilde{I} - \tilde{P})\ldots$$
$$\times (\tilde{I} - \tilde{P})\tilde{H}^{SR}(t_n)(\tilde{I} - \tilde{P})\tilde{H}^{SR}\tilde{A}]. \tag{7.179}$$

$K_n(s|t_1, \ldots, t_n)$ depends on the multi-time correlation functions

$$\omega^R(V_{j_1}^R(t_{i_1}) \cdots V_{j_k}^R(t_{i_k}))$$

of the reservoir operators appearing in the interaction Hamiltonian, up to order $k = n + 2$, the relevant times being $t_0 = s, t_1, \ldots, t_n, t_{n+1} = 0$. The Born approximation of (7.172) amounts to keeping only the term $e^{-i\tilde{H}^S s}K_0(s)$ in the expansion (7.177). The term $\lambda \sum_j \omega^R(V_j^R)V_j^S$ is a modification of the free Hamiltonian due to the interaction with the reservoir. It vanishes when $\omega^R(V_j^R) = 0$, and this will be assumed for the simplicity in what follows.

An evolution equation for the density operator of a subsystem S of a closed system $S + R$ is said to be *Markovian* if it has the form, cf. (7.6),

$$\frac{d\rho(t)}{dt} = L\rho(t). \tag{7.180}$$

However, it is, in general, impossible to derive Markovian master equation as an exact consequence of (7.172).

One does not expect that (7.172) can be reduced to the form (7.180) unless the global dynamics U_t of $S + R$ has some "singular" character, corresponding to a limiting situation in which the memory effects which are present on (7.172) become negligible. Since the completely positive maps of C^*-algebra form a convex cone which is closed in the bounded-weak topology, cf. [220], the property of complete positivity of the Heisenberg reduced dynamics is not destroyed by any limiting procedures which are employed to obtain a Markovian master equation. From the structure of the kernel $K(s)$ in (7.172), one expects that the situation mentioned above will take place if the typical variation time τ_S of $\rho(t)$ is much longer than the decay time τ_R of the correlation functions of the reservoir, i.e., Markovian approximation becomes rigorous in the limit $\tau_S/\tau_R \to \infty$.

Two possible limits can be taken.

(1) The *weak coupling limit* $\lambda \to 0$, with *rescaling time* $\tau = \lambda^2 t$, [330], [331]. In this case τ_R remains constant, while τ_S tends to infinity. Rigorous models thereof were studied by Davis, [67], [68], and Pulé, [270], and a general theory has been given by Davis, [69], [71], [70]. Roughly speaking, it turns out that in the limit $\lambda \to 0$ the expansion (7.177) reduces to the Born approximation and the integral extends to infinity due to the change in the time scale.

The simplest version of the Davis method is the following. Let us assume that $\tilde{H}^{SR}$ is bounded and $\mathrm{tr}(\tilde{H}^{SR}A) = 0$, so that $\tilde{H}_{\mathrm{eff}} = \tilde{H}^S$. In order to avoid domain problems, it is convenient to work with an integral form of (7.172). An elementary change of variables in the double integration yields

$$\rho(t) = U_t^S \rho(0) + \lambda^2 \int_0^t du \int_0^{t-u} U_{t-x-u}^S K(x)\rho(u), \qquad (7.181)$$

where $U_t^S = \exp[-it\tilde{H}^S]$.

One goes over to the interaction picture and rescales the time, setting $\tau = \lambda^2 t$ and $\sigma = \lambda^2 u$, with the purpose of letting λ go to zero. The use of the interaction picture can be interpreted as an averaging over "fast microscopic oscillations". One defines

$$\rho_I(\tau) = \lim_{\lambda \to 0} U_{-\lambda^{-2}\tau}^S \rho(\lambda^{-2}\tau) \qquad (7.182)$$

and from (7.181) one obtains

$$\rho_I(\tau) = \rho_I(0) + \lim_{\lambda \to 0} \int_0^\tau d\sigma U_{-\lambda^{-2}\sigma}^S \left[\int_0^{\lambda^{-2}(\tau-\sigma)} dx U_{-x}^S K(x) \right] U_{\lambda^{-2}\sigma}^S \rho_I(\sigma). \qquad (7.183)$$

One expects that in the limit $\lambda \to 0$ only the term K_0 in the expansion (7.177) of K will give a nonvanishing contribution, so that

$$\rho_I(\tau) = \rho_I(0) + \lim_{\lambda \to 0} \int_0^\tau d\sigma U^S_{-\lambda-2\sigma} K U^S_{\lambda-2\sigma} \rho_I(\sigma), \tag{7.184}$$

where

$$K = \int_0^\infty dx K_0(x). \tag{7.185}$$

Davis has proved that this is indeed the case under the following conditions on K_0, K_n $(n = 1, 2, \ldots)$:

$$\int_0^\infty dt \, \| K_0(t) \| < \infty, \tag{7.186}$$

$$\int_{0 \geq t_n \geq t_{n-1} \geq \cdots \geq t_1 \geq t_0 \geq t} dt_0 dt_1 \ldots dt_n \, \| K_n(t_0 | t_1, t_2, \ldots, t_n) \| \geq a_n(t), \tag{7.187}$$

with

$$a_n(t) \geq c_n t^{n/2} \quad \text{for all} \quad t \geq 0, \tag{7.188}$$

where $\sum_{n \geq 1} c_n z^n$ has an infinite radius of convergence, and

$$a_n(t) \geq d_n t^{n/2 - \epsilon} \quad \text{for some} \quad \epsilon > 0, \quad \text{and all} \quad t \geq 0. \tag{7.189}$$

Since one deals with a spatially confined system $\mathcal{S}$, the spectrum of $\tilde{H}^S$ is purely discrete and the limit $\lambda \to 0$ in (7.184) can be easily performed, to yield

$$\rho_I(\tau) = \rho_I(0) + \bar{K} \int_0^\tau d\sigma \rho_I(\sigma), \tag{7.190}$$

where

$$\bar{K} = \sum_\alpha Q_\alpha K Q_\alpha, \tag{7.191}$$

Q_α being the spectral projections of $\tilde{H}^S$ corresponding to distinct eigenvalues ω_α. Hence equivalently

$$\bar{K} = \lim_{a \to \infty} \frac{1}{2a} \int_{-a}^a U^S_{-x} K U^S_x. \tag{7.192}$$

The rigorous result in [69] is

$$\lim_{\lambda \to 0, \, \lambda^2 t = \tau} \| U^S_{-t} \mathrm{tr}_R[U_t^{(\lambda)} A\rho] - \exp(\tau \bar{K} \rho) \|_1 = 0 \tag{7.193}$$

uniformly on each interval $0 \geq \tau \geq \tau_1$.

If conditions (7.185) – (7.189) are satisfied, then using (7.156) with $\omega^R(V_\alpha^R) = 0$, the reduced dynamics in the Schrödinger picture is a semigroup whose generator $\bar{K}$ is given by [125],

$$
\begin{aligned}
\bar{K} = \sum_{\omega \in \mathrm{Sp\,\widehat{H}^S}} \sum_{\alpha,\beta} &\{-is_{\alpha\beta}(\omega)[V_\beta(\omega)^* V^\alpha(\omega), \rho] \\
&+ \frac{1}{2}\widehat{h}_{\alpha\beta}(\omega)([V_\alpha(\omega)\rho, V_\beta^*(\omega)] + [V_\alpha(\omega), \rho V_\beta^*(\omega)])\},
\end{aligned}
\tag{7.194}
$$

where $(2\pi)^{-1}\widehat{h}_{\alpha\beta}(\omega)$ is the Fourier transform of

$$
h_{\alpha\beta}(t) = \omega^R\left(V_\beta^R V_\alpha^R(t)\right),
\tag{7.195}
$$

$$
\begin{aligned}
S_{\alpha\beta}(\omega) &= i\int_0^\infty dt\, e^{-i\omega t} h_{\alpha\beta}(t) - \frac{1}{2}\widehat{h}_{\alpha\beta}(\omega) \\
&= \frac{1}{2\pi}\mathrm{P}\int_{-\infty}^\infty \frac{\widehat{h}_{\alpha\beta}(\lambda)}{\lambda - \omega}d\lambda,
\end{aligned}
\tag{7.196}
$$

P denoting the principal part, and

$$
V_\alpha(\omega) = \sum_{\epsilon_m - \epsilon_n = \omega} P_{nn} V_\alpha^S P_{mm} = V_\alpha(-\omega)^*,
$$

$$
P_{nm} = x_n\langle x_m, \cdot\rangle, \quad H^S x_m = \epsilon_m x_m.
\tag{7.197}
$$

Sufficient conditions for (7.185) – (7.189) to be satisfied are:

(i) $\sum_\alpha \| V_\alpha^S \| < \infty$;

(ii) $\int_0^\infty |h_{\alpha\beta}(t)|(1+t)^\epsilon dt < a$, with a independent of α, β;

(iii) the reservoir is quasi-free, i.e., all truncated functions

$$
\omega^R\left(V_{\alpha_1}^R(t_1), \ldots, V_{\alpha_k}^R(t_k)\right)^T
$$

of order k greater than two vanish.

If the reference state of the reservoir is KMS at inverse temperature β, then the canonical state

$$
\rho_\beta = \frac{e^{-\beta H^S}}{\mathrm{tr}\, e^{-\beta H^S}}
\tag{7.198}
$$

is a stationary state for the reduced dynamics, as follows from the KMS condition on Fourier transforms

$$
\bar{h}_{\alpha\beta}(-\omega) = e^{-\beta\omega} h_{\beta\alpha}(\omega).
\tag{7.199}
$$

Moreover, the reduced dynamics satisfies a quantum detailed balance condition as a consequence of (7.199).

(2) Models of *singular reservoirs*.

The singular reservoir limit corresponds to a limiting case in which the correlation functions $h_{\alpha\beta}(t)$, cf. (7.195), of the operators V_α^R appearing in the interaction tend to $c_{\alpha\beta}\delta(t)$, i.e., $\tau^R \to 0$. One has to observe that a correlation function (7.195) cannot tend to a δ-function if ω^R is KMS at some $\beta \neq 0$. In fact, by the continuity of the Fourier transform, $h_{\alpha\beta}(t)$ tends to a δ-function if and only if the Fourier transform $\hat{h}^{\alpha\beta}(\omega)$ tends to a constant almost everywhere. On the other hand, the KMS condition on Fourier transform (7.199) forbids $\hat{h}_{\alpha\beta}(\omega)$ to approach a constant unless $\beta = 0$. For a similar reason, if the reservoir is chosen to be a quasi-free Bose or Fermi gas and ω^R is the vacuum state, the limit of singular reservoir can be performed if one-particle energy is the whole real line. In [109] and [100] models have been studied which can account for all dynamical semigroups of an n-level system. The system S is coupled to a quasi-free boson or fermion reservoir in the vacuum state by a linear coupling of the form

$$II^{SR} = \sum_{\alpha=1}^{n^2} V_\alpha^S \otimes V_\alpha^R, \tag{7.200}$$

where $\{V_\alpha^S, \ \alpha = 1, 2, \ldots, n^2\}$ can be chosen as a hermitian CONS in M^n with $V_{n^2}^S = \frac{1}{\sqrt{n}}$;

$$V_\alpha^R = \varphi_\alpha(f^c) = \sum_{\beta=1}^{n^2} \left[\bar{\mu}_\beta \lambda_\alpha^\beta a_\beta(f^c) + \mu_\beta \lambda_\alpha^\beta a_\beta(f^c)^* \right], \tag{7.201}$$

where $a_\alpha(f^c)$, $a_\alpha^*(f^c)$ are independent Bose and Fermi creation and annihilation opeators, and

$$\sum_{\alpha=1}^{n^2} \bar{\lambda}_\alpha^\rho \lambda_\alpha^\sigma = \delta_{\rho\sigma}, \tag{7.202}$$

$$f^c(\omega) = (2\pi)^{-1/2} \exp\left[-\frac{\epsilon^2 \omega^2}{\rho} \right]. \tag{7.203}$$

The corresponding Fourier transform of the correlation function $h_{\alpha\beta}^{(c)}(t)$ has the form

$$\hat{h}_{\alpha\beta}^{(c)}(\omega) = c_{\alpha\beta}(2\pi)^{-1/2} \exp\left[-\frac{\epsilon^2 \omega^2}{4} \right], \tag{7.204}$$

where

$$c_{\alpha\beta} = \sum_{\gamma=1}^{n^2} |\mu_\gamma|^2 \lambda_\alpha^\gamma \bar{\lambda}_\beta^\gamma \qquad (7.205)$$

is the general form of a positive $(n^2 \times n^2)$-matrix. As $\epsilon \to 0$ (7.204) tends to the Fourier transform of $c_{\alpha\beta}\delta(t)$.

The generator of the reduced dynamical semigroup is given by

$$\mathcal{L}\rho = -[H^S + H^1, \rho] + \frac{1}{2} \sum_{i,j=1}^{n^2-1} c_{ij} \left\{ [V_i^S \rho, V_j^S] + [V_i^S, \rho V_j^S] \right\}, \qquad (7.206)$$

where

$$H^1 = -n^{-1/2} \sum_{k=1}^{n^2-1} \operatorname{Im} c_{k,n^2} V_k^S. \qquad (7.207)$$

7.7. Remarks

The ideas of information estimation developed in Chapter 6 can be easily extended to open systems. It simply consists in taking Markovian master equation instead of Von Neumann equation [155]. However, it is rather difficult to find the general features of such time evolution, since it strongly depends on the specific properties of Markovian master equation. This approach has been applied to laser theory [151], [152]. Finally, the question concerning the possible physical significance of decomposable dynamical semigroups will be shortly discussed.

A positive linear map ϕ of a C^*-algebra $\mathcal{A}$ into $\mathcal{B}(\mathcal{H})$ is said to be *decomposable* if there are a Hilbert space $\mathcal{K}$, a bounded operator $V: \mathcal{H} \to \mathcal{K}$, and a Jordan homomorphism π of $\mathcal{A}$ into $\mathcal{B}(\mathcal{K})$ such that $\phi(x) = V^* \pi(x) V$ for all $x \in \mathcal{A}$. Such maps have been studied in [347], [314] [54], [28], [313] and are the natural symmetrization of the completely positive ones, defined as those ϕ as above with π being a homomorphism. If $M_n(\mathcal{B})$ denotes the $n \times n$ matrices over a subspace $\mathcal{B}$ of a C^*-algebra and $M_n(\mathcal{B})^+$ the positive cone of $M_n(\mathcal{B})$, the Stinespring [314] states that a map $\phi: \mathcal{A} \to \mathcal{B}(\mathcal{H})$ is completely positive if and only if for all positive integers n, whenever $(x_{ij}) \in M_n(\mathcal{A})^+$ then $(\phi(x_{ij})) \in M_n(\mathcal{B}(\mathcal{H}))^+$.

An analogous characterization of decomposable maps has been given by Størmer [314]:

THEOREM 7.11. *Let $\mathcal{A}$ be a C^*-algebra and ϕ a linear map of $\mathcal{A}$ into $\mathcal{B}(\mathcal{H})$. Then ϕ is decomposable if and only if for all positive integers n whenever (x_{ij}) and (x_{ji}) belong to $M_n(\mathcal{A})^+$ then $(\phi(x_{ij})) \in M_n(\mathcal{B}(\mathcal{H})^+)$.*

From the proof of the above theorem one can conclude the following.

Let $(\cdot)^T$ denote the transpose map on $B(\mathcal{H})$ with respect to some CONS. Let

$$\mathfrak{U} = \left\{ \begin{pmatrix} x & 0 \\ 0 & x^T \end{pmatrix} \in M_2(B(\mathcal{H})) : \quad x = x^* \right\}. \qquad (7.208)$$

Then $\mathfrak{U}$ is a selfadjoint subspace of $M_2(B(\mathcal{H}))$ containing the identity. Moreover $\mathfrak{U}$ is a Jordan algebra.

Let ϕ be a selfadjoint map of $B(\mathcal{H})$ into itself such that

$$\phi(x^T) = (\phi(x))^T \qquad (7.209)$$

for all $x \in B(\mathcal{H})$ and $\hat{\phi}: \mathfrak{U} \to \mathfrak{U}$ is defined as

$$\hat{\phi}\begin{pmatrix} x & 0 \\ 0 & x^T \end{pmatrix} = \begin{pmatrix} \phi(x) & 0 \\ 0 & \phi(x)^T \end{pmatrix}. \qquad (7.210)$$

Then ϕ is decomposable if and only if $\hat{\phi}$ is completely positive. This means that if a system is described in terms of Jordan algebra $\mathfrak{U}$ then decomposable maps satisfying (7.209) can have almost the same physical interpretation as completely positive ones.

Chapter 8

Fractals with Information

"Fractal" is a word proposed by B. Mandelbrot, originating from "fraction", "fractional" and so on (in the context: fractal = a set with a fractional dimension). Mathematical formulation or theory of fractals is now on the road of development [225]. Therefore, the usefulness of this concept is not yet clear. However, the idea of Mandelbrot is interesting from the point of view of scientific recognition. We explain here the mathematical foundations of fractals, with some modifications, and we introduce a new concept of the fractal dimension for state spaces, not for geometrical sets.

8.1. What is a fractal?

Observe the curves in Figures 8.1 and 8.2.

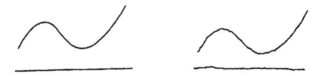

Figure 8.1   *Figure 8.2*

In conventional mathematics, the "curves in Figure 8.1" and the "curves in Figure 8.2" are considered to be topologically isomorphic and to be of dimension 1. Comparing these curves, we observe:

(1) The curves in Figure 8.1 are simpler than those in Figure 8.2.
(2) The curves in Figure 8.1 are differentiable, but those in Figure 8.2 are not differentiable at almost every point on them.
(3) The length of the curves in Figure 8.1 can be easily measured, but that of the curves in Figure 8.2 is difficult to measure.

Now, when we try to measure the length of the curves (like coastlines) in Figure 8.2, the results will be:

Measuring device		Results
by foot	(unit of 1 meter)	a meters
by a ruler	(unit of 1 centimeter)	b meters
by an atomic scale	(unit of 10^{-7} centimeter)	c meters

Comparing the results of each measurement, we may have

$$a < b < c = \infty.$$

When we measure the length of a sea coastline in an atomic scale, we must measure the length of sand, particle by particle, so that the result will be (practically) infinite. That is, the length of an object depends on a measuring device.

The mathematically abstracted "length" of a curve should not depend on a choice of a unit of measurement, so that the length of a curve has to be considered as an absolute quantity independent from a mean of measurement. However, it seems to us that this is not always true. Therefore, it may be natural for us to ask whether we can distinguish mathematically the curves in Figure 8.1 and those in Figure 8.2 by using a proper measure. For this purpose, we might need several new concepts of dimension different from the originally Euclidean dimension. This is the starting point of fractal theory.

Let us consider the middle "line" in Figure 8.3. If we briefly glance at the middle line, this line seems to be the same as the left straight one. However, a closer look at the middle line reveals its similarity with the right zigzag line. In our real world, a perfect line obviously does not exist. As in the case of simple geometric figures such as lines, the understanding of objects depends strongly on how we observe them.

rough   fine
observation   observation

Figure 8.3

This observation tells us that a complex set shows itself differently depending on how we observe it. The purpose of fractal theory is to grasp the properties of complex geometrical sets, droped out by the conventional manner of simplifying and abstracting the properties of objects. In the present stage, however, only a special complex set consisting of the copies of a basic pattern can be classified by this theory. Namely, we can treat only the sets independent of the scale of measurement on the microscopic level. Such a set is called the *scaling invariant set* (*self-similar set*). We give two typical examples of such sets.

Example 1.   The Koch curve: The basic pattern is 1° below and it appears repeatedly and makes the whole curve.

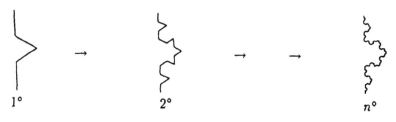

Figure 8.4

The Koch curve $n°$ is roughly the same as the curve 1°. The curve obtained in the limit $n \to \infty$ is non-differentiable at every point and any part is similar to the whole curve.

Example 2.   The Sierpinski gasket: Repeatedly take off the hatched part of the set 1°. This resulting set 1° is a basic pattern of the gasket. The limitting $n \to \infty$ set is a complex object consisting of many lines.

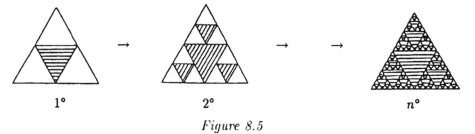

Figure 8.5

Then we have the following naive question: How can we characterize these sets? For instance, we like to find a suitable measure to distinguish the Koch curve from the straight line. In order to answer this question, we need a new concept of dimension different from the Euclidean one.

8.2. Fractal dimensions and their properties

We introduce several fractal dimensions to answer the above question.
⟨ 1 ⟩ **Scaling dimension**   We consider an object (complex set)
constructed by copies of a basic pattern. Suppose that there exist $N(1)$
basic patterns when we observe the object in a rough scale, say "scale 1",
and that there exist $N(r)$ basic patterns when we observe the object in
"scale r". Then we call

$$d_S = \frac{\log\left(N(r)/N(1)\right)}{\log 1/r} \tag{8.1}$$

the *scaling dimension* of the object. The reason for this definition is that
our experience tells us that the relation

$$N(r) \cong r^{-d_S} N(1) \tag{8.2}$$

holds in many experimental observations. Every theory in science should be
constructed by abstraction from our experience of the real world. The above
definition of the scaling dimension also comes from such an experience.

   Now, we compute the scaling dimension of a few geometrical sets.

   (i) Straight line

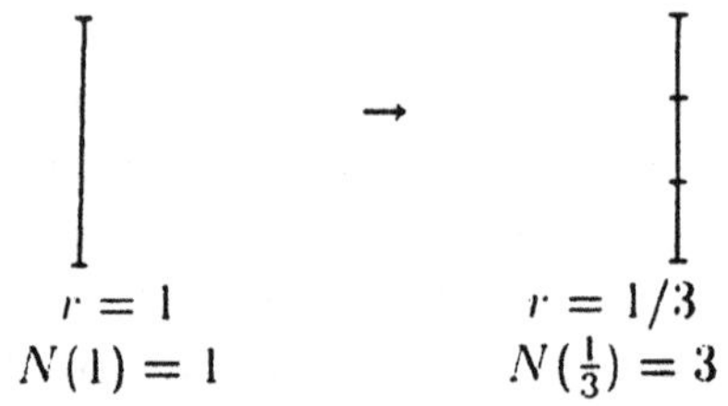

$$r = 1 \qquad\qquad r = 1/3$$
$$N(1) = 1 \qquad\qquad N(\tfrac{1}{3}) = 3$$

Figure 8.6

$$d_S = \frac{\log 3/1}{\log 1/(1/3)} = 1. \tag{8.3}$$

(ii) Koch's curve

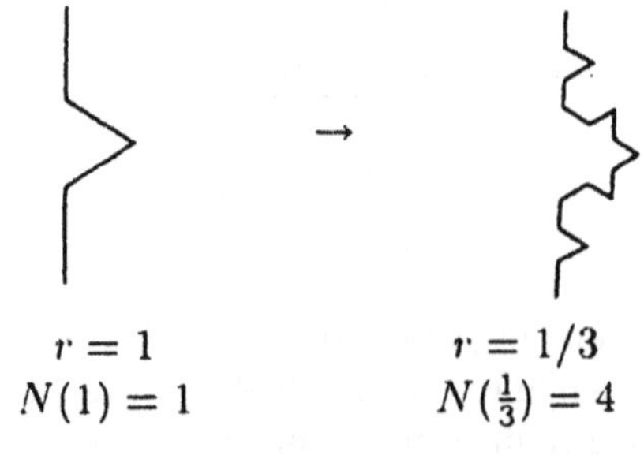

$$r = 1 \qquad\qquad r = 1/3$$
$$N(1) = 1 \qquad\qquad N(\tfrac{1}{3}) = 4$$

Figure 8.7

$$d_S = \frac{\log 4/1}{\log 1/(1/3)} = \frac{\log 4}{\log 3}. \tag{8.4}$$

As above, we can distinguish the straight line and the Koch curve by use of the scaling dimension d_S.

(iii) Square

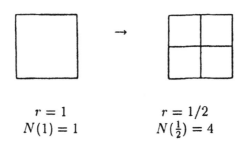

$r = 1$
$N(1) = 1$

$r = 1/2$
$N(\tfrac{1}{2}) = 4$

Figure 8.8

$$d_S = \frac{\log 4/1}{\log 1/(1/2)} = 2. \tag{8.5}$$

(iv) Sierpinski's gasket

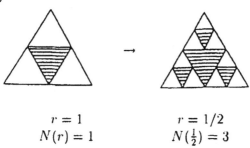

$r = 1$
$N(r) = 1$

$r = 1/2$
$N(\tfrac{1}{2}) = 3$

Figure 8.9

$$d_S = \frac{\log 3/1}{\log 1/(1/2)} = \frac{\log 3}{\log 2}. \tag{8.6}$$

From the values of (8.5) and (8.6), we can distinguish two plane figures in a quantitative way. The above examples show that the scaling dimension d_S can be considered as a dimension characterizing a certain class of geometrical sets.

In addition to the scaling dimension, there exist other dimensions classifying complex sets. We shall explain some of these notions.

⟨ 1a ⟩ **Generalized scaling dimension**

When we measure a physical quantity of an object and we get the value $Q(1)$ in the measurement of scale 1 and the value $Q(r)$ in the measurement of scale r, we call the quantity

$$d_S = \frac{\log (Q(r)/Q(1))}{\log 1/r} \tag{8.7}$$

the *generalized scaling dimension*.

⟨ 2 ⟩ **Hausdorff dimension**

For a subset X of $\mathbb{R}^d$ (d-dimensional Euclidean space) and a positive real number $\varepsilon > 0$, we denote by B a closed ball in $\mathbb{R}^d$, and by $\{B_j\}$ a set of closed balls in $\mathbb{R}^d$ covering X. Moreover

$$\mathcal{B}_\varepsilon(X) \equiv \{\{B_i\} : |B_i| \le \varepsilon, \ X \subset \bigcup B_i\}, \tag{8.8}$$

where $|B_i|$ is the diameter of B_i. Then we define

$$h_\alpha(X;\varepsilon) \equiv \inf\{\Sigma_i |B_i|^\alpha : \{B_i\} \in \mathcal{B}_\varepsilon(X)\}. \tag{8.9}$$

where $\alpha \in \mathbb{R}^1_+$.

This $h_\alpha(X;\varepsilon)$ satisfies

(i) $0 < \varepsilon_1 < \varepsilon_2 \Rightarrow$ $h_\alpha(X;\varepsilon_1) \ge h_\alpha(X;\varepsilon_2) \ge 0$
         $\Rightarrow$ $h_\alpha(X;\varepsilon)$ is decreasing for ε
         $\Rightarrow$ $\exists\, h_\alpha(X) = \lim_{\varepsilon \to 0} h_\alpha(X;\varepsilon)$ including $+\infty$.

(ii) $X_1 \subset X_2 \Rightarrow$ $h_\alpha(X_1;\varepsilon) \le h_\alpha(X_2;\varepsilon)$.

THEOREM 8.1. $h_\alpha(X)$ *is an outer measure on* $\mathbb{R}^d$, *i.e.*,
(i) $h_\alpha(\emptyset) = 0$,
(ii) $X_1 \subset X_2 \Rightarrow h_\alpha(X1) \le h_\alpha(X_2)$,
(iii) $\{X_n\} \subset 2^{\mathbb{R}^d}$ (power set of $\mathbb{R}^d$) $\Rightarrow h_\alpha(\bigcup_{n=1}^{\infty} X_n) \le \sum_{n=1}^{\infty} h_\alpha(X_n)$.

Proof. (i) By the definition of h_α, we have $h_\alpha \ge 0$. For any $0 < \delta \le \varepsilon$, denote a closed ball with the diameter less than δ by $\mathcal{B}_\delta$. Since $\emptyset \subset \mathcal{B}_\delta$, we have

$$\begin{aligned} h_\alpha(\emptyset) &= \lim_{\varepsilon \to 0} \inf\{\Sigma_i |B_i|^\alpha : \{B_i\} \in \mathcal{B}_\varepsilon(\emptyset)\} \\ &\le \lim_{\varepsilon \to 0} \inf_{\delta \le \varepsilon} \delta^\alpha \\ &= 0 \end{aligned}$$

(ii) Since $X_1 \subset X_2, \{B_i\} \in \mathcal{B}_\varepsilon(X_2)$ implies $\{B_i\} \in \mathcal{B}_\varepsilon(X_1)$, and hence we have

$$\mathcal{B}_\varepsilon(X_2) \subset \mathcal{B}_\varepsilon(X_1),$$

which reduces to

$$\inf\{\Sigma_i|B_i|^\alpha : \{B_i\} \in \mathcal{B}_\varepsilon(X_1)\} \le \inf\{\Sigma_i|B_i|^\alpha : \{B_i\} \in \mathcal{B}_\varepsilon(X_2)\}.$$

Taking the limit $\varepsilon \to 0$, we get

$$h_\alpha(X_1) \le h_\alpha(X_2).$$

(iii) For each X_n and any $\delta > 0$, we can choose a covering $\{C_n^{(i)}\} \in \mathcal{B}_\varepsilon(X_n)$ of X_n such that

$$\begin{aligned}\Sigma_i|C_n^{(i)}|^\alpha &< \inf\{\Sigma_i|B_i|^\alpha : B_i \in \mathcal{B}_\varepsilon(X_n)\} + \delta/2^n \\ &= h_\alpha(X_n) + \delta/2^n.\end{aligned}$$

Since

$$\bigcup_{n=1}^\infty X_n \subset \bigcup_{n=1}^\infty \bigcup_i C_n^{(i)},$$

we have

$$\begin{aligned}h_\alpha\left(\bigcup_{n=1}^\infty X_n\right) &\le \sum_{n=1}^\infty |C_n^{(i)}|^\alpha \\ &\le \sum_{n=1}^\infty \{h_\alpha(X_n) + \delta/2^n\} \\ &= \sum_{n=1}^\infty h_\alpha(X_n) + \delta.\end{aligned}$$

Since δ is arbitrary, we get

$$h_\alpha\left(\bigcup_{n=1}^\infty X_n\right) \le \sum_{n=1}^\infty h_\alpha(X_n).$$

From (i), (ii), (iii), $h_\alpha(X)$ is an *outer measure (Hausdorff measure)* on $\mathbb{R}^d$.
$\square$

The *Hausdorff dimension* $d_H(X)$ of a set X is given by

$$d_H(X) = \inf\{\alpha : h_\alpha(X) = 0\}.$$

THEOREM 8.2. *For $X \subset \mathbb{R}^d$, the following statements hold.*
(1) $0 \le \alpha < d_H(X) \Rightarrow h_\alpha(X) = \infty$.
(2) $d_H(X) < \alpha \Rightarrow h_\alpha(X) = 0$.

Proof. (1) We have

$$
\begin{aligned}
h_\alpha(X) &= \lim \inf_{\varepsilon \to 0} \Sigma_i |B_i|^\alpha \\
&= \lim \inf_{\varepsilon \to 0} \Sigma_i |B_i|^{d_H} |B_i|^{\alpha - d_H} \\
&\geq \lim \inf_{\varepsilon \to 0} \Sigma_i |B_i|^{d_H} (1/\varepsilon)^{\alpha - d_H} \to \infty
\end{aligned}
$$

(2) Put $h_{d_H}(X) = M < \infty$. Choose $\{C_i\} \in \mathcal{B}_\varepsilon(X)$ such that

$$
\begin{aligned}
\Sigma_i |C_i|^{d_H} &\leq \inf \Sigma_i |B_i|^{d_H} + 1 \\
&= h_{d_H}(X; \varepsilon) + 1 \\
&\leq M + 1 < \infty.
\end{aligned}
$$

Then we have

$$
\begin{aligned}
\inf \Sigma_i |B_i|^\alpha &\leq \Sigma_i |C_i|^\alpha \\
&= \Sigma_i |C_i|^{d_H + (\alpha - d_H)} \\
&\leq \varepsilon^{\alpha - d_H} \Sigma_i |C_i|^{d_H} \\
&\leq \varepsilon^{\alpha - d_H} (M + 1).
\end{aligned}
$$

By $d_H(X) < \alpha$, the limit $\varepsilon \to 0$ implies $\varepsilon^{\alpha - d_H}(M + 1) \to 0$. Therefore, we have

$$
h_\alpha(X) = \lim_{\varepsilon \to 0} h_\alpha(X; \varepsilon) = 0.
$$

$\square$

THEOREM 8.3. *Any Borel set is h_α -measurable.*

Proof. Since the Borel field is the σ-field generated by all open sets, it is also generated by all closed sets. So, we have only to show that any closed set F is h_α-measurable. We shall show

$$
h_\alpha(P \cup Q) = h_\alpha(P) + h_\alpha(Q),
$$

for all $P \subset F$ and for all $Q \subset F^c$. Here the inequality "$\leq$" obviously holds because h_α is the outer measure from Theorem 8.1. Therefore, we have to show the converse inequality "$\geq$". Define a distance ρ as follows:

$$
\begin{aligned}
\forall x, y \in X; \quad \rho(x, y) &= \max_i |x_i - y_i|, \\
\forall A, B \subset X; \quad \rho(A, B) &= \inf \{\rho(x, y); x \in A, y \in B\}.
\end{aligned}
$$

Since F is a closed set, we can take $\{Q_n\}$ such that $Q_n \downarrow Q$

$$
\rho(Q_n, F) \geq 1/n,
$$

$$
\lim_{n \to \infty} h_\alpha(Q_n) = h_\alpha(Q).
$$

Since $\rho(Q_n, P) \geq \rho(Q, F) > 0$, we get

$$h_\alpha(P \cup Q) \geq h_\alpha(P \cup Q_n) = h_\alpha(P) + h_\alpha(Q_n)$$

from the property of a metric outer measure. Taking the limit $n \to \infty$, we have

$$h_\alpha(P \cup Q) \geq h_\alpha(P) + h_\alpha(Q).$$

Therefore, any Borel set is h_α-measurable.   □

By this theorem, h_α becomes a measure over the Borel field.

The Hausdorff dimension is not so convenient for studying complex geometrical sets, that is, it is not so easy to calculate directly this dimension. As the Theorem 8.6 below shows, Hausdorff dimension is identical to the scaling dimension for some particular sets.

DEFINITION 8.4. (1) A mapping ϕ from $\mathbb{R}^d$ to $\mathbb{R}^d$ is called *similar contractive* with a contraction rate r w.r.t a metric $|\cdot|$ if $|\phi(x) - \phi(y)| = r|x - y|$ holds for any $x, y \in \mathbb{R}^d$.
(2) A set X in R^d is said to be *self-similar* if there exist similar contractive maps $\phi_1, \phi_2, \cdots, \phi_n$ with a common contraction rate r satisfying

$$X = \phi_1(X) \cup \phi_2(X) \cup \cdots \cup \phi_n(X),$$
$$\phi_j(X) \cap \phi_k(X) = \emptyset, \quad (j \neq k).$$

LEMMA 8.1. *Let* $X \subset \mathbb{R}^d$ *be a self-simillar set with a similar map* ψ *and contraction rate* r. *Then the Hausdorff measure satisfies*

$$h_\alpha(\psi(X)) = r^\alpha h_\alpha(X).$$

Proof. Put $rX = \{rx : x \in X\}$. Since "$X \subset \bigcup_i B_i$" is equivalent to "$\psi(X) \subset \bigcup_i rB_i$", we have

$$
\begin{aligned}
h_\alpha(\psi(X)) &= \liminf_{\varepsilon \to 0} \{\Sigma_i |rB_i|^\alpha : rX \subset \bigcup_i rB_i, \ |rB_i| < \varepsilon\} \\
&= \liminf_{\varepsilon \to 0} \{r^\alpha \Sigma_i |B_i|^\alpha : rX \subset \bigcup_i rB_i, \ |rB_i| < \varepsilon\} \\
&= r^\alpha \liminf_{\varepsilon/r \to 0} \{\Sigma_i |B_i|^\alpha \ X \subset \bigcup_i B_i, \ |B_i| < \varepsilon\} \\
&= r^\alpha h_\alpha(X). \qquad\qquad □
\end{aligned}
$$

By this lemma we have the following theorem.

THEOREM 8.5. *Let* X *be a self-similar set such that* $0 < h_\alpha(X) < \infty$. *Then the Hausdorff dimension of* X *equals the scaling dimension of* X.

Proof. Put $d_H = \alpha$. Let ψ be a similar map of contraction rate r. By lemma 7.5, we have

$$h_\alpha(\psi(X)) = r^\alpha h_\alpha(X).$$

Thus

$$
\begin{aligned}
\alpha &= \frac{\log h_\alpha(\psi(X))/h_\alpha(X)}{\log r} \\
&= \frac{\log h_\alpha(X)/h_\alpha(\psi(X))}{\log 1/r}.
\end{aligned}
$$

The original set X is reconstructed by some similar maps ψ_i with the contraction rate r. Here the index i indicates positions and ψ_i equals ψ for all i. Reconstructing X, we need $N(r)/N(1)$ similar maps, where $N(r)$ is the number of subsets repeating the a basic pattern in the scale r. That is

$$
\begin{aligned}
X &= \psi_1(X) \cup \psi_2(X) \cup \cdots \cup \psi_N(X), \\
N &= N(r)/N(1),
\end{aligned}
$$

$$\psi_j(X) \cap \psi_k(X) = \emptyset, \quad (j \neq k).$$

Since Borel sets are h_α-measurable, we obtain

$$
\begin{aligned}
h_\alpha(X) &= h_\alpha(\psi_1(X) \cup \psi_2(X) \cup \cdots \cup \psi_N(X)) \\
&= h_\alpha(\psi_1(X)) + h_\alpha(\psi_2(X)) + \cdots h_\alpha(\psi_N(X)) \\
&= N \cdot h_\alpha(\psi(X)),
\end{aligned}
$$

which implies

$$\frac{h_\alpha(X)}{h_\alpha(\psi(X))} = \frac{N(r)}{N(1)}.$$

Hence

$$\alpha = \frac{\log N(r)/\log N(1)}{\log 1/r} = d_S(X).$$

$\square$

Remark: Using the covering of X by some d-dimensional cubes with edge length ε instead of the balls $\{B_i\}$ in the definition of the Hausdorff dimension, we define the following measure $h'_\alpha(X;\varepsilon)$,

$$h'_\alpha(X;\varepsilon) = \sum_{i=1}^{N(\varepsilon)} \varepsilon^\alpha, \tag{8.10}$$

where $N(\varepsilon)$ is the minimum number of cubes necessary to cover X. This $h'_\alpha(X;\varepsilon)$ posesses the same properties as $h_\alpha(X;\varepsilon)$.

⟨ 3 ⟩ Topological dimension

This dimension characterizes a set of lattice points in a space of interest. This dimension is not related to scaling transformations. The topological dimension of a set consisting of lattice points is entirely determined by topological properties of the set. Let X be a set consisting of lattice points. The set of all lattice points in X reached in r steps from the origin O is called a *lattice sphere*. Let S_r be the number of lattice points on this sphere and V_r be the number of lattice points inside this sphere. Then we call

$$d_T(X) \equiv \lim_{r \to \infty} \frac{\log V_r}{\log(V_r/S_r)}. \tag{8.11}$$

the *topological dimension* of X.

For instance, the numbers S_r and V_r for the Koch curve are

$$S_r = 2, \quad V_r = 2r + 1.$$

Therefore, the topological dimension of the Koch curve is

$$\begin{aligned} d_T &= \lim_{r \to \infty} \frac{\log(2r+1)}{\log\{(2r+1)/2\}} \\ &= \lim_{r \to \infty} 1 + \frac{\log 2}{\log\{(2r+1)/2\}} = 1 \end{aligned}$$

⟨ 4 ⟩ Capacity dimension

Let $N(\varepsilon)$ be the minimum number of convex sets (e.g., cubes, balls,...) of the diameter ε necessary to cover a compact set $X \subset \mathbb{R}^d$. Then we call

$$d_C(X) \equiv \lim_{\varepsilon \to 0} \frac{\log N(\varepsilon)}{\log 1/\varepsilon} \tag{8.12}$$

the *capacity dimension* of the set X.

(i) <u>Koch's curve</u> : Let us cover the Koch curve by the discs of diameter $\varepsilon = (1/3)^n$. For $\varepsilon = 1/3$ we need $N(\varepsilon) = 4$ discs. For $\varepsilon = (1/3)^2$ we need $N(\varepsilon) = 4^2$ discs. In general, when we cover the curve by the discs of diameter $(1/3)^n$, we need $N(\varepsilon) = 4^n$ discs. Hence

$$\begin{aligned} d_C &= \lim_{\varepsilon \to 0} \frac{\log N(\varepsilon)}{\log(1/\varepsilon)} \\ &= \lim_{n \to \infty} \frac{\log 4^n}{\log(1/(1/3)^n)} \\ &= \frac{\log 4}{\log 3} \end{aligned}$$

(ii) Sierpinski's gasket : Let us cover the Sierpinski gasket by the discs of diameter $(2/\sqrt{3}) \cdot (1/2)^n$. Note that the diameter of an outer tangent circle of the equilateral triangle is $2/\sqrt{3}$. For $\varepsilon = (2/\sqrt{3}) \cdot (1/2)$ we need $N(\varepsilon) = 3$ discs, for $\varepsilon = (2/\sqrt{3}) \cdot (1/2)^2$, $N(\varepsilon) = 3^2$, etc. In general, when we cover the gasket by the discs of diameter $(2/\sqrt{3}) \cdot (1/2)^n$, we need $N(\varepsilon) = 3^n$ discs. Thus

$$
\begin{aligned}
d_C &= \lim_{\varepsilon \to 0} \frac{\log N(\varepsilon)}{\log 1/\varepsilon} \\
&= \lim_{n \to \infty} \frac{\log 3^n}{\log \frac{1}{(2\sqrt{3}/3) \cdot (1/2)^n}} \\
&= \lim_{n \to \infty} \frac{\log 3^n}{\log (\sqrt{3} \cdot 2^n/2)} \\
&= \frac{\log 3}{\log 2}
\end{aligned}
$$

THEOREM 8.6. *For any* $X \subset \mathbb{R}^d$

$$d_H(X) \leq d_C(X).$$

Proof. We have only to consider the case that $d_C(X) < \infty$ because the theorem is clear when $d_C(X) = \infty$.

$$
\begin{aligned}
h_{d_C}(X; \varepsilon) &= \inf\{\Sigma_i |B_i|^{d_C} : \{B_i\} \in \mathcal{B}_\varepsilon\} \\
&\leq \sum_{i=1}^{N(\varepsilon)} \varepsilon^{d_C} = N(\varepsilon) \varepsilon^{d_C},
\end{aligned}
$$

where we write d_C instead of $d_C(X)$ for simplicity. Since $\{B_i : |B_i| \leq \varepsilon, i = 1, 2, \ldots, N(\varepsilon)\} \in \mathcal{B}_\varepsilon$, we have

$$
\begin{aligned}
h_{d_C}(X) &= \lim_{\varepsilon \to 0} h_{d_C}(X; \varepsilon) \\
&\leq \lim_{\varepsilon \to 0} N(\varepsilon) \varepsilon^{d_C} \\
&= \lim_{\varepsilon \to 0} N(\varepsilon) \varepsilon^{\log N(\varepsilon)/\log(1/\varepsilon)} \\
&= \lim_{\varepsilon \to 0} N(\varepsilon) \varepsilon^{-\log_\varepsilon N(\varepsilon)} \\
&= 1 < \infty.
\end{aligned}
$$

Therefore, we obtain

$$d_H(X) \leq d_C(X)$$

from Theorem 8.2 and the definition of d_H.

□

Example 3. Trema endpoints and rational numbers ("trema" is from the Greek *to trēma* = 'a perforation, hole, aperture, orifice'):

Figure 8.10

The set of the points marked "o" on Fig. 8.10 is:

$$T = \{1/3, 2/3\} \cup \{1/9, 2/9, 7/9, 8/9\} \cup \{1/27, 2/27, \cdots\} \cup \cdots$$

Since T is a countable set, each point of T can be covered by an infinitely small interval for any $\varepsilon > 0$. Hence

$$h_\alpha(T; \varepsilon) = 0,$$

and

$$d_{II}(T) = 0.$$

Since $N(1) = 1$ and $N(1/3) = 2$, the scaling dimension of T is

$$d_S(T) = \frac{\log\{N(1/3)/N(1)\}}{\log\{1/(1/3)\}} = \frac{\log 2}{\log 3}.$$

In the n-th step defining T, T can be covered by 2^n closed intervals of the length $\varepsilon = 3^{-n}$. So we have $N(3^{-n}) = 2^n$, hence

$$\begin{aligned}
d_C(T) &= \lim_{n \to \infty} \frac{\log N(3^{-n})}{\log(1/3^{-n})} \\
&= \lim_{n \to \infty} \frac{\log 2^n}{\log 3^n} = \frac{\log 2}{\log 3}.
\end{aligned}$$

Therefore,

$$d_S(T) = d_C(T) > d_{II}(T) = 0.$$

By the way, for the set Q of all rational numbers in the interval $[0, 1]$, the following inclusion holds

$$T \subset Q.$$

Since Q is a countable set, we have

$$d_H(Q) = 0.$$

Hence

$$d_H(T) = d_H(Q) = 0.$$

How can we distinguish two sets Q and T? Let us try to compute the capacity dimension of Q.

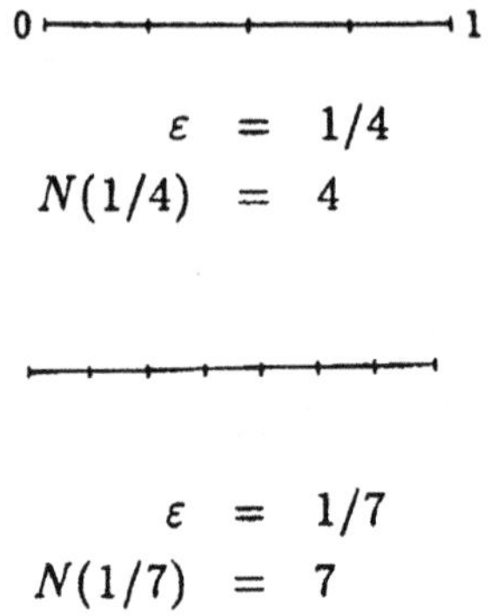

$$0 \longmapsto\!\!\!-\!\!\!-\!\!\!-\!\!\!-\!\!\!\dashv 1$$

$$\varepsilon = 1/4$$
$$N(1/4) = 4$$

$$\varepsilon = 1/7$$
$$N(1/7) = 7$$

Figure 8.11

If we take $\varepsilon = 1/\sqrt{65} < 1/7$, then we have $N(\varepsilon) = 7 + 1 = 8$ and

$$N(\varepsilon) = \begin{cases} \dfrac{1}{\varepsilon} & 1/\varepsilon : \text{integer}, \\[2mm] \left[\dfrac{1}{\varepsilon}\right] & 1/\varepsilon : \text{not integer}, \end{cases}$$

where [ ] is the Gauss symbol, i.e., the integer part of the argument. Thus

$$1/\varepsilon \le N(\varepsilon) < 1/\varepsilon + 1.$$

We have only to consider the case $\log(1/\varepsilon) > 0$ because we take the limit $\varepsilon \to 0$ for the capacity dimension. So we have

$$\lim_{\varepsilon \to 0} \frac{\log(1/\varepsilon)}{\log(1/\varepsilon)} \le d_C(Q)$$

$$= \lim_{\varepsilon \to 0} \frac{\log N(\varepsilon)}{\log(1/\varepsilon)}$$

$$\le \lim_{\varepsilon \to 0} \frac{\log[(1/\varepsilon) + 1]}{\log(1/\varepsilon)}.$$

Hence

$$1 \le d_C(Q)$$

$$\leq \lim_{\varepsilon \to 0} \frac{\log[(1/\varepsilon) + 1]}{\log(1/\varepsilon)}$$

$$= \lim_{\varepsilon \to 0} \frac{\log(1/\varepsilon) + \log(1 + \varepsilon)}{\log(1/\varepsilon)}$$

$$= 1 + \lim_{\varepsilon \to 0} \frac{\log 1 + \varepsilon}{\log 1/\varepsilon} = 1.$$

Therefore,

$$d_C(Q) = 1.$$

Thus,

$$d_C(Q) > d_C(T).$$

As a consequence, by the capacity dimension, we see that Q is denser than T, which supports the intuitive fact that Q has more elements than T.

Example 4. Cantor Set: Cantor set is constructed in the following way (Fig. 8.13): For the unit interval $[0, 1]$, we take sets E_n as

$$
\begin{aligned}
E_0 &= [0, 1] \\
E_1 &= [0, 1/3] \cup [2/3, 1] \\
E_2 &= [0, 1/9] \cup [2/9, 1/3] \cup [2/3.7/9] \cup [8/9, 1].
\end{aligned}
$$

That is, E_{n+1} is a set obtained from E_n by taking off the open middle third part of this interval. So the set E_n consisits of 2^n intervals with the length 3^{-n}. The *Cantor set* is defined by

$$E = \bigcap_{n=1}^{\infty} E_n$$

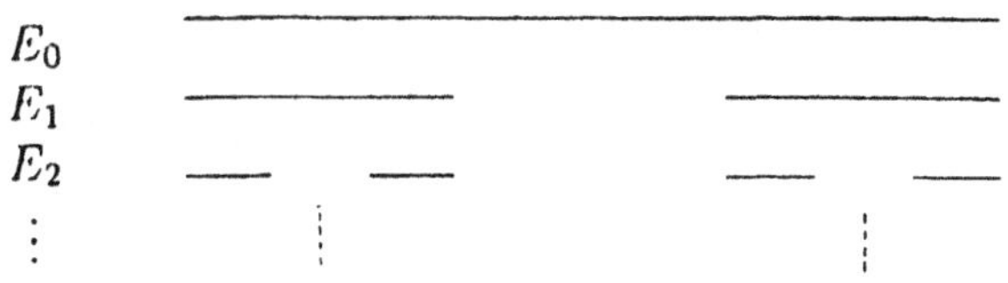

E_0
E_1
E_2
$\vdots$

Figure 8.13. Construction of the Cantor set

We compute the capacity dimension for this model. Let us divide the space by $N(\varepsilon)$ intervals of the length $(1/3)^n$.

When $\varepsilon = 1/3$, we have $N(\varepsilon) = 2$.
When $\varepsilon = (1/3)^2$, we have $N(\varepsilon) = 2^2$.
When $\varepsilon = (1/3)^3$, we have $N(\varepsilon) = 2^3$.

In general,

when $\varepsilon = (1/3)^n$,   we have $N(\varepsilon) = 2^n$.

Therefore,

$$
\begin{aligned}
d_C(X) &= \lim_{\varepsilon \to 0} \frac{\log N(\varepsilon)}{\log(1/\varepsilon)} \\
&= \lim_{n \to \infty} \frac{2^n}{\log \frac{1}{(1/3)^n}} \\
&= \lim_{n \to \infty} \frac{\log 2^n}{\log 3^n} = \frac{\log 2}{\log 3}.
\end{aligned}
$$

Finally, we collect the fractal dimensions of the sets discussed here in Table 8.1.

d	d_S	d_C	d_H	d_T
Straight line	1	1	1	
Set of rationals	0	1	0	0
Cantor set	$\log_3 2$	$\log_3 2$	$\log_3 2$	0
Trema endpoints	$\log_3 2$	$\log_3 2$	0	0
Koch's curve	$\log_3 4$	$\log_3 4$	$\log_3 4$	1
Sierpinski's gasket	$\log_2 3$	$\log_2 3$	$\log_2 3$	
Square	2	2	2	

TABLE 8.1

8.3. Fractal dimensions of states

8.3.1. GENERAL FORMULATION OF THE FRACTAL DIMENSION OF STATES

The fractal dimensions introduced in Sec. 8.2 are used to characterize geometrical sets. They cannot be applied directly to the physical systems which are not expressed by geometrical sets. Here we discuss a new fractal dimension introduced in [253], [255], [227] to characterize the general state of a physical system.

In order to define the fractal dimension of a state for a general quantum dynamical system (GQDS), including both classical and quantum systems, we need the ε-entropy for GQDS.

The ε-entropy in classical dynamical systems has been widely studied, [14], [286], in various manners. Kolmogorov [189] defined the ε-entropy for classical random variables (cf. below in Subsect. 8.3.3. where we remind this classical definition), which enables us to define the ε-entropy in GQDS, by which in turn we can introduce fractal dimensions of a state in GQDS mimicking the capacity dimension.

Let $(\mathcal{A}, \mathfrak{S})$ be a C^*-algebra and $\mathfrak{S}(\varphi; \varepsilon)$ be the set of all states $\psi \in \mathfrak{S}$ satisfying $\|\varphi - \psi\| \le \varepsilon$. Then the ε-entropy $S(\varphi; \varepsilon)$ of a state φ is defined by

$$S(\varphi; \varepsilon) = \inf\{I^0(\varphi, \psi); \psi \in \mathfrak{S}(\varphi; \varepsilon)\}, \qquad (8.13)$$

where $I^0(\varphi, \psi)$ is the *quasimutual entropy* defined by

$$I^0(\varphi, \psi) = \sup\{S(\Psi, \Psi_0); \Psi \in \mathfrak{S}_{qc}, \; \Psi \ll \Psi_0, \; \Psi_0 \equiv \varphi \otimes \psi\},$$

where $\mathfrak{S}_{qc}$ is the set of all quasicompound states Ψ on $\mathcal{A} \otimes \mathcal{A}$ satisfying $\Psi(A \otimes I) = \varphi(A)$ and $\Psi(I \otimes B) = \psi(B)$ for any $A, B \in \mathcal{A}$, and $\ll$ means 'absolutely continuous with respect to', [247]. This definition is given in [252] and is directly related to the ε-entropy of Kolmogorov, but it is somehow too general to apply in quantum systems. Therefore we provide a little more restricted definition of the ε-entropy in GQDS. This definition of the ε-entropy is given by the mutual entropy discussed in Chap. 3 as follows: Let C be the set of all channels Λ^* of interest and $C(\varphi; \varepsilon)$ the set of all channels satisfying $\|\varphi - \Lambda^*\varphi\| \leq \varepsilon$. Then the *$\varepsilon$-entropy of a state φ with respect to S*, a weak * convex subset of $\mathfrak{S}$, is defined by

$$S^S(\varphi; \varepsilon) = \inf\{J^S(\varphi; \Lambda^*) \; ; \; \Lambda^* \in C(\varphi; \varepsilon)\}, \qquad (8.14)$$

$$J^S(\varphi; \Lambda^*) = \sup\{I^S(\varphi; \Gamma^*) \; ; \; \Gamma^* \in C, \; \Gamma^*\varphi = \Lambda^*\varphi\},$$

where $I^S(\varphi, \Gamma^*)$ is the mutual entropy discussed in Chap. 3. We here call this (8.14) the *ε-entropy in GQDS* and one given by (8.13) the *generalized ε-entropy*. These two definitions of the ε-entropy equal to that of Kolmogorov when $\mathcal{A}$ is a certain commutative algebra [262] and the norm $\| \cdot \|$ is the random variable norm (see below).

The ε-entropy can be considered as a quantity representing the degree of information transmission, so that we may expect to measure the complexity of the state φ by using this entropy.

Using the above ε-entropy of a state φ, we define two fractal dimensions of φ à la capacity dimension (8.12).

The first one is

$$d_C^S(\varphi) = \lim_{\varepsilon \to 0} \frac{S^S(\varphi; \varepsilon)}{\log \frac{1}{\varepsilon}}, \qquad (8.15)$$

which is a direct generalization of the capacity dimension (8.12), so it is called the *capacity dimension of a state φ with respect to S* for GQDS. Another dimension is

$$d_I^S(\varphi; \varepsilon) = \frac{S^S(\varphi; \varepsilon)}{I(\varepsilon)}, \qquad (8.16)$$

where $I(\varepsilon)$ is a normalization constant in the limit $\varepsilon \to 0$, namely,

$$\lim_{\varepsilon \to 0} d_I^S(\varphi; \varepsilon) = 1.$$

This constant very often becomes the entropy $S^S(\varphi)$ itself. We call $d_I^S(\varphi; \varepsilon)$ the *information dimension of a state φ* in the order ε w.r.t S for GQDS.

In all cases, we drop the index "S" such as $d_C(\varphi) \equiv d_{\tilde{C}}^{S}(\varphi)$ when $S = \mathfrak{S}$. From the definition we have the following results [253]:

THEOREM 8.7. *If every decomposition measure μ of a state φ is orthogonal and has a discrete support, and if the entropy $S^{S}(\varphi)$ is finite, then $d_{C}^{S}(\varphi) = 0$. In particular, $d_C(\rho) = 0$ for a density operator ρ when $S(\rho) < +\infty$.*

THEOREM 8.8. *$S(\varphi)$ is the complexity and $S(\varphi; \varepsilon)$ is the transmitted complexity (of Chap. 4).*

The fractal dimensions introduced above describe complexity of a state, so that we may characterize states by these dimensions.

So far, we used the mutual entropy I to define the ε-entropy and fractal dimensions of a state. These quantities can be recognized as T of Chap. 4 [257].

8.3.2. FRACTAL DIMENSIONS OF STATES IN DISCRETE CDS

We turn to the discussion of the fractal dimensions of states in discrete CDS and an application [255], [261]. This application stems from an investigation of *genetic sequences*. Let us consider sequences with the length n composed of m basic symbols. In order to measure complexity of a state (sequence), we can use our fractal dimensions.

For a set X composed of n events, $X = \{x_1, x_2, \ldots, x_n\}$, let $P = \{p_1, p_2, \ldots, p_n\}$ be the state (probability distribution) of the event X. The entropy of this event system is

$$S(X)(= S(P)) = - \sum_i p_i \log p_i$$

Now, we take two event systems $(X = \{x_j : j = 1, \ldots, n\}, P = \{p_j : j = 1, \ldots, n\})$, $(Y = \{y_k : k = 1, \ldots, m\}, Q = \{q_k : k = 1, \ldots, m\})$, and we denote the compound system by $(X \odot Y, \Phi = \{r(j,k)\})$ such that $\sum_k r(j,k) = p_j$, $\sum_j r(j,k) = q_k$. The mutual entropy w.r.t the above X and Y is given by

$$I(X, Y) = \sum_{i=1}^{n} \sum_{j=1}^{m} r(i,j) \log \frac{r(i,j)}{p_i q_j} .$$

Moreover, this information can be written as

$$I(X, Y) = I(P, Q; \Phi).$$

Let $\mathcal{P}(P,Q)$ be the set of all compound states like Φ above. Then the maximum mutual entropy is

$$I^0(P, Q) = \sup\{I(P, Q; \Phi) : \Phi \in \mathcal{P}(P, Q)\} .$$

When the system is finite (like a finite sequence), it becomes

$$I^0(P,Q) = S(P) + S(Q) - \inf\{S(\Phi) : \Phi \in \mathcal{P}(P,Q)\}.$$

LEMMA 8.2. *If two states P and P' are equivalent in the sense that the components of P are equal to those of P' as a set, then $I^0(P,Q) = I^0(P',Q)$.*

Proof. Let $P = \{p_1, p_2, \ldots, p_n\}$ and $P' = \{p'_1, p'_2, \ldots, p'_n\}$. Then there exists a permutation π such that $p'_k = p_{\pi(k)}$ and the compound state $\Phi = \{r'(i,j)\}$ giving the maximum value of $I^0(P,Q)$ becomes $\{r(\pi(i),j)\}$ which is one of the compound state of P and Q, so that we have $I^0(P,Q) \geq I^0(P',Q)$. The converse inequality $I^0(P',Q) \geq I^0(P,Q)$ is similarly proved by taking a different permutation. $\square$

In this case, the ε-entropy (8.13) becomes

$$S(P;\varepsilon) = \inf\{I^0(P,Q) : Q \in \mathfrak{S}(P;\varepsilon)\}, \qquad (8.17)$$

where

$$\mathfrak{S}(P;\varepsilon) = \{Q; \|P - Q\| \leq \varepsilon\}$$

with a state space norm such as $\|P - Q\| \equiv \sum_i |p_i - q_i|$.

Now, we consider a marginal set of distributions such as

$$\mathfrak{S}_m(P;\varepsilon) = \{Q : \|P - Q\| = \varepsilon\},$$

and we define the marginal ε-entropy by

$$S_m(P;\varepsilon) = \inf\{I^0(P,Q) : Q \in \mathfrak{S}_m(P;\varepsilon)\}.$$

THEOREM 8.9. *(1) $S(P;\varepsilon) = \inf\{S_m(P;\delta) : 0 < \delta \leq \varepsilon\}$.*
(2) $0 < \varepsilon_1 \leq \varepsilon_2$ implies $S(P;\varepsilon_1) \geq S(P;\varepsilon_2)$.
(3) $S(P;0) = S(P)$.

Proof. (1) Since $\bigcup\{\mathfrak{S}_m(P;\delta) : \delta \leq \varepsilon\} = \mathfrak{S}(P;\varepsilon)$, we have

$$\begin{aligned}
\inf\{S_m(P;\delta) &: 0 < \delta \leq \varepsilon\} \\
&= \inf\{I^0(P,Q) : Q \in \mathfrak{S}_m(P;\varepsilon), \delta \leq \varepsilon\} \\
&= \inf\{I^0(P,Q) : Q \in \mathfrak{S}(P;\varepsilon)\} \\
&= S(P;\varepsilon).
\end{aligned}$$

(2) $\varepsilon_1 \leq \varepsilon_2$ follows $\mathfrak{S}(P;\varepsilon_1) \subset \mathfrak{S}(P;\varepsilon_2)$, which implies

$$\begin{aligned}
S(P;\varepsilon_1) &= \inf\{I^0(P,Q) : Q \in \mathfrak{S}(P;\varepsilon_1)\} \\
&\geq \inf\{I^0(P,Q) : Q \in \mathfrak{S}(P;\varepsilon_2)\} \\
&= S(P;\varepsilon_2).
\end{aligned}$$

(3) Take $\varepsilon = 0$. Then $\mathfrak{S}(P;0) = \{P\}$, so that

$$S(P;0) = I^0(P,P) = S(P).\qquad\square$$

There is another definition of the ε-entropy which is written, in our terminology of classical discrete systems as follows: A channel Λ^* transfering a state to another state is regarded as a transition probability $\{p(j|i)\}$ such as

$$q_j = (\Lambda^*P)_j = \sum_i p(j|i)p_i.$$

Then the mutual entropy is written by

$$I(P;\Lambda^*) = \sum_{i,j} p(j|i)p_i \log \frac{p(j|i)}{q_j}.$$

Thus the ε-entropy (8.14) is

$$S'(P;\varepsilon) = \inf\{J(P;\Lambda^*) : \Lambda^* \in \mathcal{C}(P;\varepsilon)\},\qquad(8.18)$$

where

$$J(P;\Lambda^*) \equiv \sup\{I(P;\Gamma^*) : \Gamma^* \in \mathcal{C}, \Gamma^*P = \Lambda^*P\},$$
$$\mathcal{C}(P;\varepsilon) \equiv \{\Lambda^* \in \mathcal{C} : \|P - \Lambda^*P\| \le \varepsilon\}.$$

THEOREM 8.10. $S'(P;\varepsilon) = S(P;\varepsilon)$.

Proof. For an input state $P = \{p_i\}$ and its output state $Q = \{q_j\}$, let the compound state defining $I^0(P,Q)$ be $\Phi = \{r(i,j); i,j = 1,\ldots,n\}$. Put

$$\Gamma_Q^* \equiv (p(j|i)),\quad p(j|i) \equiv \frac{r(i,j)}{p_i}.$$

Then $Q = \Gamma_Q^*P$ and $I^0(P,Q) = I(P;\Gamma_Q^*)$. Since $\mathcal{C}(P;\varepsilon) = \{\Lambda^* \in \mathcal{C}; \Lambda^*P \in \mathfrak{S}(P;\varepsilon)\}$, we have

$$\begin{aligned}
S(P;\varepsilon) &= \inf\{I^0(P,Q) : Q \in \mathfrak{S}(P;\varepsilon)\}\\
&= \inf\{I(P;\Gamma_Q^*) : Q \in \mathfrak{S}(P;\varepsilon)\}\\
&\le \inf\{J(P;\Lambda^*) : \Lambda^*P = Q \in \mathfrak{S}(P;\varepsilon)\}\\
&= \inf\{J(P;\Lambda^*) : \Lambda^* \in \mathcal{C}(P;\varepsilon)\}\\
&= S'(P;\varepsilon).
\end{aligned}$$

Conversely, let the channel attaining $J(P;\Lambda^*)$ be Γ^* and the transition probabilty representing Γ^* be $\{p(j|i)\}$. Then for a state $Q = \Gamma^*P$, the state $\{p(j|i)p_i\}$ is one of the compound states of P and Q, so

$$J(P;\Lambda^*) = I(P;\Gamma^*) \le I^0(P,Q).$$

Therefore,

$$
\begin{aligned}
S'(P;\varepsilon) &= \inf\{J(P;\Lambda^*) : \Lambda^* \in C(P;\varepsilon)\} \\
&\leq \inf\{I^0(P,Q) : Q = \Lambda^* P \in \mathfrak{S}(P;\varepsilon)\} \\
&= S(P;\varepsilon).
\end{aligned}
$$

This implies

$$
S'(P;\varepsilon) = S(P;\varepsilon). \qquad \square
$$

This proposition tells us that in classical discrete systems we can use any one of the two difinitions of ε-entropy. In the sequel of this section, we denote both ε-entropies by the same symbol $S(P;\varepsilon)$.

In CDS, the notation becomes a bit simpler and we denote the fractal dimensions by

$$
\begin{aligned}
\alpha(P;\varepsilon) &\equiv \frac{S(P;\varepsilon)}{\log\frac{1}{\varepsilon}}, \\
\beta(P;\varepsilon) &\equiv \frac{S(P;\varepsilon)}{S(P)} \qquad (\text{ i.e., } I(\varepsilon) = S(P))
\end{aligned}
$$

for any state (probability distribution) P, [255]. Note that the smallest value of ε, except $\varepsilon = 0$, is $2/n$ for this case. In our case, α and β are essentially the same because $\varepsilon = 0$ is not interesting. So we compute only β here. Some examples for the computation of fractal dimension β are listed in Table 8.2.

From Table 8.2 we can conclude that the dimension β provides a measure of the complexity (diversity) of a state P. In particular, when $S(P) = S(Q)$, β gives us a new order of the complexity between P and Q.

These types of fractal dimensions α, β can be used to classify, e.g., the shapes of rivers and those of craters on the Moon [261].

8.3.3. FRACTAL DIMENSIONS OF STATES IN CONTINUOUS CDS

In this section, we first recall the definition of *Kolmogorov's ε-entropy* for random variables, cf. [189]. Let $(\Omega, \mathcal{F}, \mu)$ be a probability space and $M(\Omega)$ be the set of all random variables, and f, g be two random variables on Ω with values in a metric space X. Let μ_f be a probability measure associated with the random variable f. Then the mutual entropy $I(f, g)$ of the random variables f and g is

$$
\begin{aligned}
I(f,g) &= S\left(\mu_{fg}, \mu_f \otimes \mu_g\right) \\
&= \begin{cases} \int_{X \times X} \dfrac{d\mu_{fg}}{d\mu_f \otimes \mu_g} \log \dfrac{d\mu_{fg}}{d\mu_f \otimes \mu_g} d\mu_f \otimes \mu_g & (\mu_{fg} \ll \mu_f \otimes \mu_g) \\ \infty & (\text{otherwise}) \end{cases},
\end{aligned}
$$

	$((P))$	$((Q))$
state(n=8) entropy $\beta(\cdot;\frac{2}{n})$	(4/8,2/8,2/8,0) 1.0397 0.6992	(5/8,1/8,1/8,1/8) 1.0735 0.6852
state(n=8) entropy $\beta(\cdot;\frac{2}{n})$	(6/8,1/8,1/8,0) 0.7356 0.5122	(5/8,3/8,0,0) 0.6616 0.4891
state(n=8) entropy $\beta(\cdot;\frac{2}{n})$	(6/8,2/8,0,0) 0.5623 0.3619	(7/8,1/8,0,0) 0.3768 0
state(n=19) entropy $\beta(\cdot;\frac{2}{n})$	(6/19,6/19,6/19,1/19) 1.2470 0.8788	(9/19,4/19,4/19,2/19) 1.2470 0.8628
state(n=20) entropy $\beta(\cdot;\frac{2}{n})$	(8/20,6/20,6/20,0) 1.0889 0.8558	(12/20,4/20,2/20,2/20) 1.0889 0.8381
state(n=20) entropy $\beta(\cdot;\frac{2}{n})$	(9/20,8/20,2/20,1/20) 1.1059 0.8530	(12/20,3/20,3/20,2/20) 1.1059 0.8406
state(n=21) entropy $\beta(\cdot;\frac{2}{n})$	(9/21,8/21,4/21,0) 1.0466 0.8521	(12/21,6/21,2/21,1/21) 1.0466 0.8396
state(n=22) entropy $\beta(\cdot;\frac{2}{n})$	(9/22,8/22,4/22,1/22) 1.1840 0.8752	(12/22,4/22,3/22,3/22) 1.1840 0.8647
state(n=23) entropy $\beta(\cdot;\frac{2}{n})$	(9/23,8/23,5/23,1/23) 1.2025 0.8825	(12/23,5/23,3/23,3/23) 1.2025 0.8725
state(n=24) entropy $\beta(\cdot;\frac{2}{n})$	(10/24,9/24,5/24,0) 1.0594 0.8682	(15/24,4/24,3/24,2/24) 1.0594 0.8529

TABLE 8.2

where $S(\cdot, \cdot)$ is the relative entropy (Kulback-Leibler information), $\mu_f \otimes \mu_g$ is the direct product probability measure of f and g, and μ_{fg} is the joint probability measure of f and g, $\frac{d\mu_{fg}}{d\mu_f \otimes \mu_g}$ is the Radon-Nikodym derivative of μ_{fg} with respect to $\mu_f \otimes \mu_g$. Moreover the entropy $S(f)$ of the random variable f is given by

$$S(f) = I(f, f).$$

$S(f)$ is often infinite in continuous case, cf. Chap. 2, so that Kolmogorov introduced the ε-entropy to avoid this inconvenience

$$S_K(f; \varepsilon) = \inf_g \left\{ I(f, g); \sqrt{\int\int_{X \times X} d(x, y)^2 d\mu_{fg}(x, y)} \leq \varepsilon \right\}, \qquad (8.19)$$

where d is the metric of X. In the sequel, we take X a Hilbert space $\mathcal{H}$, then

$$d(x, y) = \sqrt{\langle x - y, x - y \rangle}.$$

A bit more general ε-entropy is defined as follows [262]. This ε-entropy is a special case of the ε-entropy of (8.14).

Let $(\Omega_1, \mathcal{F}_1)$ be an input space, $(\Omega_2, \mathcal{F}_2)$ be an output space and $P(\Omega_k)$ be the set of all probability measures on $(\Omega_k, \mathcal{F}_k)$, $(k = 1, 2)$. We call a channel (i.e., linear map) Λ^* from $P(\Omega_1)$ to $P(\Omega_2)$ a *Markov kernel* when Λ^* is given by

$$\bar{\mu}(Q) = \Lambda^*\mu(Q) = \int_{\Omega_1} \lambda(\omega, Q) \, d\mu(\omega), \quad \mu \in P(\Omega_1),$$

where λ is a mapping from $\Omega_1 \times \mathcal{F}_2$ to $[0, 1]$ satisfying the following conditions:

(1) $\lambda(\cdot, Q)$ is a measurable function on Ω_1 for each $Q \in \mathcal{F}_2$
(2) $\lambda(\omega, \cdot) \in P(\Omega_2)$ for each $\omega \in \Omega_1$

The compound state Φ of μ and $\bar{\mu}$ is given by

$$\Phi(Q_1 \times Q_2) = \int_{Q_1} \lambda(\omega, Q_2) \, d\mu(\omega)$$

for any $Q_1 \in \mathcal{F}_1, Q_2 \in \mathcal{F}_2$. The mutual entropy in classical continuous system can be expressed by the relative entropy $S(\cdot, \cdot)$ of the compound state Φ and the direct product state $\Phi_0 = \mu \otimes \Lambda^*\mu$:

$$I(\mu; \Lambda^*) = S(\Phi, \Phi_0)$$
$$= \begin{cases} \int_{\Omega_1 \times \Omega_2} \frac{d\Phi}{d\Phi_0} \log \frac{d\Phi}{d\Phi_0} \, d\Phi_0 & (\Phi \ll \Phi_0) \\ \infty & (\text{otherwise}) \end{cases},$$

where $\frac{d\Phi}{d\Phi_0}$ is the Radon-Nikodym derivative of Φ with respect to Φ_0.

In the following, we consider the case of $(\Omega_1, \mathcal{F}_1) = (\Omega_2, \mathcal{F}_2) \equiv (\Omega, \mathcal{F})$ for simplicity. The ε- entropy (8.16) of a state $\mu \in P(\Omega)$ is written as follows in continuous CDS. The *(Ohya) ε-entropy of a state $\mu \in P(\Omega)$* is

$$S_O(\mu; \varepsilon) = \inf_{\Lambda^*} \left\{ J(\mu; \Lambda^*); \|\mu - \Lambda^* \mu\| \leq \varepsilon \right\}, \qquad (8.20)$$

where $\|\cdot\|$ is a certain norm on $P(\Omega)$ and

$$J(\mu; \Lambda^*) = \sup_{\Gamma^*} \left\{ I(\mu; \Lambda^*); \Gamma^* \mu = \Lambda^* \mu \right\}.$$

Here $J(\mu; \Lambda^*)$ is the *maximum mutual entropy* w.r.t. μ and Λ^*.

When all compound measures of an input state μ and its associated output state $\bar{\mu}(\equiv \Lambda^* \mu)$ are absolutely continuous with respect to $\mu \otimes \Lambda^* \mu$, we can rewrite the above ε-entropy as follows:

$$S_O(\mu; \varepsilon) = \inf_{\bar{\mu}} \left\{ J(\mu, \bar{\mu}); \|\mu - \bar{\mu}\| \leq \varepsilon \right\},$$

where

$$J(\mu, \bar{\mu}) = \sup_{\Phi} \left\{ I(\mu, \bar{\mu}; \Phi); \Phi \ll \Phi_0 \right\}, \quad I(\mu, \bar{\mu}; \Phi) \equiv I(\mu; \Lambda^*).$$

The fractal dimension of a state μ for a classical continuous system is defined by the above ε-entropy. The *capacity dimension of a state $\mu \in P(\Omega)$* is

$$d_C^O(\mu) = \lim_{\varepsilon \to 0} \frac{S_O(\mu; \varepsilon)}{\log \frac{1}{\varepsilon}}. \qquad (8.21)$$

Although Gaussian communication processes are treated in Chap. 3, we briefly review them again.

Let $\mathcal{B}$ be the Borel σ-field of a real separable Hilbert space $\mathcal{H}$ and μ be a Borel probability measure on $\mathcal{B}$ satisfying

$$\int_{\mathcal{H}} \|x\|^2 d\mu(x) < \infty.$$

Further, we denote the set of all positive self-adjoint trace class operators on $\mathcal{H}$ by $T(\mathcal{H})_+$ ($\equiv \{ R \in \mathcal{B}(\mathcal{H}) ; R \geq 0, R = R^*, \text{tr} R < \infty \}$) and define the mean vector $m_\mu \in \mathcal{H}$ and the covariance operator $R_\mu \in T(\mathcal{H})_+$ of μ such as

$$\langle x_1, m_\mu \rangle = \int_{\mathcal{H}} \langle x_1, y \rangle d\mu(y),$$

$$\langle x_1, R_\mu x_2 \rangle = \int_{\mathcal{H}} \langle x_1, y - m_\mu \rangle \langle y - m_\mu, x_2 \rangle d\mu(y),$$

for any $x_1, x_2 \in \mathcal{H}$. A *Gaussian measure* μ in $\mathcal{H}$ is a Borel measure such that for each $x \in \mathcal{H}$, there exist real numbers m_x and $\sigma_x(> 0)$ satisfying

$$\mu\{y \in \mathcal{H} : \langle y, x \rangle \leq a\} = \int_{-\infty}^{a} \frac{1}{\sqrt{2\pi}\sigma_x} \exp\left\{\frac{-(t - m_x)^2}{2\sigma_x}\right\} dt .$$

The notation $\mu = [m, R]$ means that μ is a Gaussian measure on $\mathcal{H}$ with a mean vector m and a covariance operator R.

Let $(\mathcal{H}_1, \mathcal{B}_1)$ be an input space, $(\mathcal{H}_2, \mathcal{B}_2)$ be an output space and $P_G^{(k)}$ be the set of all Gaussian probability measures on $(\mathcal{H}_k, \mathcal{B}_k)$ $(k = 1, 2)$. We consider the case of $(\mathcal{H}_1, \mathcal{B}_1) = (\mathcal{H}_2, \mathcal{B}_2) \equiv (\mathcal{H}, \mathcal{B})$ for simplicity. Moreover, let $\mu \in P(\mathcal{H})$ be a Gaussian measure of the input space and $\mu_0 \in P(\mathcal{H})$ be a Gaussian measure indicating noise of the channel. Then, the *Gaussian channel* Λ^* from $P(\mathcal{H})$ to $P(\mathcal{H})$ is defined by the following mapping $\lambda : \mathcal{H} \times \mathcal{B} \to [0, 1]$ such as

$$\bar{\mu}(Q) = \Lambda^*\mu(Q) \equiv \int_{\mathcal{H}} \lambda(x, Q) d\mu(x) ,$$

$$\lambda(x, Q) \equiv \mu_0(Q^x) ,$$

$$Q^x \equiv \{y \in \mathcal{H} : Ax + y \in Q\} , \quad x \in \mathcal{H}, \quad Q \in \mathcal{B} ,$$

where A is a linear transformation from $\mathcal{H}$ to $\mathcal{H}$ and λ satisfies the following conditions:

(1) $\lambda(x, \cdot) \in P(\mathcal{H})$ for each fixed $x \in \mathcal{H}$,
(2) $\lambda(\cdot, Q)$ is a measurable function on $(\mathcal{H}, \mathcal{B})$ for each fixed $Q \in \mathcal{B}$.

The compound measure Φ derived form the input measure μ and the output measure $\bar{\mu}$ is given by

$$\Phi(Q_1 \times Q_2) = \int_{Q_1} \lambda(x, Q_2) d\mu(x)$$

for any $Q_1, Q_2 \in \mathcal{B}$.

In particular, let μ be $[0, R] \in P(\mathcal{H})$ and μ_0 be $[0, R_0] \in P(\mathcal{H})$. Then, the output measure $\Lambda^*\mu = \bar{\mu}$ can be expressed as

$$\Lambda^*\mu = [0, ARA^* + R_0] .$$

When the dimension of $\mathcal{H}$ is finite, the mutual entropy (information) with respect to μ and Λ^* becomes [104], cf. Chap. 2,

$$I(\mu; \Lambda^*) = \frac{1}{2} \log \frac{|ARA^* + R_0|}{|R_0|} ,$$

where $|ARA^* + R_0|, |R_0|$ are the determinants of $ARA^* + R_0, R_0$.

The (Ohya) ε-entropy (8.20) of a state μ is different from Kolmogorov's definition (8.19) of the ε-entropy for a classical random variable. First, we show that the two definitions coincide when $\mathcal{H} = \mathbb{R}^n$ and the norm of a state μ_f is taken as

$$\|\mu_f\| = \sqrt{\frac{1}{n}\sum_{i=1}^{n}\int_{\Omega}|f_i|^2 d\mu}.$$

Then the distance between two states μ_f and μ_g induced by the above norm leads to

$$\|\mu_f - \mu_g\| = \sqrt{\frac{1}{n}\sum_{i=1}^{n}\int_{\Omega}|f_i - g_i|^2 d\mu}.$$

We call this norm the *random variable norm* (RV-norm for short). Here we consider only Gaussian measures with the mean 0 and Gaussian channels.

Let an input state $\mu_f = [0, R]$ be induced from an n-dimensional random vector $f = (f_1, \ldots, f_n)$ and its output state $\Lambda^*\mu_f$ be denoted by μ_g, where g is the random vector $g = (g_1, \ldots, g_n)$ induced from Λ^* and f.

LEMMA 8.3. *If the distance of two states is given by the above RV-norm, then*

$$J(\mu_f; \Lambda^*) = I(\mu_f; \Lambda^*).$$

Proof. From the assumption, the Gaussian channel Λ^* is represented by a conditional probability density of g with respect to f as

$$p(y|x) = \frac{1}{(2\pi)^{\frac{n}{2}}\sqrt{|R_0|}}\exp\left\{-\frac{1}{2}(y - Ax)R_0^{-1}(y - Ax)^t\right\} \quad x, y \in \mathbb{R}^n,$$

where R_0 is the covariance matrix associated to the channel Λ^*. Then the compound state Φ of μ_f and $\Lambda^*\mu_f = \mu_g$ is equal to the joint probability measure $\mu_{fg} = [0, C]$ of f and g such as

$$\mu_{fg}(Q_1 \times Q_2) = \int_{Q_1 \times Q_2}\frac{1}{(2\pi)^n\sqrt{|C|}}\exp\left\{-\frac{1}{2}zC^{-1}z^t\right\}dz \quad Q_1, Q_2 \in \mathcal{B}(\mathbb{R}^n),$$

where z is the $2n$-dimensional random vector $(x, y) = (x_1, \ldots, x_n, y_1, \ldots, y_n)$ and C is the following covariance matrix of μ_{fg} :

$$C = \begin{pmatrix} R & RA^t \\ AR & ARA^t + R_0 \end{pmatrix}, \tag{8.22}$$

where R, RA^t, AR, $ARA^t + R_0$ are $n \times n$ matrices and $(R)_{ij} = E(f_i f_j)$, $(RA^t)_{ij} = E(f_i g_j)$, $(AR)_{ij} = E(g_i f_j)$, $(ARA^t + R_0)_{ij} = E(g_i g_j)$ for each (i, j) $(i, j = 1, \ldots, n)$.

For the channel Γ^* satisfying $\Lambda^*\mu_f = \Gamma^*\mu_f$, $\Gamma^*\mu_f$ is a n-dimensional Gaussian measure μ_h induced from n-dimensional random vector $h = (h_1, \ldots, h_n)$, so that we have

$$
\begin{aligned}
J(\mu_f; \Lambda^*) &= \sup_{\Gamma^*} \{I(\mu_f; \Gamma^*) : \Lambda^*\mu_f = \Gamma^*\mu_f\} \\
&= \sup_{\Gamma^*} \{I(\mu_f; \Gamma^*) : \|\Lambda^*\mu_f - \Gamma^*\mu_f\| = 0\} \\
&= \sup_{h} \{I(f, h) : \|\mu_g - \mu_h\| = 0\} \\
&= \sup_{h} \left\{I(f, h) : E[d(g, h)^2] = 0\right\} \\
&= \sup_{h} \left\{I(f, h) : E[d(g_i, h_i)^2] = 0 \ (i = 1, \ldots, n)\right\} \\
&= \sup_{h} \{I(f, h) : g_i = h_i \quad a.e. \quad (i = 1, \ldots, n)\} \ .
\end{aligned}
$$

From (8.22), the joint probability measure μ_{fh} is a Gaussian measure $[0, \bar{C}]$, where $\bar{C}$ is the covariance matrix of μ_{fh} such as

$$
\bar{C} = \begin{pmatrix} R & RB^t \\ BR & BRB^t + \bar{R}_0 \end{pmatrix} ,
$$

where R, RB^t, BR, $BRB^t + \bar{R}_0$ are $n \times n$ matrices satisfying $B \neq A$ and $\bar{R}_0 \neq R_0$, and $(R)_{ij} = E(f_i f_j)$, $(RB^t)_{ij} = E(f_i h_j)$, $(BR)_{ij} = E(h_i f_j)$, $(BRB^t + \bar{R}_0)_{ij} = E(h_i h_j)$ for each (i, j) $(i, j = 1, \ldots, n)$. Therefore, if the equations $E(f_i g_j) = E(f_i h_j)$, $E(g_i f_j) = E(h_i f_j)$, $E(h_i h_j) = E(g_i g_j)$ for each (i, j) $(i, j = 1, \ldots, n)$ hold, then $\mu_{fg} = \mu_{fh}$. We have

$$
E(f_i h_j) = -\frac{1}{2} \left\{ E\left[d(f_i, h_j)^2\right] - E(f_i^2) - E(h_j^2) \right\}
$$

for each (i, j) $(i, j = 1, \ldots, n)$, and the following triangle inequalities for each (i, j) $(i, j = 1, \ldots, n)$

$$
\begin{aligned}
\sqrt{E[d(f_i, h_j)^2]} &\leq \sqrt{E[d(f_i, g_j)^2]} + \sqrt{E[d(g_j, h_j)^2]}, \\
\sqrt{E[d(f_i, g_j)^2]} &\leq \sqrt{E[d(f_i, h_j)^2]} + \sqrt{E[d(h_j, g_j)^2]}. \quad (8.23)
\end{aligned}
$$

When $\|\Lambda^*\mu_f - \Gamma^*\mu_f\| = 0$, since $E[d(g_j, h_j)^2] = E[d(h_j, g_j)^2] = 0$, we have

$$
E\left[d(f_i, h_j)^2\right] = E\left[d(f_i, g_j)^2\right]
$$

from (8.23). Moreover, since $h_j = g_j$ a.e. for each j $(j = 1, \ldots, n)$ holds in the case of $\|\Lambda^*\mu_f - \Gamma^*\mu_f\| = 0$, we obtain

$$
E(h_j^2) = E(g_j^2) .
$$

Therefore,

$$E(f_ih_j) = -\frac{1}{2}\left\{E\left[d(f_i,g_j)^2\right] - E(f_i^2) - E(g_j^2)\right\} = E(f_ig_j)$$

holds for each (i,j) $(i,j=1,\ldots,n)$.

It can be shown similarly that $E(g_jf_j) = E(h_if_j)$, $E(g_ig_j) = E(h_ih_j)$ for each (i,j), so that we obtain

$$\mu_{fg} = \mu_{fh}.$$

Therefore,

$$I(\mu_f;\Lambda^*) = I(f,h) = I(\mu_f;\Gamma^*),$$

which implies

$$J(\mu_f;\Lambda^*) = I(\mu_f;\Lambda^*).$$

Using the above lemma, the following theorem holds.

THEOREM 8.11. *Under the same assumption as Lemma 8.14,*

$$(1)\ S_O(\mu_f;\varepsilon) = S_K(f;\varepsilon) = \frac{1}{2}\sum_{i=1}^{n}\log\max\left(\frac{\lambda_i}{\theta^2},1\right),$$

where $\lambda_1,\ldots,\lambda_n$ are the eigenvalues of R and θ^2 is a constant uniquely determined by the equation $\sum_{i=1}^{n}\min(\lambda_i,\theta^2) = \varepsilon^2$,

(2) $d_C^O(\mu_f) = d_C^K(\mu_f) = n$.

Proof. (1) Let $\bar{C}$ be the set of all Gaussian channels from $B(\mathbb{R}^n)$ to $B(\mathbb{R}^n)$ and $\hat{C}(\mu_f;\varepsilon)$ be the set of all Gaussian channels from $B(\mathbb{R}^n)$ to $B(\mathbb{R}^n)$ satisfying $\|\mu_f - \Lambda^*\mu_f\| \le \varepsilon$. According to Lemma 8.3, we obtain

$$S_O(\mu_f;\varepsilon) = \inf\{J(\mu_f:\Lambda^*);\Lambda^*\in\bar{C}(\mu_f;\varepsilon)\}$$
$$= \inf\{I(\mu_f;\Lambda^*):\Lambda^*\in\hat{C}(\mu_f;\varepsilon)\}.$$

From (8.24), we have

$$S_O(\mu_f;\varepsilon) = \inf\{I(\mu_f;\Lambda^*):\Lambda^*\in\bar{C}(\mu_f;\varepsilon)\}$$
$$= \inf\{I(f,g):\mu_{fg}\in\bar{S}(\mu_f;\varepsilon)\}$$
$$= S_K(f;\varepsilon),$$

where $\bar{S}(\mu_f:\varepsilon) = \left\{\mu_{fg};\sqrt{\iint_{R^n\times R^n}d(f,g)^2d\mu_{fg}}\le\varepsilon\right\}.$

The expression for the ε-entropy $\dfrac{1}{2}\displaystyle\sum_{n=1}^{n}\log\left\{\max\left(\dfrac{\lambda_i}{\theta^2},1\right)\right\}$ was obtained by Pinsker [276].

(2) Since $S_O(\mu_f;\varepsilon)=\dfrac{1}{2}\displaystyle\sum_{n=1}^{n}\log\left\{\max\left(\dfrac{\lambda_i}{\theta^2},1\right)\right\}$, we have

$$
\begin{aligned}
d_C^O(\mu_f)=d_C^K(\mu_f) &= \lim_{\varepsilon\to0}\frac{S_O(\mu_f;\varepsilon)}{\log\dfrac{1}{\varepsilon}}\\[2ex]
&= \lim_{\varepsilon\to0}\frac{\dfrac{1}{2}\displaystyle\sum_{i=1}^{n}\log\left\{\max\left(\dfrac{\lambda_i}{\theta^2},1\right)\right\}}{\log\dfrac{1}{\varepsilon}}\qquad\left(i.e.\ \sum_{i=1}^{n}\min(\lambda_i,\theta^2)=\varepsilon^2\right)\\[2ex]
&= \lim_{\varepsilon\to0}\frac{\dfrac{1}{2}\displaystyle\sum_{i=1}^{n}\log\dfrac{\lambda_i}{\theta^2}}{\log\dfrac{1}{\varepsilon}}\qquad\left(\sum_{i=1}^{n}\theta^2=\varepsilon^2\right)\\[2ex]
&= \lim_{\varepsilon\to0}\frac{\dfrac{1}{2}\displaystyle\sum_{i=1}^{n}\log\dfrac{\lambda_i n}{\varepsilon^2}}{\log\dfrac{1}{\varepsilon}}=n.\qquad\square
\end{aligned}
$$

In the case of Theorem 8.14, Ohya ε-entropy coincides with Kolmogorov's ε-entropy and the fractal dimension of the state μ_f is identical to the dimension of the Hilbert space.

In the following discussion, we only consider the case of $\dim(\mathcal{H})=1$, that is, $\mathcal{H}=\mathbb{R}^1$, while the state $\mu=[0,\sigma^2]$ is one-dimensional Guassian measure (distribution) and the distance of two states is given by the total variation norm. In this subsection, we show that Ohya's ε-entropy leads to the fractal property of a Gaussian measure, but Kolmogorov's one does not. The difference between S_O and S_K comes from the assumed norm on measures. We take the norm as the total variation, namely,

$$\|\mu\|=|\mu|(\mathbb{R}^1).$$

Let $\mathcal{H}_1=\mathcal{H}_2=\mathbb{R}^1$. Then A becomes a real number β and the noise of the channel is represented by the one-dimensional Gaussian measure $\mu_0=[0,\sigma_0^2]\in P(\mathcal{H})$, so that the output state $A^*\mu$ is represented by $[0,\beta^2\sigma^2+\sigma_0^2]$. We calculate the maximum mutual entropy in the following two cases:

(1) $\beta^2\sigma^2+\sigma_0^2\geq\sigma^2$,
(2) $\beta^2\sigma^2+\sigma_0^2<\sigma^2$.

Since the channel Λ^* depends on β and σ_0^2, we write $\Lambda^* = \Lambda^*_{(\beta,\sigma_0^2)}$.

LEMMA 8.4. *If* $\beta^2\sigma^2 + \sigma_0^2 \geq \sigma^2$ *and* $\|\mu - \Lambda^*_{(\beta,\sigma_0^2)}\mu\| = |\mu - \Lambda^*_{(\beta,\sigma_0^2)}\mu|(R) = \delta$, *then*

(1) $\dfrac{4}{\sqrt{2\pi}} \dfrac{\sqrt{\beta^2\sigma^2 + \sigma_0^2} - \sigma}{\sigma} = \delta + o(\delta),$

(2) $\left\{\Lambda^*_{(\beta,\sigma_0^2)} : \|\mu - \Lambda^*_{(\beta,\sigma_0^2)}\mu\| \leq \varepsilon\right\}$

$= \left\{\Lambda^*_{(\beta,\sigma_0^2)} : \dfrac{4}{\sqrt{2\pi}} \dfrac{\sqrt{\beta^2\sigma^2 + \sigma_0^2} - \sigma}{\sigma} = \delta + o(\delta)\ \delta \in M(\varepsilon)\right\},$

where $o(\delta)$ *is the order of* δ: $\lim\limits_{\delta\to 0} o(\delta) = 0$ *and* $M(\varepsilon) = \{\delta \in R : 0 \leq \delta \leq \varepsilon\}$.

   Proof. (1) Let p_1, p_2 be the density functions of $\mu, \Lambda^*\mu$, respectively. Since $\mu = [0, \sigma^2]$ and Λ^* is a Gaussian channel from $\mathcal{B}(\mathbb{R}^1)$ to $\mathcal{B}(\mathbb{R}^1)$, we have

$$\|\mu - \Lambda^*_{(\beta,\sigma_0^2)}\mu\| = \int_R |p_1(x) - p_2(x)|\,dx$$

$$= 4\int_0^a \left|\frac{1}{\sqrt{2\pi}\sigma}\exp\left\{-\frac{x^2}{2\sigma^2}\right\} - \frac{1}{\sqrt{2\pi}\sqrt{\beta^2\sigma^2 + \sigma_0^2}}\exp\left\{-\frac{x^2}{2(\beta^2\sigma^2 + \sigma_0^2)}\right\}\right|\,d \tag{8.2}$$

$$\left(a = \sqrt{\left(\frac{1}{\sigma^2} - \frac{1}{\beta^2\sigma^2 + \sigma_0^2}\right)^{-1}\log\frac{\beta^2\sigma^2 + \sigma_0^2}{\sigma^2}}\right)$$

$$\leq 4\left(\frac{1}{\sqrt{2\pi}\sigma} - \frac{1}{\sqrt{2\pi}\sqrt{\beta^2\sigma^2 + \sigma_0^2}}\right)\sqrt{\left(\frac{1}{\sigma^2} - \frac{1}{\beta^2\sigma^2 + \sigma_0^2}\right)^{-1}\log\frac{\beta^2\sigma^2 + \sigma_0^2}{\sigma^2}}$$

$$\leq \frac{4}{\sqrt{2\pi}}\frac{\sqrt{\beta^2\sigma^2 + \sigma_0^2} - \sigma}{\sigma},$$

where the first inequality is the result of a geometrical approximation of the equation, while the second inequality is obtained from the inequality $\log x \leq x - 1$ for any positive number x.
   Since $\lim\limits_{\|\mu - \Lambda^*\mu\| \to 0} \beta^2\sigma^2 + \sigma_0^2 = \sigma^2$, the above inequality implies

$$\|\mu - \Lambda^*\mu_{(\beta,\sigma_0^2)}\mu\| = \delta \implies \frac{4}{\sqrt{2\pi}}\frac{\sqrt{\beta^2\sigma^2 + \sigma_0^2} - \sigma}{\sigma} = \delta + o(\delta).$$

(2) From (1), for any Gaussain channel $\Lambda^*_{(\bar{\beta},\bar{\sigma}_0^2)}$ satisfying $\|\mu - \Lambda^*_{(\bar{\beta},\bar{\sigma}_0^2)}\mu\| = \delta$, it is clear that

$$\Lambda^*_{(\bar{\beta},\bar{\sigma}_0^2)} \in \left\{ \Lambda^*_{(\beta,\sigma_0^2)} : \frac{4}{\sqrt{2\pi}} \frac{\sqrt{\beta^2\sigma^2 + \sigma_0^2} - \sigma}{\sigma} = \delta + o(\delta) \right\}.$$

Take a channel $\Lambda^*_{(\bar{\beta},\bar{\sigma}_0^2)}$ satisfying

$$\frac{4}{\sqrt{2\pi}} \frac{\sqrt{\bar{\beta}^2\sigma^2 + \bar{\sigma}_0^2} - \sigma}{\sigma} = \delta + o(\delta). \tag{8.25}$$

Then the covariance of $\Lambda^*_{(\bar{\beta},\bar{\sigma}_0^2)}\mu$ is uniquely given by (8.25). If $\Lambda^*_{(\bar{\beta},\bar{\sigma}_0^2)}$ satisfies $\|\mu - \Lambda^*_{(\bar{\beta},\bar{\sigma}_0^2)}\mu\| \neq \delta$, the covariance of $\Lambda^*_{(\beta,\sigma_0^2)}\mu$ doesn't satisfy the equality (8.25). Hence

$$\Lambda^*_{(\bar{\beta},\bar{\sigma}_0^2)} \in \left\{ \Lambda^*_{(\beta,\sigma_0^2)} : \|\mu - \Lambda^*_{(\beta,\sigma_0^2)}\mu\| = \delta \right\}.$$

Therefore,

$$\left\{ \Lambda^*_{(\beta,\sigma_0^2)} : \|\mu - \Lambda^*_{(\beta,\sigma_0^2)}\mu\| = \delta \right\}$$

$$= \left\{ \Lambda^*_{(\beta,\sigma_0^2)} : \frac{4}{\sqrt{2\pi}} \frac{\sqrt{\beta^2\sigma^2 + \sigma_0^2} - \sigma}{\sigma} = \delta + o(\delta) \right\}. \tag{8.26}$$

Let $M(\varepsilon)$ be the set of all $\delta \in \mathbb{R}^1$ satisfying $0 \leq \delta \leq \varepsilon$. From (8.26), it is clear that

$$\left\{ \Lambda^*_{(\beta,\sigma_0^2)} : \|\mu - \Lambda^*_{(\beta,\sigma_0^2)}\mu\| \leq \varepsilon \right\}$$

$$= \left\{ \Lambda^*_{(\beta,\sigma_0^2)} : \frac{4}{\sqrt{2\pi}} \frac{\sqrt{\beta^2\sigma^2 + \sigma_0^2} - \sigma}{\sigma} = \delta + o(\delta) \quad \delta \in M(\varepsilon) \right\}. \quad \square$$

LEMMA 8.5. *Let* $\Lambda^*_{\delta(\beta,\sigma_0^2)}$ *be a channel satisfying* $\|\mu - \Lambda^*_{\delta(\beta,\sigma_0^2)}\mu\| = \delta$ *for any* $\delta \in M(\varepsilon)$. *If a Gaussian channel* $\Lambda^*_{\delta(\beta,\sigma_0^2)}$ *satisfies the condition* $\beta^2 \leq \dfrac{C_\delta - \delta}{\sigma^2}$, *then we have*

$$J(\mu; \Lambda^*_{\delta(\beta,\sigma_0^2)}) = \frac{1}{2}\log\frac{1}{\delta} + \frac{1}{2}\log\sigma^2\left(1 + \frac{\sqrt{2\pi}}{4}(\delta + o(\delta))\right),$$

where $C_\delta = \beta^2\sigma^2 + \sigma_0^2$ *is a constant determined by* $\|\mu - \Lambda^*_{\delta(\beta,\sigma_0^2)}\mu\| = \delta$.

Proof. The mutual entropy of μ with respect to channel Λ^* is

$$I(\mu; \Lambda^*) = \frac{1}{2} \log \frac{\beta^2 \sigma^2 + \sigma_0^2}{\sigma_0^2}.$$

Thus, if $\Lambda^*_{\delta(\beta, \sigma_0^2)}$ is any channel satisfying $\|\mu - \Lambda^*_{\delta(\beta, \sigma_0^2)}\mu\| = \delta$, then we have from the above lemma 8.4

$$I(\mu; \Lambda^*_{\delta(\beta, \sigma_0^2)}) = \frac{1}{2} \log \frac{1}{C_\delta - \beta^2 \sigma^2} + \frac{1}{2} \log \sigma^2 \left(1 + \frac{\sqrt{2\pi}}{4}(\delta + o(\delta))\right),$$

The assumption implies,

$$
\begin{aligned}
J(\mu; \Lambda^*_{\delta(\beta, \sigma_0^2)}) &= \sup_{\Lambda^*_{\delta(\beta, \sigma_0^2)}} \left\{ I(\mu; \Lambda^*_{\delta(\bar\beta, \bar\sigma_0^2)}) : \Lambda^*_{\delta(\beta, \sigma_0^2)}\mu = \Lambda^*_{\delta(\bar\beta, \bar\sigma_0^2)}\mu \right\} \\
&= \sup_{\Lambda^*_{\delta(\beta, \sigma_0^2)}} \left\{ I(\mu; \Lambda^*_{\delta(\bar\beta, \bar\sigma_0^2)}) : \|\Lambda^*_{\delta(\beta, \sigma_0^2)}\mu - \Lambda^*_{\delta(\beta, \sigma_0^2)}\mu\| = 0 \right\} \\
&= \sup_{\Lambda^*_{\delta(\beta, \sigma_0^2)}} \left\{ I(\mu; \Lambda^*_{\delta(\bar\beta, \bar\sigma_0^2)}) : \beta^2 \sigma^2 + \sigma_0^2 = \bar\beta^2 \sigma^2 + \bar\sigma_0^2 \right\} \\
&= \sup_{\bar\beta} \left\{ \frac{1}{2} \log \frac{1}{C_\delta' - \bar\beta^2 \sigma^2} + \frac{1}{2} \log \sigma^2 \cdot f(\delta) : \bar\beta^2 \leq \frac{C_\delta - \delta}{\sigma^2} \right\} \\
&= \frac{1}{2} \log \frac{1}{\delta} + \frac{1}{2} \log \sigma^2 \cdot f(\delta),
\end{aligned}
$$

where $\bar\beta^2 = \dfrac{C_\delta - \delta}{\sigma^2}$ and $f(\delta) = \left(1 + \dfrac{\sqrt{2\pi}}{4}(\delta + o(\delta))\right)$. $\square$

LEMMA 8.6. *If $\beta^2 \sigma^2 + \sigma_0^2 < \sigma^2$ and a Gaussian channel $\Lambda^*_{\delta(\beta, \sigma_0^2)}$ satisfies the condition $\beta^2 \leq \dfrac{\bar C_\delta - \delta}{\sigma^2}$, with $\bar C_\delta = \beta^2 \sigma^2 + \sigma_0^2$, we have*

$$J(\mu; \Lambda^*_{(\beta, \sigma_0^2)}) = \frac{1}{2} \log \frac{1}{\delta} + \frac{1}{2} \log \frac{\sigma^2}{\left(1 + \frac{\sqrt{2\pi}}{4}(\delta + o(\delta))\right)}.$$

Proof. Similarly as Lemma 8.5. $\square$

Using the above lemmas, we obtain the following theorem.

THEOREM 8.12. *Under the same conditions of Lemma 8.5 and 8.6, we have*

(1) $S_O(\mu; \varepsilon) = \dfrac{1}{2} \log \dfrac{1}{\varepsilon} + \dfrac{1}{2} \log \dfrac{\sigma^2}{\left(1 + \frac{\sqrt{2\pi}}{4}(\varepsilon + o(\varepsilon))\right)} > S_K(\mu; \varepsilon) = 0,$

(2) $d_C^O(\mu) = \dfrac{1}{2}.$

Proof. (1) From Lemma 8.5 and 8.6, we have

$$
\begin{aligned}
S_O(\mu; \varepsilon) &= \inf_{\Lambda^*} \left\{ J(\mu; \Lambda^*); \|\mu - \Lambda^* \mu\| \le \varepsilon \right\} \\
&= \inf_\delta \left\{
\begin{array}{c}
\dfrac{1}{2} \log \dfrac{1}{\delta} + \dfrac{1}{2} \log \sigma^2 \left(1 + \dfrac{\sqrt{2\pi}}{4}(\delta + o(\delta))\right) \\
\dfrac{1}{2} \log \dfrac{1}{\delta} + \dfrac{1}{2} \log \dfrac{\sigma^2}{\left(1 + \frac{\sqrt{2\pi}}{4}(\delta + o(\delta))\right)}
\end{array}
; \quad \delta \in M(\varepsilon) \right\} \\
&= \inf_\delta \left\{ \dfrac{1}{2} \log \dfrac{1}{\delta} + \dfrac{1}{2} \log \dfrac{\sigma^2}{\left(1 + \frac{\sqrt{2\pi}}{4}(\delta + o(\delta))\right)} ; \quad \delta \in M(\varepsilon) \right\} \\
&= \dfrac{1}{2} \log \dfrac{1}{\varepsilon} + \dfrac{1}{2} \log \dfrac{\sigma^2}{\left(1 + \frac{\sqrt{2\pi}}{4}(\varepsilon + o(\varepsilon))\right)},
\end{aligned}
$$

since $\dfrac{1}{2} \log \dfrac{1}{\delta} + \dfrac{1}{2} \log \dfrac{\sigma^2}{\left(1 + \frac{\sqrt{2\pi}}{4}(\delta + o(\delta))\right)}$ is monotone decreasing with respect to δ .

(2) From (1), we obtain

$$
\begin{aligned}
d_C^O(\mu) &= \lim_{\varepsilon \to 0} \dfrac{S_O(\mu; \varepsilon)}{\log \frac{1}{\varepsilon}} \\
&= \lim_{\varepsilon \to 0} \dfrac{\dfrac{1}{2} \log \dfrac{1}{\varepsilon} + \dfrac{1}{2} \log \dfrac{\sigma^2}{\left(1 + \frac{\sqrt{2\pi}}{4}(\varepsilon + o(\varepsilon))\right)}}{\log \dfrac{1}{\varepsilon}} = \dfrac{1}{2}. \quad \square
\end{aligned}
$$

In the total variation norm, the fractal dimension of a state μ becomes a non-integer, which shows a fractal aspect of Gaussian measure. As one of the case studied, Mandelbrot analyzed time series of price data by the R/S-analysis (rescaled range analysis) [225], [273]. He showed that the fractal dimension of the time series of data depending on a Gaussian measure is $\frac{1}{2}$. His result coincides with our fractal dimension of a Gaussian measure.

THEOREM 8.13. *If the noise of a channel is completly neglected, then*

$$S_O(\mu;\varepsilon) = \infty.$$

Proof. For every Gaussian channels satisfying $\|\mu - \Lambda^*\mu\| \leq \varepsilon$ and, for any $\delta \in M(\varepsilon)$ with $\|\mu - \Lambda^*\mu\| = \delta$,

$$
J(\mu;\Lambda^*) = \sup_{\sigma_0^2} \left\{ \begin{array}{l} \dfrac{1}{2}\log\dfrac{1}{\sigma_0^2} + \dfrac{1}{2}\log\sigma^2\left(1 + \dfrac{\sqrt{2\pi}}{4}(\delta + o(\delta))\right) \\[2ex] \dfrac{1}{2}\log\dfrac{1}{\sigma_0^2} + \dfrac{1}{2}\log\dfrac{\sigma^2}{\left(1 + \frac{\sqrt{2\pi}}{4}(\delta + o(\delta))\right)} \end{array} \right\}
$$

$$
= \lim_{\sigma_0^2 \to 0} \left\{ \begin{array}{l} \dfrac{1}{2}\log\dfrac{1}{\sigma_0^2} + \dfrac{1}{2}\log\sigma^2\left(1 + \dfrac{\sqrt{2\pi}}{4}(\delta + o(\delta))\right) \\[2ex] \dfrac{1}{2}\log\dfrac{1}{\sigma_0^2} + \dfrac{1}{2}\log\dfrac{\sigma^2}{\left(1 + \frac{\sqrt{2\pi}}{4}(\delta + o(\delta))\right)} \end{array} \right\}
$$

$$
= \infty.
$$

Therefore,

$$S_O(\mu;\varepsilon) = \infty. \qquad \square$$

When the noise is completely neglected, "sup" = " $\lim\limits_{\sigma_0^2 \to 0}$ ", so that the ε-entropy of a state is infinite. The result $S_O(\mu;\varepsilon) = S(\mu) = \infty$ is quite natural from the point of view of the information transmission since the entropy itself is infinite in a continuous system in Shannon's formulation.

The fractal dimension of state can be used to analyse chaoitic properties of systems.

Appendix 1

A.1 Elements of Hilbert Spaces

In this Section we explain the fundamental facts about Hilbert spaces to facilitate the understanding of this book.

DEFINITION A.1.1. Let $\mathbb{R}$ (resp. $\mathbb{C}$) be a field consisting of all real numbers (resp. complex numbers), denoted by $\alpha, \beta, \ldots$, and X be a real (resp. complex) linear space with elements denoted by $x, y, \ldots$. Then, an *inner product* denoted by $\langle \cdot, \cdot \rangle$ on $X \times X$ with values in $\mathbb{R}$ (resp. $\mathbb{C}$) is defined by the mapping satisfying

$$
\begin{aligned}
\langle y, x \rangle &= \overline{\langle x, y \rangle}, \\
\langle x + y, z \rangle &= \langle x, z \rangle + \langle y, z \rangle, \\
\langle x, \alpha y \rangle &= \alpha \langle x, y \rangle, \\
\langle x, x \rangle &\geq 0, \qquad \langle x, x \rangle = 0 \Leftrightarrow x = 0.
\end{aligned}
$$

X with an inner product defined in it is called an *inner product space*.

Let $\|x\| := \sqrt{\langle x, x \rangle}$. Then, the *Schwarz inequality* is expressed by the following statement:

THEOREM A.1.2.

$$
|\langle x, y \rangle| \leq \|x\| \cdot \|y\|, \quad x, y \in X
$$

holds, where the left hand side term is exactly equal to the right hand side if and only if there exists a real (resp. complex) constant λ satisfying $y = \lambda x$ or $x = 0$.

This inequality shows that the functional $\|\cdot\|$ satisfies the following conditions of a *norm*:

$$
\begin{aligned}
\|x\| &\geq 0, \qquad \|x\| = 0 \Leftrightarrow x = 0, \\
\|\alpha x\| &= |\alpha| \cdot \|x\|, \\
\|x + y\| &\leq \|x\| + \|y\|.
\end{aligned}
$$

If an inner product space X is a *Banach space*, that is, a complete normed space whose norm arises from the inner product, this space is said to be a *Hilbert space* H.

THEOREM A.1.3 (Parallelogram law).

$$\|x + y\|^2 + \|x - y\|^2 = 2(\|x\|^2 + \|y\|^2), \quad x, y \in \mathcal{H}.$$

THEOREM A.1.4 (Polar identity).

$$\langle x, y \rangle = \frac{1}{4} \left[\|x + y\|^2 - \|x - y\|^2 + i\|x + iy\|^2 - i\|x - iy\|^2 \right], \quad x, y \in \mathcal{H}.$$

Let X be a Banach space with the norm $\|\cdot\|$, and the functional $\ll \cdot, \cdot \gg$ on $X \times X$ be defined by the following equation:

$$\ll x, y \gg = \frac{1}{4} \left[\|x + y\|^2 - \|x - y\|^2 + i\|x + iy\|^2 - i\|x - iy\|^2 \right], \quad x, y \in X.$$

If the norm $\|\cdot\|$ satisfies the parallelogram law, it can be proved that $\ll \cdot, \cdot \gg$ satisfies the conditions for inner product, and $(X, \ll \cdot, \cdot \gg)$ has the same metric structure as Hilbert spaces.

Example 1: $\mathbb{R}^n$ equipped with the inner product defined by

$$\langle x, y \rangle = \sum_{i=1}^{n} x_i y_i.$$

is a real Hilbert space.

Example 2: $\mathbb{C}^n$ equipped with the inner product defined by

$$\langle x, y \rangle = \sum_{i=1}^{n} \bar{x}_i y_i.$$

is a complex Hilbert space.

Example 3: Let ℓ^2 be the linear space consisting of the square summable complex sequences, and $\langle \cdot, \cdot \rangle$ be the fuctional on $\ell^2 \times \ell^2$ defined by

$$\langle \{x_n\}, \{y_n\} \rangle = \sum_{n=1}^{\infty} \bar{x}_n y_n, \quad \{x_n\}, \{y_n\} \in \ell^2.$$

Then, $(\ell^2, \langle \cdot, \cdot \rangle)$ is a Hilbert space.

Example 4: Let $(\Omega, \mathcal{F}, \mu)$ be a σ-finite measure space, $L^2(\Omega)$ be the linear space consisting of all equivalence classes of square integrable complex-valued functions on Ω, and $\langle \cdot, \cdot \rangle$ be the functional on $L^2(\Omega) \times L^2(\Omega)$ defined by

$$\langle x(\cdot), y(\cdot) \rangle = \int_\Omega \overline{x(\omega)} y(\omega) d\omega, \quad x, y \in L^2(\Omega).$$

Then, $(L^2(\Omega, \langle \cdot, \cdot \rangle)$ is a Hilbert space.

DEFINITION A.1.5. Let $\mathcal{H}$ be a Hilbert space with the inner product $\langle \cdot, \cdot \rangle$, and x and y be elements of $\mathcal{H}$. Then x is said to be *orthogonal* to y, if $\langle x, y \rangle = 0$ holds. A subset $\{x_j : j \in J\}$ is said to be an *orthogonal system* if it satisfies $\langle x_i, x_j \rangle = 0$ for any distinct $i, j \in J$. An orthogonal system $\{x_j : j \in J\}$ is said to be an *orthonormal system*, in short ONS, if it satisfies $\|x_i\| = 1$ for all $i \in J$. Moreover, an orthonormal system $\{x_j; j \in J\}$ is said to be *complete*, in short CONS, if $\langle y, x_i \rangle = 0$ for all $j \in J$ implies $y = 0$.

Zorn's lemma assures that any Hilbert space has a complete orthonormal system. In particular, a Hilbert space having a countable orthonormal system is said to be *separable*.

THEOREM A.1.6 (Bessel's inequality). *If $\{x_j\}$ is an orthonormal system, then*

$$\sum_{j \in J} |\langle x, x_j \rangle|^2 \leq \|x\|^2, \quad x \in \mathcal{H},$$

where the equality holds iff $\{x_j\}$ is complete.

THEOREM A.1.7 (Fourier's expansion formula). *If $\{x_j\}$ is a complete orthonormal system, then*

$$x = \sum_{j \in J} \langle x, x_j \rangle x_j, \quad x \in \mathcal{H}.$$

$\{\langle x, x_j \rangle : j \in J\}$ *is said to be a set of Fourier coefficients of x with respect to $\{x_i : i \in J\}$.*

THEOREM A.1.8 (Parseval's equality). $\{x_j\}$ *is a CONS iff*

$$\langle x, y \rangle = \sum_{j \in J} \langle x, x_j \rangle \langle x_j, y \rangle, \quad x, y \in \mathcal{H}.$$

DEFINITION A.1.9. Let $\mathcal{L}$ be a subspace of $\mathcal{H}$. Then, $\mathcal{L}$ is said to be *closed* if it is a closed subset of $\mathcal{H}$. The *orthogonal subspace* of $\mathcal{L}$, denoted by $\mathcal{L}^\perp$, is defined by $\{x \in \mathcal{H} : \langle x, y \rangle = 0, y \in \mathcal{L}\}$.

It is clear that $\mathcal{L}^{\perp}$ is also a closed subspace. For any two subspaces $\mathcal{L}$ and $\mathcal{K}$, $\mathcal{L}$ is said to be *orthogonal* to $\mathcal{K}$ if for any $x \in \mathcal{L}$ and any $y \in \mathcal{K}$, $\langle x, y \rangle = 0$ holds. For such an orthogonal pair $\mathcal{L}$ and $\mathcal{K}$, the direct sum $\mathcal{L} \oplus \mathcal{K}$ is a closed subspace of $\mathcal{H}$. For any subset $\mathcal{L}$ of $\mathcal{H}$, $\mathcal{L}^{\perp\perp}$ defined by $(\mathcal{L}^{\perp})^{\perp}$ is the smallest closed subspace including $\mathcal{L}$. Moreover, if $\mathcal{L}$ is a complete orthonormal system of $\mathcal{H}$, then $\mathcal{L}^{\perp}$ is equal to $\{0\}$ and $\mathcal{L}^{\perp\perp}$ is equal to $\mathcal{L}$.

THEOREM A.1.10 (Least distance theorem). *Let $\mathcal{L}$ be a closed subspace. Then, for any $x \in \mathcal{H}$, there uniquely exists an element z of $\mathcal{L}$ satisfying the following equation:*

$$\|x - z\| = \inf\{\|x - y\| : y \in \mathcal{L}\}.$$

Let $\mathcal{P}_{\mathcal{L}}$ be the mapping on $\mathcal{H}$ with values in $\mathcal{L}$ defined by $\mathcal{P}_{\mathcal{L}} x = z$, where x belongs to $\mathcal{H}$ and z is the unique element of $\mathcal{L}$ satisfying the above equality. Then, it is known that the mapping $\mathcal{P}_{\mathcal{L}}$ is a bounded linear operator. This operator is usually called a *projection*. We have the following two important results:

THEOREM A.1.11 (Projection theorem). *Let $\mathcal{L}$ be a closed subspace of $\mathcal{H}$. Then,*

$$\mathcal{H} = \mathcal{L} \oplus \mathcal{L}^{\perp}.$$

THEOREM A.1.12 (Riesz's theorem). *For any bounded linear functional f, there uniquely exists $y \in \mathcal{H}$ such that*

$$f(x) = \langle y, x \rangle, \quad x \in \mathcal{H},$$

and the following equality holds:

$$\sup\{|f(x)| : x \in \mathcal{H}, \|x\| = 1\} = \|y\|.$$

DEFINITION A.1.13. For any two Hilbert spaces $\mathcal{H}$ and $\mathcal{K}$, $\mathcal{H}$ is said to be *isomorphic* to $\mathcal{K}$, denoted $\mathcal{H} \cong \mathcal{K}$, if there exists a bounded linear operator U on $\mathcal{H}$ with values in $\mathcal{K}$ satisfying

$$\langle x, y \rangle_{\mathcal{H}} = \langle Ux, Uy \rangle_{\mathcal{K}}, \quad x, y \in \mathcal{H}.$$

It is known that if $\dim \mathcal{H} = n$, then $\mathcal{H}$ is isomorphic to the n-dimensional vector space $\mathbb{R}^n$ or $\mathbb{C}^n$, and moreover, if $\mathcal{H}$ is a separable infinite dimensional Hilbert space, then $\mathcal{H}$ is isomorphic to ℓ^2.

Let x (resp. y) be an element of $\mathcal{H}$ (resp. $\mathcal{K}$) and $x \otimes y$ be the *conjugate bilinear functional* on $\mathcal{H} \times \mathcal{K}$ defined by

$$x \otimes y(u, v) = \langle x, u \rangle_{\mathcal{H}} \langle y, v \rangle_{\mathcal{K}}, \quad u \in \mathcal{H}, v \in \mathcal{K}.$$

Let $\mathcal{E}$ denote the linear space consisting of conjugate bilinear functionals constructed by forming all finite linear combinations of the elements of $\{x \otimes y : x \in \mathcal{H}, y \in \mathcal{K}\}$. It is clear that the *sesquilinear functional* (from Latin *sesqui* = 'one and a half') $\langle \cdot, \cdot \rangle$ defined by

$$\langle x \otimes y, u \otimes v \rangle = \langle x, u \rangle_{\mathcal{H}} \langle y, v \rangle_{\mathcal{K}}, \quad x, u \in \mathcal{H}, \quad y, v \in \mathcal{K}$$

satisfies the conditions for the inner product. Now, the *tensor product Hilbert space*, denoted by $\mathcal{H} \otimes \mathcal{K}$, is defined as the completion of $\mathcal{E}$ in the norm constructed by the inner product stated above. It is known that if $\{x_j\}$ (resp. $\{y_k\}$) is a complete orthonormal system of $\mathcal{H}$ (resp. $\mathcal{K}$), then $\{x_j \otimes y_k\}$ is also a complete orthonormal system of $\mathcal{H} \otimes \mathcal{K}$.

Let (X, μ) and (Y, ν) be two σ-finite measure spaces, $\mathcal{H}$ be a separable Hilbert space and $\mathcal{E}(X, \mu; \mathcal{H})$ denote the linear space consisting of the Hilbert space valued functions constructed by taking finite linear combinations of the elements of $\{f(\cdot)x : f(\cdot) \in L^2(X, \mu), x \in \mathcal{H}\}$. It is clear that the sesqui-linear functional $\ll \cdot, \cdot \gg$ defined by

$$\ll \sum_{j=1}^{m} f(\cdot)x_j, \sum_{k=1}^{n} g(\cdot)y_k \gg = \sum_{j=1}^{m} \sum_{k=1}^{n} \int_X \overline{f_j} g_k d\mu \langle x_j, y_k \rangle$$

satisfies the conditions for the inner product. Now, the Hilbert space consisting of all square summable Hilbert space valued functions, denoted by $L^2(X, \mu; \mathcal{H})$, is defined by the completion of $\mathcal{E}(X, \mu; \mathcal{H})$ in the norm constructed by the inner product $\ll \cdot, \cdot \gg$.

THEOREM A.1.14. $L^2(X, \mu) \otimes L^2(Y, \nu)$ *is isomorphic to* $L^2(X \times Y, \mu \otimes \nu)$, *and* $L^2(X, \mu) \otimes \mathcal{H}$ *is isomorphic to* $L^2(X, \mu; \mathcal{H})$.

Let $\mathcal{D}$ be a subspace of $\mathcal{H}$ and A be a mapping on $\mathcal{D}$ with values in $\mathcal{K}$. If $A(\alpha x + \beta y) = \alpha A x + \beta B y$ holds for all $\alpha, \beta \in C$ and $x, y \in \mathcal{H}$, A is called a *linear operator*, $\mathcal{D}$ is called the *domain* of A and denoted by dom A and $\{Ax : x \in \text{dom} A\}$ is called the *range* of A and denoted by ran A. The *graph* of A, which is denoted by $\mathcal{G}(A)$ is defined by $\{(x, Ax) : x \in \text{dom } A\}$. If $\mathcal{G}(A)$ is a closed subset of $\mathcal{H} \oplus \mathcal{K}$, then A is called a *closed operator*.

THEOREM A.1.15. $\mathcal{G}(A)$ *is closed iff* $\|x_n - x\| \to 0$ *and* $\|A x_n - y\| \to 0$ *imply that both* $x \in \text{dom} A$ *and* $Ax = y$ *hold*.

Let A and B be linear operators. Then, A is said to *include* B, denoted by $A \subset B$, if $\mathcal{G}(A) \subset \mathcal{G}(B)$ holds. A is called a *closable operator* if there exists a closed operator including A. Especially, the smallest operator including A is called the *closure* of A, which is denoted by $\overline{A}$.

Let A be an operator on $\mathcal{H}$ with values in $\mathcal{H}$ satisfying $\overline{\mathrm{dom}A} = \mathcal{H}$. Then, by Riesz's theorem, for any $y \in \mathrm{dom}A$, there uniquely exists $z \in \mathcal{H}$ satisfying

$$\langle y, Ax \rangle = \langle z, x \rangle, \quad x \in \mathcal{H}.$$

With the use of this result, the *adjoint operator* A^* of A is defined by $A^* y = z$, where $y \in \mathcal{H}$ and z is the unique vector satisfying the above equality.

DEFINITION A.1.16. Let A be an operator with $\overline{\mathrm{dom}A} = \mathcal{H}$. A is said to be *symmetric* if $A \subset A^*$ holds. If a symmetric operator A satisfies $\mathrm{dom}A = \mathrm{dom}A^*$, then A is said to be *selfadjoint*, and if the closure of a symmetric operator A is self adjoint, then A is said to be *essentially selfadjoint*.

It can be easily proved that the closure of a symmetric operator A is equal to A^{**}. The *kernel* of a linear operator A on $\mathcal{H}$, denoted by $\ker A$, is the linear subspace of $\mathcal{H}$ defined as $\{x \in \mathcal{H}; Ax = 0\}$.

THEOREM A.1.17. *The necessary and sufficient conditions which a symmetric operator A should satisfy to be a selfadjoint operator is:*

$$\ker(A^* + iI) = \ker(A^* - iI) = \{0\},$$
$$\mathrm{ran}(A + iI) = \mathrm{ran}(A - iI) = \mathcal{H}.$$

DEFINITION A.1.18. Let A be a linear operator on $\mathcal{H}$. If there exists a positive constant M satisfying $\|Ax\| \leq M\|x\|$ for all $x \in \mathcal{H}$, then A is said to be *bounded*. The set of all bounded linear operators on $\mathcal{H}$ is denoted by $B(\mathcal{H})$. The norm of A, denoted by $\|A\|$, is defined by

$$\|A\| = \sup\{\|Ax\| : \|x\| = 1\}.$$

It can be proved that $B(\mathcal{H})$ equipped with the norm $\|\cdot\|$ is a Banach space.

A bounded operator A is *positive* if it satisfies

$$\langle x, Ax \rangle \geq 0, \quad x \in \mathrm{dom}A.$$

It follows from applying Gelfand's representation theorem (Appendix 2) to the smallest commutative algebra generated by $\{A, I\}$ that there uniquely exists a positive operator B satisfying $A = B^2$. According to the unique existence of B, B is denoted by $\sqrt{A}$ and is called the *square root* of A. Especially, $|A|$ denotes $\sqrt{A^*A}$ and is called the *absolute value* of A.

It should be mentioned that not all positive operators on a real Hilbert space are selfadjoint, but all positive operators on a complex Hilbert space are selfadjoint.

Let A be a bounded operator. Then, A is called a *normal operator* if $A^*A = AA^*$. A is called a *projection* if $A = A^* = A^2$. A is called an *isometry* if $\|Ax\| = \|x\|$ holds for all $x \in \mathcal{H}$. A is called a *unitary operator* if $\mathrm{ran}A = \mathcal{H}$ and $\langle Ax, Ay \rangle = \langle x, y \rangle$ holds for all $x, y \in \mathcal{H}$. A is a *partial isometry* if A is an isometry on $(\ker A)^\perp$. $(\ker A)^\perp$ is called the *initial space* asociated with A and $\mathrm{ran}A$ is called the *final space* associated with A.

For any partial isometry A, it is known that A^*A is the projection whose range is $(\ker A)^\perp$ and AA^* is the projection whose range is $\mathrm{ran}A$.

THEOREM A.1.19. *For any bounded operator A, there uniquely exists a partial isometry W satisfying the following equations:*

$$
\begin{aligned}
A &= W|A|, \\
|A| &= W^*A, \\
A^* &= W^*|A^*|, \\
|A^*| &= W|A|W^*.
\end{aligned}
$$

Moreover, the initial space associated with the partial isometry W is $\overline{\mathrm{ran}|A|}$ and the final space associated with W is $\overline{\mathrm{ran}A}$. If either A is normal or $\mathcal{H}$ is finite-dimensional, then W is unitary.

DEFINITION A.1.20. Let A be a linear operator on $\mathrm{dom}A$ with the values in $\mathrm{ran}A$. Then, A is said to be *invertible* if A is injective. A^{-1} denotes the inverse operator of A on $\mathrm{ran}A$ with the values in $\mathrm{dom}A$. Here, the set of all *resolvents* of A is defined by

$$\mathrm{Re}(A) = \{\lambda \in \mathbb{C}: \ (A - \lambda I)^{-1} \text{ is bounded}\}.$$

The set of all spectral points, called the *spectrum*, is defined by

$$\mathrm{Sp}(A) = \mathbb{C} - \mathrm{Re}(A).$$

Moreover, the spectrum can be devided into the following three sets:

(1) The *point spectrum* of A is defined by

$$\mathrm{Sp}^{(p)}(A) = \{\lambda \in \mathbb{C}: (A - \lambda I)^{-1} \text{ does not exist}\}.$$

(2) The *continuous spectrum* of A is defined by

$$\mathrm{Sp}^{(c)} = \{\lambda \in \mathbb{C}: (A - \lambda I)^{-1} \text{ is unbounded}, \ \overline{\mathrm{ran}(A - \lambda I)} = \mathcal{H}\}.$$

(3) The *residual spectrum* of A is defined by

$$\mathrm{Sp}^{(r)} = \{\lambda \in \mathbb{C} : (A - \lambda I)^{-1} \text{ is unbounded, } \overline{\mathrm{ran}(A - \lambda I)} \neq \mathcal{H}\}.$$

It is clear that the following equalities hold:

$$\mathrm{ran}(A - \lambda I) = \mathrm{dom}(A - \lambda I)^{-1},$$
$$\mathrm{Sp}(A) = \mathrm{Sp}^{(p)}(A) \cup \mathrm{Sp}^{(c)}(A) \cup \mathrm{Sp}^{(r)}(A).$$

DEFINITION A.1.21. Let A be an operator on $\mathcal{H}$. Then, a complex number λ is called an *approximate spectral point* if there exists a sequence $\{x_n\}$ such that

$$\forall \|x_n\| \geq 1, \quad n \in \mathbb{N}, \lim_{n \to \infty} \|(A - \lambda I)x_n\| = 0.$$

Let $\mathrm{Sp}^{(ap)}(A)$ denotes the set of all approximate spectral points. Then, it is easily proved that

$$\mathrm{Sp}^{(p)}(A) \cup \mathrm{Sp}^{(c)}(A) \subset \mathrm{Sp}^{(ap)}(A) \subset \mathrm{Sp}(A),$$

and if $\mathrm{Sp}^{(r)}(A)$ is empty, then $\mathrm{Sp}^{(ap)}(A) = \mathrm{Sp}(A)$ holds.

An element of $\mathrm{Sp}^{(p)}(A)$ is called a *proper value (eigenvalue)* of A, and a non-zero vector x satisfying $Ax = \lambda x$ is called a *proper vector (eigenvector)* associated with λ. The subspace spanned by all proper vectors associated with λ is called the *proper space (eigenspace)* associated with λ, and the dimension of this space is called the *multiplicity* of λ.

DEFINITION A.1.22. Let $\mathcal{H}$ be a Hilbert space, and $\{E_\lambda : -\infty \leq \lambda \leq \infty\}$ be a family of projections on $\mathcal{H}$ indicated by real numbers. Then, $\{E_\lambda : -\infty \leq \lambda \leq \infty\}$ is called a *spectral measure* if it satisfies

$$E_\lambda E_\mu = E_\mu E_\lambda = E_\mu, \quad \mu \leq \lambda,$$
$$\lim_{\lambda \to -\infty} \|E_\lambda x\| = 0, \quad x \in \mathcal{H},$$
$$\lim_{\lambda \to \infty} \|E_\lambda x - x\| = 0, \quad x \in \mathcal{H}.$$

THEOREM A.1.23. *A spectral measure is strongly right continuous and has a left strong limit if $E_{\lambda+\varepsilon} \downarrow E_\lambda$, as ε tends to 0 decreasingly, and there exists a projection denoted by $E_{\lambda-}$ satisfying $E_{\lambda-\varepsilon} \uparrow E_{\lambda-0} \leq E_\lambda$, as ε tends to 0 decreasingly. Moreover, let*

$$\overline{\left\{x \in \mathcal{H} : \int_{-\infty}^{\infty} \lambda d \|E_\lambda x\|^2 < \infty\right\}} = \mathcal{H}$$

is satisfied. If A is the operator defined by

$$\langle x, Ay \rangle = \int_{-\infty}^{\infty} \lambda d \langle x, E_\lambda y \rangle, \quad x, y \in \mathcal{H},$$

then A is selfadjoint.

THEOREM A.1.24 (spectral decomposition theorem). *For any selfadjoint operator A, there uniquely exists a spectral measure E_λ satisfying*

$$\langle x, Ay \rangle = \int_{-\infty}^{\infty} \lambda d \langle x, E_\lambda y \rangle, \quad x, y \in \mathcal{H}.$$

THEOREM A.1.25. $\lambda \in \mathrm{Sp}^{(p)}(A)$ *holds if and only if if $E_{\lambda-0} \neq E_\lambda$ holds, and $\mathrm{ran}(E_\lambda - E_{\lambda-0})$ is exactly equal to the proper space of λ. Moreover, if $\mathrm{Sp}(A) = \mathrm{Sp}^{(p)}(A)$ holds, then $\mathrm{Sp}(A)$ is countable and A can be represented by*

$$A = \sum_{\lambda \in \mathrm{Sp}(A)} \lambda P_\lambda,$$

where P_λ is the projection whose range is $\mathrm{ran}(E_\lambda - E_{\lambda-0})$.

DEFINITION A.1.26. For any $x, y \in \mathcal{H}$, let $x \odot y$ be the operator defined by

$$(x \odot \bar{y})z = \langle y, z \rangle x, \quad z \in \mathcal{H}.$$

Then, $x \otimes \bar{y}$ is bounded and $\mathrm{ran}(x \odot \bar{y}) = \{\lambda x : \lambda \in \mathbb{C}\}$ holds. Moreover, the following equalities hold:

$$
\begin{aligned}
\|x \odot \bar{y}\| &= \|x\| \|y\|, \\
(x \odot \bar{y})^* &= y \odot \bar{x}, \\
(\alpha x) \odot \overline{(\beta y)} &= \alpha \bar{\beta}(x \odot \bar{y}), \\
(x + y) \odot \bar{z} &= x \odot \bar{z} + y \odot \bar{z}, \\
(x \odot \bar{y})(z \odot \bar{w}) &= \langle y, z \rangle x \odot \bar{w}, \\
A(x \odot \bar{y}) &= (Ax) \odot \bar{y}, \quad A \in B(\mathcal{H}), \\
(x \odot \bar{y})A &= x \odot \overline{A^* y}, \quad A \in B(\mathcal{H}).
\end{aligned}
$$

THEOREM A.1.27. *(1) Let A be the operator defined by $\sum_{j \in J} \lambda_j x_j \odot \bar{y_j}$. Then the following equalities hold:*

$$A^* = \sum_{j \in J} \overline{\lambda_j} y_j \otimes \bar{x_j},$$

$$A^*A = \sum_{j \in J} |\lambda_j|^2 y_j \otimes \overline{y_j},$$

$$|A| = \sum_{j \in J} |\lambda_j| y_j \otimes \overline{y_j},$$

$$\|A\| = \|A^*\| = \|A^*A\|^{1/2} = \| \, |A| \, \|.$$

(2) For A of (1), let c and d be positive numbers satisfying $0 < c \leq d$. If for any $j \in J$ $c \leq |\lambda_j| \leq d$ holds, then A is invertible and A^{-1} can be represented by

$$A^{-1} = \sum_{j \in J} \lambda_j^{-1} y_j \otimes \overline{x_j}.$$

THEOREM A.1.28. *For an operator A on $\mathcal{H}$,*

(1) A is a projection operator $\Leftrightarrow$ $\forall$ CONS$\{x_j\} \subset \overline{\mathrm{ran}A}$, $A = \sum_j x_j \otimes \bar{x}_j$.

(2) A is a unitary operator $\Leftrightarrow$ $\exists$ CONS $\{x_j\}$, $\{y_j\} \subset \mathcal{H}$, $A = \sum_j x_j \otimes \bar{y}_j$.

(3) A is an isometric operator $\Leftrightarrow$ $\exists$ ONS $\{x_j\}$, CONS $\{y_j\} \subset \mathcal{H}$, $A = \sum_j x_j \otimes \bar{y}_j$.

(4) A is a semi-isometric operator $\Leftrightarrow$ $\exists$ ONS $\{x_j\}$, $\{y_j\} \subset \mathcal{H}$, $A = \sum_j x_j \otimes \bar{y}_j$.

An operator $A \in B(\mathcal{H})$ satisfying $\dim(\mathrm{ran}A) < \infty$ is called *finite rank*. We denote the set of all finite rank operators by $F(\mathcal{H})$: $A \in F(\mathcal{H}) \Leftrightarrow \exists n \in \mathbb{N}$, $\{\lambda_j\} \subset \mathbb{R}$, CONS $\{x_j\}$, $\{y_j\}$, $A = \sum_{j=1}^{n} \lambda_j x_j \otimes \bar{y}_j$.

An operator $A : \mathcal{H}_1 \to \mathcal{H}_2$ between two Hilbert spaces $\mathcal{H}_1$ and $\mathcal{H}_2$ is *completely continuous* (c.c. for short) or *compact* if the image $A\mathcal{L}$ of a bounded set $\mathcal{L} \subset \mathcal{H}$ through A is totally bounded. We denote a set of all c.c. operators by $C(\mathcal{H}_1, \mathcal{H}_2)$. Then the following statements are satisfied.

1° A is c.c. $\Leftrightarrow$ if $x_n \overset{w}{\to} x$ (i.e., $\langle y, x_n \rangle \to \langle y, x \rangle$, $\forall y \in \mathcal{H}$, *weak convergence*), implies $Ax_n \overset{s}{\to} Ax$ (i.e., $\|Ax_n - Ax\| \to 0$, *strong convergence*).

2° $\{A_n\} \subset C(\mathcal{H}_1, \mathcal{H}_2)$ and $A_n \overset{u}{\to} A$ (i.e., *operator (uniform) norm* $\|Q\| \equiv \sup \{\|Qx\| \, ; \, x \in \mathcal{H}_1, \|x\| = 1\}$, $\|A_n - A\| \to 0$, *uniform convergence*, or *norm convergence*) and if $A \in B(\mathcal{H}_1, \mathcal{H}_2)$ (i.e., the set of all bounded linear operators from $\mathcal{H}_1$ to $\mathcal{H}_2$), then $A \in C(\mathcal{H}_1, \mathcal{H}_2)$.

3° $C(\mathcal{H}_1, \mathcal{H}_2)$ is a closed subspace in Banach space $B(\mathcal{H}_1, \mathcal{H}_2)$ w.r.t. the operator norm, therefore $C(\mathcal{H}_1, \mathcal{H}_2)$ is a Banach space itself.

In the following statements, we assume $\mathcal{H}_1 = \mathcal{H}_2 = \mathcal{H}$, hence, $C(\mathcal{H}_1, \mathcal{H}_2) = C(\mathcal{H}, \mathcal{H}) \equiv C(\mathcal{H})$.

4° $C(\mathcal{H})$ is a two-sided ideal of $B(\mathcal{H})$ (i.e., $\forall A \in C(\mathcal{H})$, $\forall B \in B(\mathcal{H}) \Rightarrow AB$, $BA \in C'(\mathcal{H})$.

5° If one of A, A^*, $|A|$ and $|A^*|$ for $A \in B(\mathcal{H})$ is c.c., then the other ones are also c.c.

6° For any sequence of points $\{\lambda_j\} \subset \mathbb{C}$ satisfying $\lambda_j \to 0$ and two ONS$\{x_j\}$, $\{y_j\}$, $A = \sum \lambda_j x_j \otimes \bar{y}_j$ is c.c.

7° If $A \in C(\mathcal{H})$, then there exists $x \in \mathcal{H}$ such that

$$|\langle x, Ax \rangle| = \max \{|\langle y, Ay \rangle| : \ y \in \mathcal{H}, \|y\| = 1\}.$$

Moreover, if $A \in C(\mathcal{H})$ is selfadjoint, then there exists $x \in \mathcal{H}$ such that $\|A\| = |\langle x, Ax \rangle|$. Here x is the eigenvector of A. That is, $Ax = \lambda x$ ($\lambda = \|A\|$, $-\|A\|$).

THEOREM A.1.29 (Eigenvalue expansion theorem). *A selfadjoint operator $A \in C(\mathcal{H})$ satisfies the following properties:*

(1) $\mathrm{Sp}(A) = \mathrm{Sp}^{(p)}(A) \cup \{0\} \subset \mathbb{R}$ *and the set* $\mathrm{Sp}(A)$ *is at most countable. Eigenvalues except 0 have finite multiplicity and can be arranged as follows:*

$$\lambda_1 : \ |\lambda_1| = \|A\| \ (\text{multiplicity} = n_1) \Rightarrow \lambda_1 = \lambda_2 = \cdots = \lambda_{n_1};$$

$$\lambda_{n_1+1} : \ |\lambda_{n_1+1}| = \max \{|\lambda| : \lambda \in \mathrm{Sp}(A) - \{\lambda_1\}\} \ (\text{multiplicity} = n_2)$$

$$\Rightarrow \lambda_{n_1+1} = \lambda_{n_1+2} = \cdots = \lambda_{n_1+n_2}$$

Then we determine λ_n recursively. The sequence $\{\lambda_1, \lambda_2, \ldots\}$ obtained is finite or $|\lambda_n| \downarrow 0$.

(2) If we can choose the eigenvectors x_n ($\|x\| = 1$) for every $\lambda_n \in \mathrm{Sp}^{(p)}(A)$ such that $x_n \perp x_m$ ($m \neq n$), then $A = \sum \lambda_n x_n \odot \bar{x}_n$. If every eigenvalue is nondegenerate, then this decomposition is unique.

(3) If $\mathcal{L} = \overline{\mathrm{span}}\{x_n : n = 1, 2, \cdots\}$ and $K = \{z \in \mathcal{H}; \ Az = 0\}$, then

$$\mathcal{L} = K^\perp, \quad y = \sum \langle x_n, y \rangle x_n + P_K y, \quad Ay = \sum \lambda_n \langle x_n, y \rangle x_n \quad (\forall y \in \mathcal{H}).$$

THEOREM A.1.30 (Mini-max theorem). *When $A \in C(\mathcal{H})$ is positive, the eigenvalues $\{\lambda_n\}$ of A can be calculated in the following mini-max form:*

$$\lambda_n = \min \{\max \{\langle Ax, x \rangle : \ \|x\| = 1, x \perp \mathcal{L}\} : \dim \mathcal{L} = n - 1\}.$$

Here $\mathcal{L}$ is an $(n-1)$-dimensional closed subspace (λ_1 is the eigenvalue calculated in 7°).

8° *When A, $B \in C(\mathcal{H})$ are selfadjoint operators, the following relation is satisfied:*

$$AB = BA \Leftrightarrow \exists \mathrm{ONS} \{x_n\} \subset \mathcal{H} \text{such that}$$

$$A = \sum_n \lambda_n x_n \odot \overline{x_n}, \quad B = \sum_n \mu_n x_n \odot \overline{x_n}.$$

$9°$ *For any $A \in C(\mathcal{H})$ there exists a unique point sequence $\{\lambda_n\}$ such that $0 \leq \lambda_n \downarrow 0$ (λ_n: eigenvalue of $|A|$). Then $A = \sum_n \lambda_n x_n \otimes \overline{y_n}$ for ONS $\{x_n\}$, $\{y_n\}$.*

$10°$ *If a selfadjoint operator A satisfies $A^n \in C(\mathcal{H})$ for a natural number $n \in \mathbb{N}$, then A is c.c. itself.*

DEFINITION A.1.31. For any two CONS $\{x_i\}$, $\{x_j\} \subset \mathcal{H}$, if one of the following three infinite series

$$\sum_j \|Ax_j\|^2, \quad \sum_j \|A^*y_j\|^2, \quad \sum_{i,j} |\langle y_j, Ax_i\rangle|^2$$

converges, then the other ones also do and they have the same limit value. Then $A \in B(\mathcal{H})$ is called a *Schmidt operator (Hilbert-Schmidt operator)* and the set of all Schmidt operators is denoted by $S(\mathcal{H})$.

For any CONS $\{x_i\}$ and $A \in S(\mathcal{H})$, $\|A\|_2 \equiv \sqrt{\sum_j \|Ax_j\|^2}$ is a norm (*Schmidt norm (Hilbert-Schmidt norm)*) and $S(\mathcal{H})$ is a Banach space w.r.t. $\|\cdot\|_2$.

The elements $A, B \in S(\mathcal{H})$ have the following properties :

$1°$ $\|x \otimes \overline{y}\|_2 = \|x\| \|y\|$, $\quad \forall x, y \in \mathcal{H}$.

$2°$ $\|A\| \leq \|A\|_2$, $\|A^*\|_2 = \|A\|_2$, $\|AB\|_2 \leq \|A\| \|B\|_2$, $\|BA\|_2 \leq \|A\| \|B\|_2 \Rightarrow S(\mathcal{H})$ is a two-sided ideal of $B(\mathcal{H})$.

$3°$ $A \in S(\mathcal{H}) \Rightarrow |A| \in S(\mathcal{H})$, $\||A|\|_2 = \|A\|_2$.

For any $A, B \in S(\mathcal{H})$ and any CONS $\{x_j\}$, $\langle\langle A, B\rangle\rangle = \sum_j \langle Ax_j, Bx_j\rangle$ absolutely converges, which is independent of a choice of $\{x_j\}$. Then $S(\mathcal{H})$ is a Hilbert space w.r.t. this inner product $\langle\langle \cdot, \cdot\rangle\rangle$.

This inner product $\langle\langle \cdot, \cdot\rangle\rangle$ has the following properties :

$4°$ $\langle\langle x \otimes \overline{y}, u \otimes \overline{v}\rangle\rangle = \langle x, u\rangle \langle y, v\rangle$, $\forall x, y, u, v \in \mathcal{H}$.

$5°$ For $A, B \in S(\mathcal{H}).C \in B(\mathcal{H})$, $\langle\langle A^*, B^*\rangle\rangle = \langle\langle B, A\rangle\rangle$, $\langle\langle CA, B\rangle\rangle = \langle\langle A, C^*B\rangle\rangle$, $\langle\langle AC, B\rangle\rangle = \langle\langle A, BC^*\rangle\rangle$.

$6°$ $S(\mathcal{H}) \subset C(\mathcal{H})$, so that $A = \sum_j \lambda_j x_j \otimes \overline{y_j}$ for any $A \in S(\mathcal{H})$. Moreover, c.c. operator $A = \sum_j \lambda_j x_j \otimes \overline{y_j} \in S(\mathcal{H}) \Leftrightarrow \sum_j |\lambda_j|^2 < \infty$.

DEFINITION A.1.32. If an infinite series $\sum_j \langle x_j, Ax_j\rangle$ absolutely converges for any CONS $\{x_j\} \subset \mathcal{H}$, then the limit value does not depend on the choice of CONS $\{x_j\}$. Then $A \in B(\mathcal{H})$ is called a *trace class operator* and the set of all trace class operators is denoted by $T(\mathcal{H})$. Note that a positive trace class operator is often called a *density operator* in physics.

$\sum_j \langle x_j, Ax_j\rangle$ is called a *trace* and denoted by $\operatorname{tr} A$.

A trace class operator has the following properties:

1° The following statements are equivalent: (i) $A \in T(\mathcal{H})$, (ii) $|A| \in$
$T(\mathcal{H})$, (iii) $|A|^{1/2} \in S(H)$, (iv) $\operatorname{tr}|A| < \infty$.
2° For any $A, B \in T(\mathcal{H})$ and $\alpha, \beta \in \mathbb{C}$, $\alpha A + \beta B \in T(\mathcal{H})$, $A^* \in T(\mathcal{H})$.

If $\|A\|_1 = \operatorname{tr}|A|$ $(\forall A \in T(\mathcal{H}))$, then $\|\cdot\|_1$ becomes a norm (*trace norm*)
and $T(\mathcal{H})$ becomes a Banach space w.r.t. $\|\cdot\|_1$.

3° $\operatorname{tr} x \otimes \overline{y} = \langle y, x \rangle$, $\|x \otimes \overline{y}\|_1 = \|x\| \|y\|$, $\forall x, y \in \mathcal{H}$.
4° $T(\mathcal{H})$ is a two sided ideal of $B(\mathcal{H})$.
5° $\|A\| \leq \|A\|_2 \leq \|A\|_1$.
6° $A \in T(\mathcal{H}) \Leftrightarrow \exists B, C \in S(\mathcal{H})$ such that $A = BC$.
7° $A = \sum_j \lambda_j x_j \otimes y_j \in C(\mathcal{H})$ $(\lambda_j \geq 0)$. $A \in T(\mathcal{H}) \Leftrightarrow \sum_j \lambda_j < \infty$. Then
$$\|A\|_1 = \sum_j \lambda_j.$$

4° and 5° imply $T(\mathcal{H}) \subset S(\mathcal{H})$

THEOREM A.1.33. $F(\mathcal{H}) \subset T(\mathcal{H}) \subset S(\mathcal{H}) \subset C(\mathcal{H}) \subset B(\mathcal{H})$.

1° $F(\mathcal{H})$ is the minimal ideal in other spaces. $C(\mathcal{H})$ is an ideal closed by
a uniform norm in $B(\mathcal{H})$. Moreover, if $\mathcal{H}$ is separable, then $C(\mathcal{H})$ is
the maximal ideal.
2° (i) $\overline{F(\mathcal{H})}^{\|\cdot\|_1} = T(\mathcal{H})$ (ii) $\overline{F(\mathcal{H})}^{\|\cdot\|_2} = S(\mathcal{H})$ (iii) $\overline{F(\mathcal{H})}^{\|\cdot\|} = C(\mathcal{H})$.
3° Each $F(\mathcal{H}), T(\mathcal{H}), S(\mathcal{H}), C(\mathcal{H})$ is dense w.r.t. the weak topology
$(A_n \xrightarrow{w} A)$ in $B(\mathcal{H})$.

Let X^* be the set of all bounded linear functionals φ (i.e., $\varphi : X \to \mathbb{C}$
is linear and $|\varphi(A)| \leq M \|A\|$, $0 < M < +\infty$) on a Banach space X. Then
the following statements are held:
(1) For any $\varphi \in C(\mathcal{H})^*$ there exsits a unique $T_\varphi \in T(\mathcal{H})$ such that $\varphi(A) = \operatorname{tr} T_\varphi A$ $(\forall A \in C(\mathcal{H}))$.
(2) If $\varphi_T(A) = \operatorname{tr} TA$ $(\forall A \in C(\mathcal{H}))$ for any $T \in T(\mathcal{H})$, then $\varphi_T \in C(\mathcal{H})^*$.
(3) $\|T\|_1 = \|\varphi_T\| \equiv \sup \{|\varphi_T(A)| : A \in C(H), \|A\| = 1\}$
(4) For any $\psi \in T(\mathcal{H})^*$ there exsits a unique $B_\psi \in B(\mathcal{H})$ such that $\psi(A) = \operatorname{tr} AB_\psi$ $(\forall A \in T(\mathcal{H}))$.
(5) If $\psi_B(A) = \operatorname{tr} AB$ $(\forall A \in T(\mathcal{H}))$ for any $B \in B(\mathcal{H})$, then $\psi_B \in T(\mathcal{H})^*$
and $\|B\| = \|\psi_B\|$.

$T \in T(\mathcal{H})$ and $B \in B(\mathcal{H})$ can be identified with $\varphi_T \in C(\mathcal{H})^*$ and
$\psi_B \in T(\mathcal{H})^*$, respectively, so that we have the following theorem:

THEOREM A.1.34. *(1)* $C(\mathcal{H})^* = T(\mathcal{H})$.
(2) $T(\mathcal{H})^* = B(\mathcal{H})$.
(3) $S(\mathcal{H})^* = S(\mathcal{H})$.
(4) $B(\mathcal{H})^* = T(\mathcal{H}) \oplus C(\mathcal{H})^\perp$, *where*
$$C(\mathcal{H})^\perp = \{\varphi \in B(\mathcal{H})^* : \varphi(A) = 0. \forall A \in C(\mathcal{H})\}.$$

Appendix 2

A.2 Elements of Operator Algebras

In Appendix 1, we discussed the Hilbert space method to describe quantum systems. Presently, a more general method, the operator algebraic approach is considered. Namely, we explain some of fundamental facts of C^*-algebras and Von Neumann algebras. The operator algebraic approach to quantum systems is particularly important when a quantum system has infinitely many degrees of freedom. In physics, one considereds several systems with an infinite number of degrees of freedom, such as quantum fields, quantum statistical systems with symmetry breaking, etc.

Let A be a linear space on the set of complex numbers $\mathbb{C}$. A is said to be an *algebra* if it satisfies: (i) $AB \in A$ for any $A, B \in A$; (ii) $(A+B)+C = A+(B+C)$, $(AB)C = A(BC)$ for any $A, B, C \in A$.

An algebra A having the *involution* $*$ from A to A such that (i) $(A^*)^* = A$, (ii) $(A + \lambda B)^* = A^* + \overline{\lambda} B^*$, (iii) $(AB)^* = B^* A^*$ for any $A, B \in A$ and $\lambda \in C$, is called a *$*$-algebra* on $\mathbb{C}$.

Moreover, the *norm* $\|\cdot\|$ on A is a mapping from A to $\mathbb{R}^+$ satisfying (i) $\|A\| \geq 0, \|A\| = 0 \Leftrightarrow A = 0$, (ii) $\|\lambda A\| = |\lambda| \|A\|$, (iii) $\|A + B\| \leq \|A\| + \|B\|$ for $A, B \in A$ and $\lambda \in \mathbb{C}$.

DEFINITION A.2.1. Let A be an algebra with norm $\|\cdot\|$.

(1) A is a *normed algebra* if $\|AB\| \leq \|A\| \|B\|$ for any $A, B \in A$.
(2) A is a *Banach algebra* if A is a complete normed algebra w.r.t. $\|\cdot\|$.
(3) A is a *Banach $*$-algebra* ($B^*-algebra$) if A is a Banach algebra with $\|A^*\| = \|A\|$ for any $A \in A$.
(4) A is a *C^*-algebra* if A is a B^*-algebra with $\|A^* A\| = \|A\|^2$ for any $A \in A$.

Let us give some examples for the above algebras.

(1) The set $M_n(\mathbb{C})$ of $n \times n$ matrices with complex entries is a C^*-algebra.
(2) The set $B(\mathcal{H})$ of all bounded linear operators on a Hilbert space $\mathcal{H}$ and the set $C(\mathcal{H})$ of all compact operators on $\mathcal{H}$ are C^*-algebras.
(3) The set $T(\mathcal{H})$ of all trace class operators on $\mathcal{H}$ and the set $S(\mathcal{H})$ of all Hilbert-Schmidt operators are B^*-algebras, but not C^*-algebras.
(4) The set $C(X)$ of all continuous functions on a locally compact Hausdorff space X is a C^*-algebra.

Now take a subset A of $B(\mathcal{H})$ and define the *commutant* of A by

$$A' \equiv \{A \in B(\mathcal{H}) : AB = BA, B \in A\},$$

and the *double commutant* A'' by

$$A'' = (A')'$$

DEFINITION A.2.2. A is a *Von Neumann algebra* (*VN-algebra* for short) if $A'' = A$.

Remark that a VN-algebra is a C^*-algebra. It is easily shown that
(1) $A'' \supset A$,
(2) $A' = A^{(3)} = A^{(5)} = \ldots$,
(3) $A'' = A^{(4)} = A^{(6)} = \ldots$.

Let A be a C^*-algebra with an identity I (i.e., $AI = IA = A, \forall A \in A$). Any C^*-algebra without I can be always extended to a C^*-algebra with I.

DEFINITION A.2.3. Let A, B be elements of A and λ, μ be elements of $\mathbb{C}$.
(1) A is *regular* (*invertible*) if there exists $A^{-1} \in A$ satisfying $AA^{-1} = A^{-1}A = I$.
(2) *Resolvent set*: $\mathrm{Re}(A) \equiv \left\{\lambda \in \mathbb{C} : \exists (A - \lambda I)^{-1} \in A\right\}$.
(3) *Spectral set*: $\mathrm{Sp}(A) \equiv \mathbb{C} - \mathrm{Re}(A)$.
(4) The mapping χ from A to $\mathbb{C}$ is a *character* if χ is a homomorphism not identically 0. The set $\mathrm{Sp}(A)$ of all characters of A is called a *character space* or a *spectral space*.
(5) The mapping $\hat{A}$ from $\mathrm{Sp}(A)$ to $\mathbb{C}$ is defined by $\hat{A}(\chi) = \chi(A)$, and the correspondence $\wedge : A \to \hat{A}$ is called a *Gelfand representation*.

Note that $\lambda \in \mathrm{Sp}(A)$ iff there exists a character χ such that $\chi(A) = \lambda$, which is a reason that we call $\mathrm{Sp}(A)$ a spectral space.

THEOREM A.2.4. *The following statements are equivalent:*
(1) A is an abelian (i.e., $AB = BA, \forall A, B \in A$) C^*-algebra.
(2) The mapping $\wedge$ from A to $C(Sp(A))$ (the set of all continuous functions on $Sp(A)$) is a $*$-isometry (i.e., $\hat{A}^* = \left(\hat{A}\right)^*$, $\|\hat{A}\| = \|A\|$).

By the above theorem, A and $C(\mathrm{Sp}(A))$ are identified. $\mathrm{Sp}(A)$ is a weak $*$-compact Hausdorff space, and it becomes locally compact when $I \in A$.

The mapping φ from A to $\mathbb{C}$ is a *linear functional* on A if $\varphi(\lambda A + \mu B) = \lambda\varphi(A) + \mu\varphi(B)$, and a linear functional φ from A to C is said to be *positive* if $\varphi(A^*A) \geq 0$ for any $A \in A$. The *norm* $\|\cdot\|$ of φ is defined by

$$\|\varphi\| \equiv \sup\{|\varphi(A)| : A \in A, \|A\| \leq 1\}.$$

For a linear functional φ, the following theorems are satisfied:

1° *Schwarz's inequality:* $|\varphi(A^*B)|^2 \leq |\varphi(A^*A)| |\varphi(B^*B)|$.

2° $A = I \Rightarrow \|\varphi\| = \varphi(I)$.

φ is *faithful* if $\varphi(A^*A) = 0 \Rightarrow A = 0$. Let A^*, called the *dual*, be the set of all continuous linear functionals φ on A (i.e., $\|A_n - A_m\| \to 0 \Rightarrow |\varphi(A_n) - \varphi(A_m)| \to 0$), and A_+^* be the set of all positive φ in A^*.

DEFINITION A.2.5. :

(1) $\varphi \in A_+^*$ is a *state* on A if $\|\varphi\| = 1$. The set of all states on A is denoted by $\mathfrak{S}(A)$ or $\mathfrak{S}$ for simplicity.

(2) For $\varphi_1, \varphi_2 \in A_+^*$, φ_1 *dominates* φ_2 (denoted by $\varphi_1 \geq \varphi_2$) if $\varphi_1 - \varphi_2 \in A_+^*$.

(3) $\varphi \in \mathfrak{S}$ is a *mixed state* if there exist $\lambda \in \mathbb{R}^+$ and $\psi \in \mathfrak{S}$ satisfying $\lambda\varphi \geq \psi$, and φ is a *pure state* if φ is not mixed state.

When A contains I, φ is a mixed state iff there exists $\lambda \in (0,1)$ and $\psi_1, \psi_2 \in \mathfrak{S}$ such that $\psi_1 \neq \psi_2$, $\varphi = \lambda\psi_1 + (1-\lambda)\psi_2$.

The topology on A^* is usually defined by one of the following two ε-neighborhoods $N_\varepsilon(\varphi)$ of φ for any $\varphi \in A^*$:

$$N_\varepsilon(\varphi) \equiv \{\psi \in A^* : \|\varphi - \psi\| < \varepsilon\},$$

and

$$N_\varepsilon(\varphi) \equiv \{\psi \in A^* : |\varphi(A_k) - \psi(A_k)| < \varepsilon, A_1, \cdots, A_n \in A\}.$$

The first topology is called the *uniform topology* and the second is the *weak *-topology* or $\sigma(A^*, A)$-*topology*.

Below we list up some fundamental properties of the state space $\mathfrak{S}$ as 1° and 2° above.

3° When $I \in A$, $\mathfrak{S}$ is weak *-compact.

4° By using the Krein-Milman theorem we obtain $\mathfrak{S} = \mathrm{w}^* \overline{\mathrm{co}} \, \mathrm{ex} \, \mathfrak{S}$, where $\mathrm{ex} \, \mathfrak{S}$ is the set of all extreme points of $\mathfrak{S}$ (an *extreme point* of $\mathfrak{S}$ is an element not represented by a convex combination of two other elements of $\mathfrak{S}$). When $I \in A$, $\mathrm{ex} \, \mathfrak{S} = \mathfrak{S}_p$, i.e., the set of all pure states. Moreover, $\mathrm{w}^* \overline{\mathrm{co}} \, \mathrm{ex} \, \mathfrak{S} \equiv \{\Sigma_n \lambda_n \varphi_n : \Sigma_n \lambda_n = 1, \lambda_n \geq 0, \varphi_n \in \mathrm{ex} \, \mathfrak{S}\}^{-\mathrm{w}^*}$, where $-w*$ means the closure of the set $\{\dots\}$ with respect to the weak*-topology.

Let α be a mapping from a one-parameter group G to $\mathrm{Aut}(A)$, which is the set of all automorphisms on A satisfying the following conditions :

(i) $\alpha_{t+s} = \alpha_t \alpha_s$,

(ii) $\lim_{t \to 0} \|\alpha_t(A) - A\| = 0$ (*strong continuity*),

(iii) $\alpha_t(A^*) = \alpha_t(A)^*$ $(\forall t \in G, \forall A \in A)$.

The *general quantum dynamical system* (GQDS) is described by a triple $(A, \mathfrak{S}, \alpha)$ or $(A, \mathfrak{S}, \alpha(G))$. This description should contain the usual Hilbert space description as a special case. This is assured by the following *GNS (Gelfand-Naimark-Segal) theorem:*

THEOREM A.2.6 (GNS). *For any $\varphi \in \mathfrak{S}$, there uniquely exist the following items up to the unitary equivalence:*

(1) GNS Hilbert space $\mathcal{H}_\varphi$;
(2) GNS representation π_φ (i.e., π_φ is a *-homomorphism, i.e., $\pi_\varphi(AB) = \pi_\varphi(A)\pi_\varphi(B)$, $\pi_\varphi(A^*) = \pi_\varphi(A)^*$) from $\mathcal{A}$ to $B(\mathcal{H})$;
(3) GNS cyclic vector $x_\varphi \in \mathcal{H}_\varphi$ (i.e., $\{\pi_\varphi(A)x_\varphi; A \in \mathcal{A}\}^- = \mathcal{H}_\varphi$) satisfying $\varphi(A) = \langle x_\varphi, \pi_\varphi(A)x_\varphi \rangle$.

THEOREM A.2.7. *For an α-invariant state φ (i.e., $\varphi(\alpha_t(A)) = \varphi(A)$, $\forall A \in \mathcal{A}$, $\forall t \in \mathbb{R}$), there exists a strongly continuous unitary one-parameter group $\{u_t : t \in \mathbb{R}\}$ satisfying the following equalities:*

(1) $u_t x_\varphi = x_\varphi$,
(2) $\pi_\varphi(\alpha_t(A)) = u_t \pi_\varphi(A) u_{-t}$.

From the above theorems, the usual Hilbert space description is reconstructed if it is needed. Moreover, if we take a sufficiently large Hilbert space, then we have

THEOREM A.2.8. *For general C^*-algebra $\mathcal{A}$, there exists a Hilbert space $\mathcal{H}$ such that a C^*-algebra B ($\subset B(\mathcal{H})$) on $\mathcal{H}$ is isomorphic to $\mathcal{A}$.*

Next we discuss some topologies on $B(\mathcal{H})$. Let $\{A_\lambda\}$ be a net of $B(\mathcal{H})$, and "$A_\lambda \xrightarrow{\tau} A$" means that A_λ converges to A as $\lambda \to$ "∞" with respect to the (operator) topology τ:

(1) *Uniform (operator) topology τ^u*: $\overset{\text{def}}{\Longleftrightarrow} \|A_\lambda - A\| \to 0$.
(2) *Strong (operator) topology τ^s*: $\overset{\text{def}}{\Longleftrightarrow} \|(A_\lambda - A)x\| \to 0$, $\forall x \in \mathcal{H}$.
(3) *Weak (operator) topology τ^w*: $\overset{\text{def}}{\Longleftrightarrow} \langle x, (A_\lambda - A)y \rangle \to 0$, $\forall x, y \in \mathcal{H}$.
(4) *Ultrastrong (operator) topology $\tau^{u.s}$*: $\overset{\text{def}}{\Longleftrightarrow}$ For any $\{x_n\} \subset \mathcal{H}_F \equiv \{\{x_n\} \subset \mathcal{H} : \sum_{n=1}^\infty \|x_n\|^2 < +\infty, \sum_{n=1}^\infty \|(A_\lambda - A)x_n\|^2 \to 0\}$.
(5) *Ultraweak (operator) topology $\tau^{u.w}$*:
$\overset{\text{def}}{\Longleftrightarrow} \sum_n |\langle x_n, (A_\lambda - A)y_n \rangle| \to 0$, $\forall \{x_n\}, \{y_n\} \in \mathcal{H}_F$.
(6) *Strong* (operator) topology τ^{s*}*:
$\overset{\text{def}}{\Longleftrightarrow} \|(A_\lambda - A)x\|^2 + \|(A_\lambda^* - A^*)x\|^2 \to 0$, $\forall x \in \mathcal{H}$.
(7) *Ultrastrong* (operator) topology τ^{us*}*:
$\overset{\text{def}}{\Longleftrightarrow} \sum_n \{\|(A_\lambda - A)x_n\|^2 + \|(A_\lambda^* - A^*)x_n\|^2\} \to 0$, $\forall \{x_n\} \subset \mathcal{H}$.

These topologies are related as follows:

$$\begin{array}{ccccc} \tau^u & > & \tau^{us*} & > & \tau^{u.s} & > & \tau^{u.w} \\ & & \vee & & \vee & & \vee \\ & & \tau^{s*} & > & \tau^s & > & \tau^w \end{array}$$

("strong" topology > "weak" topology). When $\dim \mathcal{H} < +\infty$, every topology coincides each other.

Denote by $\mathcal{B}$ a uniformly bounded subset of $B(\mathcal{H})$, for example, the unit ball $\mathcal{B}_1 \equiv \{A \in B(\mathcal{H}) : \|A\| \leq 1\}$. On this ball, $\tau^{u.w} = \tau^w, \tau^{u.s} = \tau^s$, $\tau^{us*} = \tau^{s*}$.

When $A_\lambda \to A$, $B_\lambda \to B$ in a certain topology τ, we shall show whether the following propositions hold or not: (a) $A_\lambda^* \to A^*$; (b) $A_\lambda Q \to AQ$ and $QA_\lambda \to QA$, $\forall Q \in B(\mathcal{H})$; (c) $A_\lambda B_\lambda \to AB$; (d) $A_\lambda B_\lambda \to AB$ for $\{A_\lambda\}, \{B_\lambda\} \subset \mathcal{B}$.

We write $\bigcirc$ if the proposition holds, and write $\times$ if it does not.

τ	(a)	(b)	(c)	(d)
τ^u	$\bigcirc$	$\bigcirc$	$\bigcirc$	$\bigcirc$
τ^s	$\times$	$\bigcirc$	$\times$	$\bigcirc$
τ^w	$\bigcirc$	$\bigcirc$	$\times$	$\times$
$\tau^{u.s}$	$\times$	$\bigcirc$	$\times$	$\bigcirc$
$\tau^{u.w}$	$\bigcirc$	$\bigcirc$	$\times$	$\times$
τ^{s*}	$\bigcirc$	$\bigcirc$	$\times$	$\bigcirc$
τ^{us*}	$\bigcirc$	$\bigcirc$	$\times$	$\bigcirc$

Note that $\tau_1 < \tau_2$ implies $\overline{\mathfrak{M}}^{\tau_2}$ (the closure of τ_2-topology) $\subset \overline{\mathfrak{M}}^{\tau_1}$. Let $\mathfrak{M}$ be a *-algebra such that $\overline{\mathfrak{M}}^{u.s} = \mathfrak{M}$ for an algebra $\mathfrak{M} \subset B(\mathcal{H})$. A linear functional φ is τ-continuous if $A_\lambda \to A(\tau)$ implies $\varphi(A_\lambda - A) \to 0$. When $\tau_1 < \tau_2$, τ_1-continuity of φ implies τ_2-continuity of φ.

Denote by $\mathfrak{M}^*$ the set of all τ^u-continuous linear functionals. Put $\mathfrak{M}^*_+ \equiv \{\varphi \in \mathfrak{M}^* : \varphi(A^*A) \geq 0\}$, $\mathfrak{S} \equiv \{\varphi \in \mathfrak{M}^*_+ : \|\varphi\| = 1\}$, $\mathfrak{M}_\sim \equiv \{\varphi \in \mathfrak{M}^* : \tau^w$-continuous$\}$, $\mathfrak{M}_* \equiv \{\varphi \in \mathfrak{M}^*; \tau^{uw}$-continuous$\}$.

We have the following properties:

5° $\overline{\mathfrak{M}_\sim} = \mathfrak{M}^*$, where the closure is taken for the norm $\|\cdot\|$ on $\mathfrak{M}^*$.

6° $\varphi \in \mathfrak{M}_\sim \Leftrightarrow \exists \{x_n\}_{n=1}, \{y_n\}_{n=1} \subset \mathcal{H}$, and $\exists N \in \mathbb{N}$ such that $\varphi(A) = \sum_{n=1}^{N} \langle x_n, Ay_n \rangle$.

7° $\varphi \in \mathfrak{M}^* \Leftrightarrow \exists \{x_n\}_{n=1}, \{y_n\}_{n=1} \subset \mathcal{H}$, and such that $\sum_{n=1}^{\infty} \|x_n\|^2 < +\infty$, $\sum_{n=1}^{\infty} \|y_n\|^2 < +\infty$, $\varphi(A) = \sum_n \langle x_n, Ay_n \rangle$.

When $\mathfrak{M}$ is identical with $B(\mathcal{H})$, we have $\mathfrak{M}_* = T(\mathcal{H})$ (trace class). This identity is particularly important.

THEOREM A.2.9. $\mathfrak{M}^*$ is a Banach space with respect to the norm of $\mathfrak{M}^*$, and $(\mathfrak{M}^*)^* = \mathfrak{M}$.

DEFINITION A.2.10. $\varphi \in \mathfrak{M}^*$ is normal if $0 \leq A_\lambda \uparrow A$ implies $\varphi(A_\lambda) \uparrow \varphi(A)$.

The following two theorems are essential.

THEOREM A.2.11. *For* $\varphi \in \mathfrak{M}_+^*$, *the following statements are equivalent:*

(1) $\varphi \in \mathfrak{M}^*$.
(2) φ *is normal.*
(3) *There exists* $\rho \in T(\mathcal{H})_+$ *such that* $\varphi(A) = \mathrm{tr}\rho A$.

THEOREM A.2.12 (Von Neumann density theorem). *When* $\mathfrak{M}$ *is a* $*$-*algebra with* I, *the following equations are equivalent:*

(1) $\mathfrak{M}'' = \mathfrak{M}$.
(2) $\overline{\mathfrak{M}}^w = \mathfrak{M}$.
(3) $\overline{\mathfrak{M}}^{us*} = \mathfrak{M}$.

In the sequel, we assume that $\mathfrak{M}$ is a Von Neumann algebra.

DEFINITION A.2.13. $\mathfrak{M}$ is said to be σ-*finite* if any family of mutual orthogonal projections is countable.

THEOREM A.2.14. *The following statements are equivalent:*

(1) $\mathfrak{M}$ *is* σ-*finite.*
(2) *There exists a faithful normal state on* $\mathfrak{M}$.
(3) $\mathfrak{M}$ *is isomorphic to a Von Neumann algebra* $\mathfrak{M}_1$ *having a separating and cyclic vector* x *(see below).*

For a Von Neumann algebra $\mathfrak{M}$ over a Hilbert space $\mathcal{K}$, a vector $x \in \mathcal{K}$ is *cyclic* if $\overline{\mathfrak{M}x} = \mathcal{K}$, and is *separating* if $Ax = 0$ implies $A = 0$.
Now one can easily prove the following two propositions:

8° If $\mathcal{H}$ is separable, then $\mathcal{H}$ is σ-finite.
9° If φ is a faithful normal state, then $\pi_\varphi(\mathfrak{M})'' = \pi_\varphi(\mathfrak{M}) \cong \mathfrak{M}, \mathcal{H}_\varphi \cong \mathcal{H}$.

8° implies that almost all Von Neumann algebras used in physics are σ-finite.
Let $\mathcal{P}(\mathfrak{M})$ be a set of all projections on $\mathfrak{M}$, and $E, F \in \mathcal{P}(\mathfrak{M})$.
E *dominates* F (denoted by $F \leq E$) if $\mathrm{ran}F \subset \mathrm{ran}E$, and E is *equivalent* to F (denoted by $E \cong F$) if there exists a partial isometry such as $E = W^*W, F = WW^*$.

DEFINITION A.2.15. (1) E is *finite* if $E \cong F \leq E$ implies $E = F$.
(2) E is *semifinite* if any $F \leq E$ has no finite subprojection.
(3) E is *infinite* if E is not finite.
(4) E is *purely infinite* if there exists no non-zero finite projection F such that $F \leq E$.

DEFINITION A.2.16. Let $\mathfrak{M}$ be a Von Neumann algebra.

(1) $\mathfrak{M}$ is *finite* $\Leftrightarrow I$ is finite.
(2) $\mathfrak{M}$ is *semifinite* $\Leftrightarrow I$ is semi-finite.
(3) $\mathfrak{M}$ is *infinite* $\Leftrightarrow I$ is infinite.
(4) $\mathfrak{M}$ is *purely infinite* $\Leftrightarrow I$ is purely infinite.

Let us introduce the "trace" on a Von Neumann algebra $\mathfrak{M}$. A *trace* τ over $\mathfrak{M}$ is a map from $\mathfrak{M}_+$ to $\overline{\mathbb{R}^+} = [0, \infty]$ satisfying:

(i) $\tau(\lambda A + B) = \lambda\tau(A) + \tau(B)$, $\forall A, B \in \mathfrak{M}_+$, $\lambda \geq 0$.
(ii) $\tau(A^*A) = \tau(AA^*)$, $\forall A \in \mathfrak{M}$.

This τ is not a linear functional on a VN-algebra $\mathfrak{M}$ in general, but it can be extended to a linear functional $\overline{\tau}$ on $\mathfrak{M}$ in the following sense: Put $\mathfrak{M}_+^f \equiv \{A \in \mathfrak{M}_+ : \tau(A) < \infty\}$ and $\mathfrak{M}^f \equiv \{A \in \mathfrak{M} : A^*A \in \mathfrak{M}_+^f\}$, $\mathfrak{M}_t \equiv \{\Sigma_{i=1}^n \lambda_i A_i B_i; A_i, B_i \in \mathfrak{M}^f, n < \infty\}$.

THEOREM A.2.17. *(1)* $\mathfrak{M}_t$ *is a two sided ideal of* $\mathfrak{M}$, *namely,* $\mathfrak{M}_t\mathfrak{M} \subset \mathfrak{M}$ *and* $\mathfrak{M}\mathfrak{M}_t \subset \mathfrak{M}$.
(2) There exists a unique linear functional $\overline{\tau}$ *on* $\mathfrak{M}_t$ *such that* $\overline{\tau} \uparrow \mathfrak{M}_t \cap \mathfrak{M}_+ = \tau$ *and* $\tau(AB) = \tau(BA)$ *($\forall A \in \mathfrak{M}_t, \forall B \in \mathfrak{M}$)*.

We denote $\overline{\tau}$ by the same symbol τ for simplicity.

DEFINITION A.2.18. Let τ be a trace,

(1) τ is *faithful* if $\tau(A) = 0, A \in \mathfrak{M}_+$ implies $A = 0$.
(2) τ is *normal* if $0 \geq A_\lambda \uparrow A$ implies $\tau(A_\lambda) \uparrow \tau(A)$.
(3) τ is *finite* if $\tau(A) \leq \infty, \forall A \in \mathfrak{M}_+$.
(4) τ is *semifinite* if for any $A(\neq 0) \in \mathfrak{M}_+$, there exists $B(\neq 0) \in \mathfrak{M}_+$ such that $0 \leq B \leq A$ and $\tau(B) < +\infty$. (Note that if τ is normal and semifinite, then $\tau(A) = \sup\{\tau(B) : 0 \leq B \leq A, \tau(B) < +\infty\}$)
(5) A family of traces $\{\tau_j : j \in J\}$ is *sufficient* if for any $A(\neq 0) \in \mathfrak{M}_+$, there exists $j \in J$ such that $\tau_j(A) \neq 0$.

THEOREM A.2.19. *(1)* $\mathfrak{M}$ *is finite if* $\mathfrak{M}$ *has a family of sufficient normal finite traces.*
(2) When $\mathfrak{M}$ *is* σ-*finite,* $\mathfrak{M}$ *is finite if* $\mathfrak{M}$ *has a faithful normal finite trace.*
(3) $\mathfrak{M}$ *is semi-finite if* $\mathfrak{M}$ *has a faithful normal semifinite trace.*

Examples of traces:

(1) When $\mathfrak{M} = M_n(C)$, $\tau(A) = \operatorname{tr} A = \Sigma_{i=1}^n a_{ii}$, a_{ii} = diagonal elements of A is a faithful normal finite trace.
(2) When $\mathfrak{M} = B(\mathcal{H})$ and $\dim \mathcal{H} = \infty, \tau(A) = \operatorname{tr} A$ (= $\Sigma_n\langle x_n, Ax_n\rangle$ for a CONS $\{x_n\}$ of $\mathcal{H}$) is a faithful normal semifinite trace.

(3) When $\mathfrak{M} = L^\infty(\Omega, \mu)$ (a finite Von Neumann algebra on $\mathcal{H} = L^2(\Omega, \mu)$), $f \in M(\Omega)$ (the set of all μ-measurable functions) and $f \geq 0$, μ-a.e., put

$$\tau(A) \equiv \int_\Omega A(\omega) f(\omega) d\mu(\omega).$$

Then

(i) $f > 0$, μ-a.e. $\Rightarrow \tau$ is a faithful normal trace.
(ii) $f \in L^1(\Omega, \mu) \Rightarrow \tau$ is a finite trace.
(iii) $f \in L^\infty(\Omega, \mu) \Rightarrow \tau$ is a semifinite trace.

We now explain some representations of commutative C^*-algebras. The fact that $\mathcal{A}$ is commutative and $I \in \mathcal{A}$ implies $\mathcal{A} = C(\Omega)$ with $\Omega = \mathrm{Sp}(\mathcal{A})$. So, for any $\varphi \in \mathcal{A}^*_{+,1} = C(\Omega)^*_{+,1}$, there exists a Baire measure μ such that

$$\varphi(A) \equiv \int_\Omega A(\omega) d\mu(\omega)$$

by the Riesz-Markov-Kakutani theorem.

Moreover, put $\mathcal{H}_\varphi = \mathcal{H}_\mu = L^2(\Omega, \mu)$ and $(\pi_\mu(f)g)(t) = f(t)g(t)$ for any $g \in \mathcal{H}_\mu$. Then

$10°$ $\pi_\mu(C(\Omega))'' = L^\infty(\Omega, \mu)$.

THEOREM A.2.20. *Let $\mathcal{H}$ be a separable Hilbert space and $\mathfrak{M}$ be a commutative Von Neumann algebra. Then there exists a locally compact Hausdorff space Ω satisfying the second countable axiom (i.e., it has a countable basis of open sets) and a probability measure μ such that $\mathfrak{M} = L^\infty(\Omega, \mu)$.*

From these facts, we recognize that the formulation of QPT (*quantum probability theory*) by Von Neumann algebras or C^*-algebras is indeed an extension of CPT (*classical probability theory*).

As explained above, there exist various Von Neumann algebras, and purely infinite Von Neumann algebras are needed in some special situations. Although it is sufficient to treat semifinite Von Neumann algebras over separable Hilbert spaces, such as $B(\mathcal{H})$, a purely infinite Von Neumann algebraic (C^*-algebraic) description is fundamental for the case when a physical system has infinite number of degrees of freedom and some kind of symmetry breaking.

We summarize the description of quantum systems by Von Neumann algebras and C^*-algebras.

⟨1⟩ Description by Von Neumann algebras: The Von Neumann algebraic description of quantum systems is given by 4-tuple $(\mathcal{H}, \mathcal{P}, \mathfrak{M}, \varphi)$:

CPT		QPT
Ω	$\Leftrightarrow$	$\mathcal{H}$
$\mathcal{F}$	$\Leftrightarrow$	$\mathcal{P}_{\mathcal{H}}$
f	$\Leftrightarrow$	A
μ	$\Leftrightarrow$	$\varphi \in \mathfrak{S}$
$L^{\infty}(\Omega)$	$\Leftrightarrow$	$\mathfrak{M}$

⟨2⟩ Description by C^{*}-algebras: The C^{*}-algebraic description of quantum systems is given by $(\mathcal{A}, \varphi)$:

CPT		QPT
Ω	$\Leftrightarrow$	Does not exist
$\mathcal{F}$	$\Leftrightarrow$	Does not exist
μ	$\Leftrightarrow$	$\varphi \in \mathfrak{S}$
$L^{\infty}(\Omega)$	$\Leftrightarrow$	$\mathcal{A}$

Once φ is given, we can construct the Hilbert space $\mathcal{H}_\varphi$ through the GNS construction theorem. Therefore the description ⟨1⟩ is essentially the same as that of ⟨2⟩.

Finally, we briefly discuss the KMS (Kubo-Martin-Schwinger) state and the *Tomita-Takesaki theory*.

Let $(\mathcal{A}, \mathfrak{S}, \alpha)$ be a C^{*}-dynamical system. A state $\varphi \in \mathfrak{S}$ is a *KMS state* with respect to a constant β and α_t if for any $A, B \in \mathcal{A}$, there exists a complex function $F_{A,B}(z)$ such that (1) $F_{A,B}(z)$ is analytic for any $z \in D_\beta \equiv \{z \in \mathbb{C};\ -\beta < \operatorname{Im} z < 0\}$ (if $\beta < 0$, $D_\beta \equiv \{z \in \mathbb{C};\ 0 < \operatorname{Im} z < -\beta\}$), (2) $F_{A,B}(z)$ is bounded and continuous for any $z \in \bar{D}_\beta \equiv \{z \in \mathbb{C};\ -\beta \leq \operatorname{Im} z \leq 0\}$, (3) $F_{A,B}(z)$ satisfies the following boundary conditions: (i) $F_{A,B}(t) = \varphi(\alpha_t(A)B)$ and (ii) $F_{A,B}(t - i\beta) = \varphi(B\alpha_t(A))$. The KMS state with respect to the constant β and α_t is called (β, α_t)-*KMS state*. We denote the set of all (β, α_t)-KMS states by $K_\beta(\alpha)$.

One can easily prove the following two propositions:

1° φ is $(0, \alpha_t)$-KMS state $\Longleftrightarrow$ φ is a tracial state.
2° φ is (β, α_t)-KMS state $\Longleftrightarrow$ φ is $(-1, \alpha_{\beta t})$-KMS state.

THEOREM A.2.21. (β, α_t)-*KMS state* φ *is* α-*invariant*.

THEOREM A.2.22. *The following statements are equivalent:*

(1) $\varphi \in K_\beta(\alpha)$.
(2) $\varphi(AB) = \varphi(B\alpha_{i\beta}(A))$, $\forall A, B \in \mathcal{A}_\alpha$.
(3) For any f whose Fourier transform is in $C_0^\infty(\mathbb{R})$ and for any $A, B \in \mathcal{A}$, the following equation is satisfied:

$$\int f(t)\varphi(\alpha_t(A)B)\,dt = \int f(t - i\beta)\varphi(B\alpha_t(A))\,dt,$$

where $f(t - i\beta) = \int \hat{f} e^{i\omega(t - i\beta)} d\omega$.

THEOREM A.2.23. *If φ is faithful, then for a certain β, there uniquely exists a one-parameter automorphism group α_t such that φ is (β, α_t)-KMS state.*

Let $\{\mathcal{H}_\varphi, \pi_\varphi, x_\varphi, u_t^\varphi\}$ be the GNS representation of $\varphi \in \mathfrak{S}$. $\tilde{\varphi}$ is defined by $\tilde{\varphi}(Q) = \langle x_\varphi, Q x_\varphi \rangle$ for any $Q \in \pi_\varphi(A)''$, which is called the *natural extension of φ* to $\pi_\varphi(A)''$. The *natural extension* $\tilde{\alpha}_t$ of α_t is defined by $\tilde{\alpha}_t(Q) = u_t^\varphi Q u_{-t}^\varphi$ for any $Q \in \pi_\varphi(A)''$.

3° For any $\varphi \in K_\beta(\alpha)$, $\tilde{\alpha}_t(Q) = Q, \forall Q \in \mathcal{Z}_\varphi$.
4° $K_\beta(\alpha) = \{\varphi\} \Longrightarrow \mathcal{Z}_\varphi = \mathbb{C}I$.
5° $\varphi \in \mathrm{ex}\, I(\alpha) \Longrightarrow \pi_\varphi(A)' \cap u_t^\varphi(\mathbb{R})' = \mathbb{C}I$.
6° $\varphi \in K_\beta(\alpha) \cap \mathrm{ex}\, I(\alpha) \Longrightarrow \varphi \in ex\, K_\beta(\alpha) \Longleftrightarrow \mathcal{Z}_\varphi = \mathbb{C}I$.
7° Since $K_\beta(\alpha)$ is compact in weak*-topology and convex $K_\beta(\alpha) = w \ *$ $\overline{\mathrm{co}}\,\mathrm{ex}\, K_\beta(\alpha)$.

Let us consider a Von Neumann algebra $\mathfrak{N}$ having a cyclic and separating vector x in a Hilbert space $\mathcal{H}$. Define conjugate linear operators S_o and F_o by

$$S_o A x = A^* x \quad (\forall A \in \mathfrak{N}),$$

$$F_o A' x = A'^* x \quad (\forall A' \in \mathfrak{N}).$$

Their domains contain $\mathfrak{N}x$ and $\mathfrak{N}'x$, respectively, and they are closable operators: $\bar{S}_o = S$, $\bar{F}_o = F$

(1) $S_o^* = F$, $F_o^* = S$,
(2) $S = J\Delta^{1/2}$ polar decomposition.

Δ is called the *modular operator* w.r.t $\{\mathfrak{N}, x\}$, which is unbounded positive selfadjoint. J is called the *modular conjugate operator*, which is conjugate unitary (i.e., $\langle Jx, Jy \rangle = \langle y, x \rangle$, $J^2 = I$ for any $x, y \in \mathcal{H}$).

(3) $\Delta = FS$, $\Delta^{-1} = SF$.
(4) $F = JSJ = \Delta^{1/2}J = J\Delta^{-1/2}$.

Remark: These operators depend on x, so that they should be denoted by S_x, F_x, etc., but we omitted x here.

THEOREM A.2.24 (Tomita). *(1) $J\mathfrak{N}J = \mathfrak{N}'$, $J\mathfrak{N}'J = \mathfrak{N}$.*
(2) $JAJ = A'$, $(\forall A \in \mathfrak{N} \cap \mathfrak{N}')$.
(3) $\Delta^{it}\mathfrak{N}\Delta^{-it} = \mathfrak{N}$, $\forall t \in \mathbb{R}$.

Take $\phi(A) = \langle x, Ax \rangle$, $A \in \mathfrak{N}$ and define $\sigma_t^\phi(A) = \Delta^{it} A \Delta^{-it}$, $A \in \mathfrak{N}$.

THEOREM A.2.25. *(Takesaki) ϕ satisfies the KMS-condition w.r.t. σ_t^ϕ at $\beta = -1$.*

Bibliography

1. Abragam, A. (1961) *The principle of nuclear magnetism*, Clarendon Press, Oxford.
2. Abramowitz, M., and Stegun, A. (1984) *Handbook of Mathematical Functions*, abridged ed.: M. Danos and J. Rafelski (eds.) *Pocketbook of Mathematical Functions*, Deutsch, Thun.
3. Accardi, L. (1974) Noncommutative Markov chains, in *International School of Mathematical Physics*, Camerino, pp. 268–295.
4. Accardi, L. (1976) Non-relativistic quantum mechanics as a non-commutative Markov process, *Adv. Math.* **20**, 329–366.
5. Accardi, L., Frigerio, A., Lewis, J. (1982): *Quantum stochastic processes*, Publications of the Research institute for Mathematical Sciences Kyoto University, **18**, pp. 97-.
6. Accardi, L. and Ohya, M. (1992) Compound channels, transition expectations and liftings, *J. Multivariate Analysis*, to appear.
7. Accardi, L., Ohya, M., and Suyari, H. (1994) Computation of mutual entropy in quantum Markov chains, *Open Sys. Information Dyn.* **2**, 337–354.
8. Accardi, L., Ohya, M. and Watanabe, N. (1996) Kolmogorov-Sinai entropy through quantum Markov chain, *Open Sys. Information Dyn.*, to appear.
9. Accardi, L., Ohya, M., Watanabe, N. : *Dynamical entropy through quantum Markov chain*, to appear in Open System and Information Dynamics.
10. Accardi, L., Ohya, M., Watanabe, N. : *Note on quantum dynamical entropies*, to appear in Rep. Math. Phys.
11. Aczel, J. (1969) On different characterizations of entropies, *Lecture Notes Math.* **89**, 1–11.
12. Aczel, J., and Daróczy, Z. (1975) *On Measures of Information and their Characterizations*, Academic Press, New York.
13. Adler, R. L., Konheim, A.G. and McAdrew, M.H. (1965) Topological entropy, *Trans. Amer. Math. Soc.* **114**, 309–319.
14. Akashi, S. (1992) A relation between Kolmogorov-Prokhorov's condition and Ohya's fractal dimensions, *IEEE Trans. Information Theory* **38**, 1567–1570.
15. Akashi, H. : "Superposition representability problems of quantum information channels", to appear in Open System and Information Dynamics.
16. Alberti, M.A., Cotta-Ramusio, P., and Ramella, G. (1975) Preliminary results in the Banach space formulation of master equation and subdynamics theory in quantum statistics, *J. Math. Phys.* **16**, 132–146.
17. Alicki, R. (1976) On the detailed balance condition for non-Hamiltonian systems, *Rep. Math. Phys.* **10**, 249–258.
18. Alicki, R. (1982) Path integrals and stationary phase approximation for quantum dynamical semigroups. Quadratic systems, *J. Math. Phys.* **23**, 1370–1375.
19. Alicki, R., and Lendi, K. (1990) *Quantum Dynamical Semigroups and Applications*, Lect. Notes in Phys. 286, Springer, Berlin.
20. Alicki, L. and Fannes, M. (1994) Defining quantum dynamical entropy, *Lett. Math. Phys.* **32**, 75–82.
21. Allen, A.B., and Kamiya, N. (eds.) (1964) *Primitive motile systems in cell biology*,

Academic Press, New York.

22. Amari, S. (1985), *Differential-Geometrical Methods in Statistics*, Springer, Berlin (Lecture Notes in Statistics **28**).

23. Antoniewicz, R. (1975) The chaotic state of the electromagnetic field coherent in the second order, Rep. Math. Phys. **7**, 289–301.

24. Antonelli, P.L., Ingarden, R.S., and Matsumoto, M. (1993) *The Theory of Sprays and Finsler Spaces with Applications in Physics and Biology*, Kluwer Academic Publishers, Dordrecht.

25. Appel, P., and Kampé de Fériet (1926) *Functions hypegéometriques et hyperspheériques. Polynomômes d'Hermite*, Gauthiers-Villars, Paris.

26. Araki, H. (1976) Relative entropy for states of Von Neumann algebras, *Publ. RIMS Kyoto Univ.* **11**, 809–833.

27. Araki, H, (1977) Relative entropy of states of Von Neumann algebras II, *Publ. RIMS Kyoto Univ.* **13**, 173–192.

28. Arverson, W.B. (1969) Subalgebras of C^*-algebras, *Acta Math.* **123**, 141–224.

29. Baker, C.R. (1978) Capacity of the Gaussain channel without feedback, *Information & Control* **37**, 70–89.

30. Bateman, H., and Erdélyi, A. (1953) *Higher Transcendental Functions*, vol.1, McGraw-Hill, New York (Russ. transl. Nauka, Moscow, 1965).

31. Bayer, W., and Ochs, W. (1973) Quantum states with maximum information I, *Z. Naturforsch.* **28a**, 693–701.

32. Bar-Hillel, Y. (1964) *Language and Information. Selected Essays on their Theory and Application*, Addison-Wesley, Reading, Mass.

33. Belavkin, V.P. (1993): "Quantum diffusion, their measurement and filtering, I", Probability Theory and its Applications, **38**, pp. 742–757.

34. Bellman, R. (1961) *Adaptive Control Processes: A Guided Tour*, Princeton Univ. Press, Princeton, N. J.

35. Belis, M., and Guiaşu, S. (1968) A quantitative-qualitative measure of information in cybernetic systems, *IEEE Trans. Inform. Theory* **IT−14**, 593–594.

36. Benatti, F. (1993) *Deterministic Chaos in Infinite Quantum Systems*, Trieste Notes in Phys., Springer, Berlin.

37. Berezin, F. A. (1986) *Methods of Second Quantization* (in Russain), Nauka, Moscow.

38. Bergmann, P. G. (1951), Generalized statistical mechanics, *Phys. Rev.* **84**, 1026–1033.

39. Berthalanffy, L. von (1971) *General System Theory. Foundations Development Applications*, Penguin Books, Harmondsworth, Middlesex.

40. Billingsley, P. (1965) *Ergodic Theory and Information*, Wiley, New York.

41. Boltzmann. L. (1923) *Vorlesungen über Gastheorie*, vol.1, 3rd ed., Barth, Leipzig. (1st ed. 1895).

42. Bongard, M. (1963) On the notion of useful information, *Problemy Kibernet.* **9**, 71–102.

43. Bratelli, O. and Robinson, D.W. (1979), *Operator Algebras and Quantum Statistical Mechanics I*, Springer, Berlin.

44. Bratelli, O. and Robinson, D.W. (1981), *Operator Algebras and Quantum Statistical Mechanics II*, Springer, Berlin.

45. Brillouin, L. (1956) *Science and Information Theory*, Academic Press, New York, 2nd ed. 1962.

46. Brillouin, L. (1964) *Scientific Uncertainty, and Information*, Academic Press, New York.

47. Brown, L.M., Pais, A., and Pippard, B. Sir (eds.) (1995) *Twentieth Century Physics*, vol.1, Inst. Phys. Publ., Bristol.

48. Callen, H.B. (1985), *Thermodynamics and an Introduction to Thermostatistics*, 2nd ed., Wiley, New York.

49. Campbell, L.L. (1965) A coding theorem and Rényi's entropy, *Information and*

Control **8**, 423-429.

50. Carathéodory, C. (1909) *Math. Ann.* **67**, 355.
51. Chaitin, G. (1966), On the length of programs for computing finite binary sequences, *J. Assoc. Comput. Machines* **13**, 547-569.
52. Chaitin, G.J. (1987): *Algorithmic Information Theory*, Cambridge Uni. Press.
53. Choi, M.D. (1972) Positive linear maps on C^*-algebras, *Can. J. Math.* **24**, 520-526.
54. Choi, M.D. (1975) Positive semidefinite biquadratic forms, *Linear Algebra and Appl.* **12**, 95-100.
55. Choquet, G. (1969) *Lecture Analysis I,II,III*, Benjamin, New York.
56. Clausius, R. (1876) *Die mechanische Wärmetheorie*, 3 vols., Vieweg, Braunschweig (3rd vol. 1891).
57. Coleman, A. J. (1963) Structure of fermion density matrices, *Rev. Mod. Phys.* **35**, 668-689.
58. Connes, A. and Størmer, E. (1975) Entropy for automorphisms of II_1 Von Neumann algebras, *Acta Math.* **134**, 289-306.
59. Connes, A., Narnhofer, H. and Thirring, W. (1987) Dynamical entropy of C^*-algebras and Von Neumann algebras, *Commun. Math. Phys.* **112**, 691-719.
60. Connes, A., Narnhofer, H. and Thirring, W. (1987) The dynamical entropy of quantum systems, in H. Mitter and L. Pittner (eds.) *Recent Developments in Mathemaical Physics*, Springer, New York, pp. 102-136.
61. Czajkowski, G.Z. (1971) An information-theoretical approach to the problem of polymolecularity of polymers (in Polish), Doctor Thesis, Preprint No. 140, Inst. Phys., N. Copernicus Univ., Toruń, 101 pp.
62. Czajkowski, G.Z. (1972) An Approximation method of calculating of Lagrange coefficients in classical information thermodynamics, Preprint No. 177, Inst. Phys., N. Copernicus Univ., Toruń, 13 pp.
63. Czajkowski, G.Z. (1973) Generalized gamma functions and their application in classical information thermodynamics, *Bull. Acad. Polon. Sci. Sér. math. astr. phys.* **21**, 759-763.
64. Davies, C. (1961) Operator-valued entropy of a quantum mechanical measurement, *Proc. Japan. Acad.* **37**, 533-538.
65. Davies, E.B. (1970) Quantum stochastic processes II, *Commun. Math. Phys.* **19**, 83-105.
66. Davies, E.B. (1972) Diffusion for weakly coupled quantum oscillators, *Commun. Math. Phys.*, 309-325.
67. Davies, E.B. (1973) The harmonic oscillator in a heat bath, *Commun. Math. Phys.* **33**, 171-186.
68. Davies, E.B. (1974) Dynamics of a multilevel Wigner-Weisskopf atom, *J. Math. Phys.* **15**, 2036-2041.
69. Davies, E.B. (1974) Markovian master equations, *Commun. Math. Phys.* **39**, 91-110.
70. Davies, E.B. (1975) Matkovian master equations, *Ann. Inst. Henri Poincaré* **11**, 265-273.
71. Davies, E.B. (1976) Markovian master equations, *Math. Ann.* **219**, 147-158.
72. Davies, E.B. (1976) *Quantum Theory of Open Systems*, Academic Press, New York.
73. DeLuca, A., Termini, S. (1972): A definition of a nonprobabilistic entropy in the setting of fuzzy set theory, Inform. Control., **20**, pp. 301-312.
74. Domotor, Z. (1970) Qualitative information and entropy structures, in J. Hintikka and P. Suppes (eds.) *Information and Inference*, Reidel, Dordrecht, 148-194.
75. Donald, M.J. (1985) On the relative entropy, *Commun. Math. Phys.* **105**, 13-34.
76. Dubikajtis, L., Ingarden, R.S., and Kossakowski, A. (1968) On regularity of information, *Bull. Acad. Polon. Sci. Sér. math. astr. phys.* **16**, 55-56.
77. Ebanks, B.R. (1983): On measures of fuzziness and their representations, J. Math. Anal. Appl., **94**, pp. 24-37.
78. Eddington, A.S (1947) *The Nature of the Physical World*, Dent, London.

79. Edwards, M. (1985) *Dictionary of Key Words*, Macmillan, London.
80. Eggleston, H.G. (1958) *Convexity*, Cambridge Univ. Press, London.
81. Ehrenfest, P., and Ehrenfest, T. (1906), Über zwei bekannte Einwände gegen das Boltzmannsche H-theorem, *Z. Phys.* **8**, 311–314.
82. Ehrenfest, P., and Ehrenfest, T. (1959) *The Conceptual Foundations of the Statistical Aproach in Mechanics*, Cornell University Press, Ithaca, N.Y. (first published in German in *Encyklopädie der math. Wissenschaften*, Vol. IV 2 II, No. 6, Teubner, Leipzig 1912).
83. Emch, G. G. (1972) *Algebraic Methods in Statistical Mechanics and Quantum Field Theory*, Wiley-Interscience, New York.
84. Emch, G.G. (1974) Positivity of the K-entropy on non-abelian K-flows, *Z. Wahrscheinlichkeitstheorie verw. Gebiete* **29**, 241–252.
85. Emch, G.G., and Sewell, G.L. (1968) Nonequilibrium statistical mechanics of open systems, *J. Math. Phys.* **9**, 946–958.
86. Evans, D.E. (1977) Irreducible quantum dynamical semigroups, *Commun. Math. Phys.* **54**, 293–297.
87. Faddeev, D.K. (1956) On the concept of entropy of a finite probabilistic scheme (in Russian), *Uspekhi Mat. Nauk* **11**, no.1(67), 227–231 (German transl. in *Math. Forschungberichte IV, Arbeiten zur Informationstheorie I*, 3rd ed. DVW, Berlin 1967, 86–90).
88. Favre, C., and Martin, P. (1968) Approach to equilibrium in an explicit quantum model, *Helv. Phys. Acta* **41**, 333–351.
89. Favre, C., and Martin, P. (1968) Dynamique quantique des systèmes amortis "non markoviens", *Helv. Phys. Acta* **41**, 333–361.
90. Feinstein, A. (1958) *Foundations of Information Theory*, McGraw-Hill, New York.
91. Feller, W. (1950) *An Introduction to Probability and its Applications*, 2 vols., Wiley, New York (2nd vol. 1966).
92. Fey, P. (1963) *Informationtheorie*, Akademie-Verlag, Berlin.
93. Fichtner, K.H., Freudenberg, W., Liebscher. V.: "Beam splitting and time evolutions of Boson systems", preprint.
94. Fikus, M. (1995) DNA, gens and patents (in Polish), *Wiedza i Życie* no.1, 10–11.
95. Fisher, R.A. (1922) On the mathematical foundations of theoretical statistics, *Phil. Trans. Roy. Soc. (London)* **A222**, 309–368 (also in R.A. Fisher, *Contributions to Mathematical Statistics*, Wiley, New York 1950, paper 10).
96. Fisher, R. A. (1925), Theory of statistical estimation, *Proc. Camb. Phil. Soc.* **22**, 700–725, (also in R.A. Fisher, *Contributions to Mathematical Statistics*, Wiley, New York 1950, paper 11).
97. Fisz, M. (1963) *Probability Theory and Mathematical Statistics*, 3rd ed., Wiley, New York.
98. Fock, V. A. (1930) Näherungsmethode zur Lösung des quantenmechanischen Mehrkörperproblems, *Z. Phys.* **611**, 126–148.
99. Frigerio, A. (1977) Quantum dynamical semigroups and approach to equilibrium, *Lett. Math. Phys.* **2**, 79–87.
100. Frigerio, A. Novellone, C., and Verri, M. (1977) Master equation treatment of the singular reservoir limit, *Rep. Math. Phys.* **12**, 279–284.
101. Frigerio, A. (1978) Stationary states of quantum dynamical semigroups, *Commun. Math. Phys.* **63**, 269–276.
102. Gallager, R. G. (1968) *Information Theory and Reliable Communication*, Wiley, New York.
103. Gelfand, I.M., and Jaglom (Yaglom), A.M. (1958) Über die Berechnung der Menge an Information über eine zufällige Funktion, die in einer anderen zufälligen Funktion enthalten ist, *Math. Forschungberichte VI, Arbeiten zur Informationstheorie II*, DVW, Berlin, 7–56 (Russ. original Usp. Math. Nauk 11 No. 73, 1957, 3–52).

104. Gelfand, I.M. and Yaglom, A.M. (1959) Calculation of the amount of information about a random function contained in another such function, *Amer. Math. Soc. Trans.* **12**, 199–246.
105. Gibbs, J. W. (1902) *Elementary Priciples in Statistical Mechanics*, Yale Univ. Press, New Haven, Conn.; Dover, New York 1960.
106. Gleason, A.M. (1957), Measures on the closed subspaces of a Hilbert space, *J. Math. Mech.* **6**, 885–893.
107. Gnedenko, B.W., and Kolmogorov, A.N. (1960) *Grenzverteilungen von Summen unabhängiger Zufallsgrössen*, Akademie-Verlag, Berlin (except of the Russian original there is an English transl. Polish transl. PWN, Warszawa, 1957).
108. Gorini, V., Kossakowski, A., Sudarshan, E.C.G. (1976) Completely positive dynamical semigroups of *n*-level systems, *J. Math. Phys.* **17**, 821–825.
109. Gorini, V. and Kossakowski, A. (1976) *N*-level system in contact with a singular reservoir, *J. Math. Phys.* **17**, 1298–1305.
110. Gorini, V., Frigerio, A., Verri, M., Kossakowski, A., and Sudarshan (1978) Properties of quantum Markovian master equations, *Rep. Math. Phys.* **13**, 149–172.
111. Grabowski, M., and Staszewski, P. (1977) On continuity properties of the entropy of an observable, *Rep. Math. Phys.* **11**, 233–237.
112. Grabowski, M. (1978) *A*-entropy for spectrally absolutely continuous observables, *Rep. Math. Phys.* **14**, 377–384.
113. Grenlander. V. (1963) *Probabilities on algebraic structures*, Wiley, New York.
114. Grzegorczyk. A. (1993). Philosophical aspects of mathematics (in Polish), in M. Skwarczyński *et al.* (eds.) *Leksykon Matematyczny (Mathematical Lexicon)*, Wiedza Powszechna, Warszawa, 206–217.
115. Gudder, S. and Merchand, J.-P. (1972) Noncommutative probability on Von Neumann algebras, *J. Math. Phys.* **13**, 799–806.
116. Guiaşu, S. (1971), Weighted entropy, *Rep. Math. Phys.* **2**, 165–179.
117. Guiaşu, S. (1977), *Information Theory with Applications*, McGraw-Hill, New York.
118. Gupta, H. C. and Sharma, S. D. (1975) Noiseless coding theorem for non-additive measures of entropy and inaccuracy, *J. Math. Sciences* **10**, 86–95.
119. Gupta, H. C. (1977) *A Study on Sub-additive and Non-additive Measures of Information*, Ph.D. Thesis, Univ. of Delhi (typescript, 280 pp., unpublished).
120. Haake, F. (1973) in *Springer Tracts in Modern Physics*, **66**, Springer, Berlin.
121. Haken, H. (1970) Laser theory, in Flüge (ed.), *Handbuch der Phys.*, vol. XXV/2c, Springer, Berlin.
122. Haken, H. (1983) *Synergetics, An Introduction*, 3rd ed., Springer, Berlin.
123. Haken, H. (1987) *Advanced Synergetics*, Springer, Berlin.
124. Haken, H. (1988), *Information and Self-Organization*, Springer, Berlin.
125. Halmos, P.R. (1950), *Measure Theory*, Van Nostrand, New York.
126. Hardy, G.H., Littlewood, J.E., and Polya, G. (1952) *Inequalities* Cambridge Univ. Press, Cambridge.
127. Hartley, R.V.L. (1928), Transmission of information, *Bell System Techn. J.* **7**, 535–563.
128. Hartree, D. R. (1928) The wave mechanics of an atom with a non-Coulomb central field, *Proc. Cambr. Phil. Soc.* **24**, 89, 111.
129. Hermann, A. (1972), *Lexicon Geschichte der Physik A-Z*, Deubner, Stuttgart.
130. Hiai, F., Ohya, M. and Tsukuda. M. (1981) Sufficiency, KMS condition and relative entropy in Von Neumann algebras. *Pacific J. Math.* bf96, 99–109.
131. Hiai, F., Ohya, M. and Tsukuda. M. (1983) Sufficiency and relative entropy in *-algebras with application to quantum systems, *Pacific J. Math.* **107**, 117–140.
132. Holevo, A.S. (1973) Statistical problems in quantum physics, in *Proc. 2nd Japan-USSR Sympos. Probability Theory*, Lecture Notes in Math. **330**, pp. 104–119.
133. Holevo, A.S. (1973) Some estimates for the amount of information transmittable by a quantum communication channel (in Russian), *Problemy Peredachi Informacii*

9, 3–11.
134. Holevo, A.S. (1991) Quantum probability and quantum statistics (in Russian), *Itogi Nauki i Techniki ser. Modern Problems of Math., Fundamental Trends* **83**, 5-132, 266–270.
135. Ihara, S. (1979) On the capacity of the discrete time Gaussian channel with feedback, in *Trans. 8th Prague Conf.* C, Czechoslovak Acad. Sci., Prague, pp. 175-186.
136. Ingarden, R.S., and Urbanik, K. (1961) Information as a fundamental notion of statistical physics, *Bull. Acad. Polon. Sci., Sér. math. astr. phys.* **9**, 313–316.
137. Ingarden, R.S., and Urbanik, K. (1962), Information without probability, *Coll. Math,* **9**, 131-150.
138. Ingarden, R.S., and Urbanik, K. (1962) Quantum informational thermodynamics, *Acta Phys. Polon.* **21**, 281–304.
139. Ingarden, R.S. (1963) A simplified axiomatic definition of information, *Bull. Acad. Polon. Sci., Sér. math. astr. phys.* **11**, 209–211.
140. Ingarden, R.S. (1963), Information theory and variational principles in statistical theories, *Bull. Acad. Polon. Sci., Sér. math. astr. phys.* **11**, 541–547.
141. Ingarden, R.S. (1964) Theoretical investigations into fundamentals of thermodynamics in Wrocław, in J. Kaczér *Physics and Techniques of Low Temperatures, Proc. 3rd Regional Conf. Prague 1963*, Czechoslov. Acad. Sci., Prague, 45–52.
142. Ingarden, R.S. (1964) Information theory and thermodynamics of light. Part I. Foundations of information theory, *Fortschr. Phys.* **12**, 567–594.
143. Ingarden, R.S. (1965) The Higher Order Temperatures and the Zeroth Principle of Thermodynamics, *Bull. Acad. Polon. Sci. Sér. math. astr. phys.* **13**, 69–72, errata 532.
144. Ingarden, R.S. (1965) Information theory and thermodynamics of light. Part II. Principles of information thermodynamics, *Fortschr. Phys.* **13**, 755–805.
145. Ingarden, R.S. and Kossakowski, A. (1965) Statistical thermodynamics with higher order temperatures for ideal gases of bosons and fermions, *Acta Phys. Polon.* **28**, 499–511.
146. Ingarden, R.S. (1965) Simplified axioms for information without probability, *Prace Mat.* **9**, 273–282.
147. Ingarden, R.S., and Kossakowski, A. (1968), An axiomatic definition of information in quantum mechanics, *Bull. Acad. Polon. Sci. Sér. math. astr. phys.* **16**, 61–65.
148. Ingarden, R.S. (1968) Notion de température et pompage optique', *Ann. Inst. Henri Poincaré, Sect. A, Phys. Théor.* **8**, 1–23, 2nd ed.: Acad. Polon. Sci., Centre Sci. à Paris, Conférences **71**, 3–20.
149. Ingarden, R.S. (1969) Generalized thermodynamics of the electromagnetic radiation in a cavity. Part I. A 5-temperature thermodynamics of one mode, *Acta Phys. Polon* **36**, 855–885.
150. Ingarden, R.S. (1971) An information-theoretical approach to the theory of lasers, *Bull. Acad. Polon. Sci., Sér. sci. math. astr. phys.* **19**, 77–82.
151. Ingarden, R.S. (1973) Generalized irreversible thermodynamics and its application to lasers. Part I. General theory, *Acta Phys. Polon.* **A43**, 3–14.
152. Ingarden, R.S. (1973) Generalized irreversible thermodynamics and its application to lasers. Part II. Thermodynamics of a laser, *Acta Phys. Polon.* **A43**, 15–35.
153. Ingarden, R.S. (1974) *Information Theory and Thermodynamics*, vol.1: Part I. Information Theory, Chap.I. Classical Theory, Preprint No. 270, Inst. Phys., N. Copernicus Univ., Toruń.
154. Ingarden, R.S. (1974) *Information Theory and Thermodynamics*, vol.2: Part I. Information Theory, Chap.II. Quantum Theory, Preprint No. 275, Inst. Phys., N. Copernicus Univ., Toruń.
155. Ingarden, R.S., and Kossakowski, A. (1975) On the connection of nonequilibrium information thermodynamics with non-Hamiltonian quantum mechanics of open

systems, *Ann. Phys. (N. Y.)* **89**, 451–485.

156. Ingarden, R.S. (1976) Quantum information theory, *Rep. Math. Phys.* **10**, 43–72.
157. Ingarden, R.S., Janyszek, H., Kossakowski, A. and Kawaguchi, T. (1979) *Tensor* **33**, 347–353.
158. Ingarden, R.S. (1980) An information-thermodynamical model of the primitive motile system in cell biology, *Bull. Acad. Polon. Sci., Sér. phys. astron.* **28**, 125–131.
159. Ingarden, R.S. (1987) Geometry of thermodynamics, in H.D. Doebner and J.D. Hennig (eds.) *Differential Geometric Methods in Theoretical Physics* (XV DGM Conference Clausthal 1986), World Scientific, Singapore, pp. 455–465.
160. Ingarden, R.S. (1987) Self-organization of the visual information channel and solitons, in H. Haken (ed.), *Computational Systems - Natural and Artificial*, Proc. Intern. Symp. on Synergetics, Schloss Elmau, Bavaria, May 4–9, 1987, Springer, Berlin, 56–64.
161. Ingarden, R.S., and Górniewicz, L. (1990) Shape as an information-thermodynamical concept and the pattern recognition problem, *Math. Nachr.* **145**, 97–109.
162. Ingarden, R.S. (1992) Towards mesoscopic thermodynamics: small systems in higher-order states, *Open Sys. Information Dyn.* **1**, 75–102, errata 309.
163. Ingarden, R.S., and Nakagomi, T. (1992) The second order extension of the Gibbs state, *Open Sys. Information Dyn.* **1**, 259–268.
164. Ingarden, R.S., and Meller, J. (1994) Temperatures in linguistics as a model of thermodynamics, *Open Sys. Information Dyn.* **2**, 211-230.
165. Izumi, S., *et al.* (1977) *Kyoritsu Mathematical Formulae and Tables of Functions* (in Japanese), 4th ed., Kyoritsu, Tokyo.
166. Jaworski, W., and Ingarden, R.S. (1980) On the partition function in information thermodynamics with higher order temperatures, *Bull. Acad. Polon. Sci. Sér. phys. astr.* **28**, 119–123.
167. Jaworski, W. (1981) On information thermodynamics with temperatures of the second order (in Polish, unpublished), Master Thesis, Inst. Phys., N. Copernicus Univ., Toruń.
168. Jaworski, W. (1981) Information thermodynamics with the second order temperatures for the simplest classical systems, *Acta Phys. Polon.* **A60**, 645–659.
169. Jaworski, W. (1983) On the thermodynamic limit in information thermodynamics with higher-order temperatures, *Acta Phys. Polon.* **A63**, 3–19.
170. Jaworski, W. (1985) Investigation of the thermodynamical limit for the states maximizing entropy under auxiliary conditions for higher-order statistical moments (in Polish, unpublished), Doctor Thesis, Inst. Phys., N. Copernicus Univ., Toruń.
171. Jaworski, W. (1985) On the microcanonical ensemble of a spin system interacting with one-mode electromagnetic field, *Z. Phys. B - Condensed Matter* **59**, 483–491.
172. Jaworski, W. (1987) Higher-order moments and the maximum entropy inference: the thermodynamical limit approach, *J. Phys. A: Math. Gen.* **20**, 915–926.
173. Jaynes, E.T. (1957) Information theory and statistical mechanics, *Phys. Rev.* **106**, 620-630.
174. Jeffreys, H. (1946), An invariant form for the prior probability in estimation problems, *Proc. Roy. Soc. (London)* **A186**, 453–461.
175. Jeffreys, H. (1948) *Theory of Probability*, 2nd ed., Oxford Univ. Press, Oxford.
176. Jensen, J.L.W.V. (1906) Sur les fonctions convexes et les inégalités entre les valeurs moyennes, *Acta Math.* **30**, 175–193.
177. Jumarie, G. (1986) *Subjectivity, Information, Systems. Introduction to a Theory of Relativistic Cybernetics*, Gordon and Breach, New York.
178. Jumarie, G. (1990) *Relative Information. Theories and Applications*, Springer, Berlin.
179. Kadison, R.V. (1952) A generalized Schwartz inequality and algebraic invariants for operator algebras, *Ann. of Math.* **56**, 494-503.

180. Kampé de Fériet, J. (1957) *Functions de physique mathématique*, CNRF, Paris.
181. Kendall, D.G. (1964), Functional equations in information theory, *Z. Wahrscheinlichkeitstheorie* **2**, 225–229.
182. Khinchin, A.J. (1957) *Mathematical Foundations of Information Theory*, Dover, New York (Russ. originals 1953 and 1956).
183. Klein, O. (1931) Zur quantenmechanischen Begründung des zweiten Hauptsatzes der Wärmelehre, *Z. Phys.* **72** 767–775.
184. Kolmogorov, A.N. (1931) Über die analytische Methoden in der Wahrscheinlichkeitsrechnung, *Math. Ann.* **104**, 415–436.
185. Kolmogorov, A.N. (1933) *Grundbegriffe der Wahrscheinlichkeitsrechnung*, Springer, Berlin.
186. Kolmogorov, A.N., Gelfand, I.M., and Yaglom, A.M. (1956) On the general definition of amount of information (in Russian), *Dokl. Akad. N. SSSR* **111**, 745–748.
187. Kolmogorov, A.N., Gelfand, I.M., and Yaglom, A.M. (1958), Amount of information and entropy of continuous distributions (in Russian), *Proc. 3rd Allunion Math. Congress*, Academy Press, Moscow, vol. 3, 300–320.
188. Kolmogorov, A.N. (1958): *A new metric invariant of transitive dynamic systems and automorphisms in Lebesgue spaces*, Dokl. Akad. Nauk, SSSR **119**, pp. 861–869.
189. Kolmogorov, A.N. (1963) Theory of transmission of information, *Amer. Math. Soc. Transl.* Ser.2, **33**, 291.
190. Kolmogorov, A.N. (1965): *Their approaches to the quantitive definition of information*, Prob. of Inform. Trans. **1**, pp.3–11.
191. Kolmogorov, A.N. (1968), Three approaches to the quantitative definition of information, *Intern. J. Computer Math.* **2**, 157–168 (Russ. original 1965).
192. Kolmogorov, A.N. (1968) Logical basis for information theory and probability theory, *IEEE Trans. Inform. Theory* **IT–14**, 778–782.
193. Kolakowski, L. (1994), *Presence of Myth* (in Polish), Wydawnictwo Dolnośląskie, Wroclaw (Germ. transl. *Die Gegenwärtigkeit des Mythos*, 1972, Polish 1st edition, Paris, 1973).
194. Kossakowski, A. (1968) On a modification of the Kullback information, *Bull. Acad. Polon. Sci. Sér. math. astr. phys.* **16**, 349–352.
195. Kossakowski, A. (1969) On the quantum informational thermodynamics, *Bull. Acad. Polon. Sci. Sér. math. astr. phys.* **17**, 263–267.
196. Kossakowski, A. (1970) Information-theoretical decision scheme in quantum statistical mechanics (in Polish), Preprint No. 122, Habilitation Dissertation, Inst. Phys., N. Copernicus Univ., Toruń.
197. Kossakowski, A. (1972) On quantum statistical mechanics of non-Hamiltonian systems, *Rep. Math. Phys.* **3**, 247–274.
198. Kossakowski, A. (1972) On necessary and sufficient conditions for a generator of a quantum dynamical semi-group, *Bull. Acad. Polon. Sci., Sér math. astr. phys.* **20**, 1021–1025.
199. Kossakowski, A. (1973) On the general form of the generator of a dynamical semigroup for the spin 1/2 system, *Bull. Acad. Polon. Sci., S'er. math. astr. phys.* **21**, 649–653.
200. Kossakowski, A. and Sudershan, E.C.G. (1976) Completely positive dynamical semigroup of N-level system, *J. Math. Phys.* **17** 821–825.
201. Kossakowski, A., Frigerio, A., Gorini, V., and Verri, M. (1977) Quantum detailed balance and KMS condition, *Commun. Math. Phys.* **57**, 97–110.
202. Kraus, K. (1971) General state changes in quantum theory, *Ann. Phys. (N.Y.)* **64**, 311–333.
203. Kraus, K. (1983) *States, Effects and Operations*, Springer, Berlin.
204. Krilov, V.I. (1967) *Approximate Evaluation of Integrals* (in Russian), 2nd ed., Nauka, Moscow.

205. Kullback, S., and Leibler, R.A. (1951) On information and sufficiency, *Ann. Math. Statist.* **22**, 79–86.
206. Kullback, S. (1959) *Information Theory and Statistics*, Wiley, New York.
207. Kummer, H. (1967) n-representability problem for reduced density matrices, *J. Math. Phys.* **8**, 2063–2081.
208. Kuratowski, C. (1958) *Topologie*, vol.1, 4th ed., PWN, Warszawa, (Engl. ed. Academic Press, New York 1966, Russ. ed. Mir, Moscow 1966).
209. Landau, L. D. and Lifshitz, E. M. (1984) *Fluid Mechanics*, Pergamon Press, Oxford.
210. Lanz, L., Lugiato, L., and Ramella, G. (1971) On the existence of independent subdynamics in quantum statistics, *Physica* **54**, 94–136.
211. Laurikainen, K.V. (1988) *Beyond the Atom. The Philosophical Thought of Wolfgang Pauli*, Springer, Berlin.
212. Lavenda, B.H. (1991) *Statisitcal Physics. A Statistical Approach*, Wiley, New York.
213. Lawson, J.L., and Uhlenbeck, G.E. (1950) *Threshold Signals*, McGraw-Hill, New York.
214. Lee, P.M. (1964), On the axioms of information theory, *Ann. Math. Statist.* **35**, 415–418.
215. Lehmann, E.L. (1950), Some principles of the theory of testing hypotheses, *Ann. Math. Statist.* **21**, 1–26.
216. Levin, R.D. (1988) Patterns of maximal entropy, in W. Güttinger and G. Dangemeyr (eds.), *The Physics of Structure Formation*, Springer, Berlin, 78–86.
217. Lieb, E.H., Ruskai, M.B. (1973) Proof of the strong subadditivity of quantum mechanical entropy, *J. Math. Phys.* **14**, 1938–1941.
218. Lindblad, G. (1973) Entropy, information and quantum measurement, *Commun. Math. Phys.* **33**, 305–322.
219. Lindblad, G. (1975) Completely positive maps and entropy inequalities, *Commun. Math. Phys.* **40**, 147–151.
220. Lindblad, G. (1977) On the generators of quantum dynamical semigroup, *Commun. Math. Phys.* **48**, 119–130.
221. Louisell, W.H. (1973) *Quantum Statistical Properties of Radiation*, Wiley, New York.
222. Lomnicki, A. (1923) Nouveaux fondements du calcul des probabilités, *Fund. Math.* **4**, 34–71.
223. Łoś, Z.M. (1970) Classical information thermodynamics of conformation of polymers (in Polish), Master Thesis, Inst. Phys., N. Copernicus Univ., Touń unpublished).
224. Majewicz, A.F. (1985), *The Grammatical Category of Aspect in Japanese and Polish in a Comparative Perspective*. A. Mickiewicz Univ. Press, Poznań.
225. Mandelbrot, B.B. (1982) *The Fractal Geometry of Nature*, Freeman, San Francisco. Linguistic Series 2, Poznań.
226. Martin, S.E. (1988) *A Reference Grammar of Japanese*, 2nd ed., Tuttle, Rutland, Vermont, Tokyo.
227. Matsuoka, T. and Ohya, M. (1995) Fractal dimensions of states and its application to Ising model, *Rep. Math. Phys.* **36**, 27–41.
228. Maurin, K. (1972) *Methods of Hilbert Spaces*, 2nd ed., PWN, Warszawa.
229. Maurin, K. (1993) Mathematics and physics (in Polish), in M. Skwarczyński *et. al.* (eds.) *Leksykon Matematyczny (Mathematical Lexicon)*, Wiedza Powszechna, Warszawa, 781–954.
230. Michalski, M. (1992) Rényi entropies, Lyapunov exponents and dimension spectra for analytically perturbed piecewise linear maps, *Open Syst. Inform. Dyn.* **1**, 409–422.
231. Morrison, D.F. (1976) *Multivariate Statistical Methods*, 2nd ed., McGraw–Hill, New York.
232. Morse, P.M, and Feshbach, H. (1953) *Methods of Theoretical Physics*, McGraw-Hill, New York (Russ. transl. Izd. Inostr. Lit., Moscow, 1st vol. 1958, 2nd vol.

1960).

233. Muraki, N., Ohya, M. and Petz, D. (1992) Note on entropy of general quantum systems, *Open Sys. Information Dyn.* **1**, 43–56.

234. Muraki, N. and Ohya M. (1996) Entropy functionals of Kolmogorov-Sinai type and their limit theorems, *Lett. Math. Phys.*, to appear.

235. Murti, T.R.V. (1987) *The Central Philosophy of Buddhism. A Study of the Madhyamika System*, Unwin Paperbacks, London.

236. Nagao, G. (1989) *The Foundational Standpoint of Madhyamika Philosophy*, transl. by J. P. Keenan, State Univ. of New York Press, Albany, N. Y.

237. Nakagomi, T. (1981) Mesoscopic thermodynamics of nonequilibrium open systems.I. Negentropy consumption and residual entropy, *J. Statist. Phys.* **26**, 567–611.

238. Nakagomi, T. (1992) Mesoscopic version of thermodynamic equilibrium condition. Another approach to higher order temperatures, *Open Sys. Information Dyn.* **1**, 233–241.

239. Nakamura, M. and Umegaki, H. (1962) On Von Neumann theory of measurements in quantum statistics, *Math. Jap.* **7**, 151–157.

240. Nettleton, R.E., and Sobolev, S.L. (1995) Applications of Extended thermodynamics to chemical, rheological, and transport processes: a special survey Part I. Approaches and scalar rate processes, *J. Non-Equilib. Thermodyn.* **20**, 205–229.

241. Nicolis, G. and Prigogine, I. (1977) *Selforganization in Non-Equilibrium Systems*, Wiley-Interscience, New York.

242. Nicolis, G. (1989) *Exploring Complexity*, Freeman, New York.

243. Ochs, W. and Bayer, W. (1973) Quantum states with maximum information entropy II, *Z. Naturforsch.* **28a**, 1571–1585.

244. Ochs, W. (1975) A new axiomatic characterization of the Von Neumann entropy, *Rep. Math. Phys.* **8**, 109–120.

245. Ochs, W. (1976) Basic properties of the generalized Boltzmann-Gibbs-Shannon entropy, *Rep. Math. Phys.* **9**, 135–155.

246. Ohya, M. (1981) Quantum ergodic channels in operator algebra, *J. Math. Anal. Appl.* **84**, 318–328.

247. Ohya, M. (1983) On compound state and mutual information in quantum information theory, *IEEE Trans. Information Theory* **29**, 770–777.

248. Ohya, M. (1983) Note on quantum probability, *Il Nuovo Cimento* **38**, 402–406.

249. Ohya, M. (1984) Entropy transmission in C^*-dynamical systems, *J. Math. Anal. Appl.* **100**, 222–235.

250. Ohya, M. (1985) State change and entropies in quantum dynamical systems, in L. Accardin and W. Von Waldenfels (eds.), *Quantum Probability and Applications II*, Lecture Notes in Math. **1136**, Springer, Berlin, pp. 397–408.

251. Ohya, M. and Watanabe, N. (1986) A new treatment of communication processes with Gaussian channels, *Jap. J. Appl. Math.* **3**, 197–206.

252. Ohya, M. (1989) Some aspects of quantum information theory and their applications to irreversible processes, *Rep. Math. Phys* **27**, 19–47.

253. Ohya, M. (1989) Fractal dimensions of general quantum states, *Proc. Symp. Appl. Func. Anal.* **11**, 45–57.

254. Ohya, M. (1991) Information dynamics and its application to optical communication processes, in *Lecture Notes in Phys.* **378**, Springer, Berlin, pp. 81–92.

255. Ohya, M. (1991) Fractal dimension of states, in *Quantum Probability and Related Topics VI*, World Scientific, Singapore, pp. 359–369.

256. Ohya, M. and Petz, D. (1993) *Quantum Entropy and its Use*, Springer, Berlin.

257. Ohya, M. (1995) State change, complexity and fractal in quantum systems, in V. P. Belavkin, O. Hirota and R.L. Hudson (eds.), *Quantum Communication and Measurement*, Plenum Press, New York, pp. 309–320.

258. Ohya, M., Suyari, H. (1995): "An application of lifting theory to optical communiation processes", Rep. Math. Phys. **36**, pp. 403–420.
259. Ohya, M. (1996) Foundation on entropy, complexity and fractal in quantum systems, in *Probability Towards the Year 2000*, to appear
260. Ohya, M. and Watanabe, N. (1996), Note on irreversible dynamics and quantum information, in *Proc. A. Frigerio Conf.*, to appear.
261. Ohya, M. and Matsuoka, T. (1996) Analysis of complexity of craters and rivers by fractal dimensions of states (in Japanese), *Trans. Inst. Electronics, Information & Communication*, to appear.
262. Ohya, M., Matsuoka, T. and Inoue, K. (1996) New approach to ε-entropy and its comparison with Kolmogorov's ε-entropy, preprint.
263. Ohya, M.: *Foundation of entropy, complexity and fractal in quantum systems*, to be published in International Congress of Probability Towards the Year 2000.
264. Ohya, M. : "Fundamentals of mutual entropy and capacity in classical and quantum systems", SUT preprint.
265. Ohya,M., Petz, D., Watanabe, N. : "On capacities on quantum channels", SUT preprint.
266. Ojima, I., Hasegawa, H. and Ichiyanagi, M. (1988) Entropy production and its positivity in nonlinear response theory of quantum dynamical systems, *J. Statist. Phys.* **50**, 633–655.
267. Ojima, I. (1989) Entropy production and nonequilibrium stationarity in quantum dyanmical systems. Physical meaning of Van Hove limit, *J. Statist. Phys.* **56**, 203–226.
268. Onicescu, O. (1966) Energia informationala (in Rumanian), *Stud. Cercet. Matem.* **18**, 1419–1420.
269. Organ, T.W. (1975) *Western Approaches to Eastern Philosophy*, Ohio Univ. Press, Athens, Ohio.
270. Palé, J. V. (1974) The Bloch equations, *Commun. Math. Phys.* **38**, 241–256.
271. Palmer, F.R. (1986) *Mood and Modality*, Cambridge Univ. Press, Cambridge.
272. Parthasarathy, K.R. (1967) *Probability measures on metric spaces*, Academic Press, New York.
273. Peters, E.E. (1988) *Chaos and Order in the Capital Markets*, Springer, New York.
274. Petz, D. (1986) Sufficient subalgebras and the relative entropy of states on a Von Neumann algebra, *Commun. Math. Phys.* **105**, 123–131.
275. Petz, D. (1990) *The Algebra of Cannonical Communication Relation*, Leuven Univ. Press, Leuven.
276. Pinsker, M.S. (1963) Gaussian sources (in Russian), *Problems Information Transmission* **14**, 59–100.
277. Pippard, A.B. (1957) *Elements of Classical Thermodynamics*, Cambridge Univ. Press, Cambridge.
278. Pötschke, D., and Sobik. F. (1980) *Mathematische Informationstheorie*, Akademie-Verlag, Berlin.
279. Powles, J.G., and Carazza, B. (1969) An information theory of line shape in nuclear magnetic resonance, Paper 112, Intern. Conf. on Magnetic Resonance, Monash, Australia, August 1969.
280. Presutti, E., Scacciatelli, E., Sewell, G.L., and Wandelingh, F. (1972) Studies in the C^*-algebra theory of nonequilibrium statistical mechanics, dynamics of open and mechanically driven systems, *J. Math. Phys.* **13**, 1085–1098.
281. Prigogine, I., George, C., Honin, F., and Rosenfeld, L. (1973) *Chemical Scripta* **4**, 5–32.
282. Prigogine, I. (1980) *From Being to Becoming, Time and Complexity in the Physical Sciences*, Freeman, San Francisco.
283. Prigogine, I., and Stengers, I. (1984) *Order out of Chaos, Man's New Dialogue with Nature*, Bantam Books, Toronto.
284. Primas, H. (1981) *Chemistry, Quantum Mechanics and Reductionism*, Springer,

Berlin, 2nd ed. 1983.
285. Primas, H. (1992) Time-asymmetric phenomena in biology, *Open Sys. Information Dyn.* 1, 3–34.
286. Prosser, R.T. and Root, W.L. (1968) The ε-entropy and ε-capacity of time-invariant channels, *J. Math. Anal. Appl.* 21, 233.
287. Pruski, S. (1967) A many-state extension of the Hartree-Fock-Slater method, *Physica* 37, 246–252.
288. Radin, C. (1971) Non-commutative mean ergodic theory, *Commun. Math. Phys.* 21, 291–302.
289. Ralston, A. (1965) *A First Course in Numerical Analysis*, McGraw-Hill, New York.
290. Rényi, A. (1961) On measures of entropy and information, *Proc. Fourth Berkeley Symp. Math. Statist. Probab.* 1, 547–561.
291. Rényi, A. (1970) *Probability Theory*, Akadémiai Kiadó, Budapest, North-Holland, Amsterdam.
292. Rényi, A. (1984) *A Diary on Information Theory*, Akadémiai Kiadó, Budapest.
293. Rényi, A., and Balatoni, J. (1967) Über den Begriff der Entropie, Math. Forschungsberichte IV, Arbeiten zur Informationstheorie I 3rd ed., DVW, Berlin, 117–134, (Hung. original 1956).
294. Reza, F. M. (1961) *An Introduction to Information Theory*, McGraw-Hill, New York.
295. Riesz, F., and Nagy, B.Sz. (1952) *Leçons d'analyse fonctionnelle*, Acad. Kiado, Budapest.
296. Ruelle, D. (1969), *Statistical Mechanics, Rigorous Results*, Benjamin, New York (Russ. transl. Mir, Moscow 1971).
297. Schneider, I. (ed.) (1989) *Die Entwicklung der Wahrscheinlichkeitstheorie von den Anfängen bis 1933. Einführungen und Texte*, Akademie, Berlin.
298. Schrödinger, E. (1945) *What is life?*, Cambridge Univ. Press, New York.
299. Schweber, S. S. (1961) *An Introduction toi Relativistic Quantum Field Theory*, Row, Peterson & Co., Evanston, Ill.
300. Shannon, C.E. (1948) A mathematical theory of communication, *Bell System Techn. J.* 27, 379–423, 623–656.
301. Shannon, C.E. (1949) Communication in the presence of noise, *Proc. IRE* 37, 10–21.
302. Shohat, J.A., and Tamarkin, J.D. (1963) *The Problem of Moments*, 3rd printing. of rev. ed., Amer. Math. Soc., Providence RI.
303. Sinai, Ya. G. (1959) On the concept of entropy for dynamical systems (in Russian), *Dokl. Akad. Nauk SSSR* 124, 768–771.
304. Sobczyk, K. (1991) *Stochastic Differential Equations with Applications to Physics and Engineering*, Kluwer Academic Publishers, Dordrecht.
305. Solomonoff, R.L. (1964) A formal theory of inductive inference, *Inform. Control* 7, 1–22.
306. Spohn, H. (1976) Approach to equilibrium for completely positive dynamical semigroups of n-level systems, *Rep. Math. Phys.* 10, 189–194.
307. Spohn, H. (1978) Entropy production for quantum dynamical semigroups, *J. Math. Phys.* 19, 1227–1230.
308. Staszewski, P. (1978) On the characterization of the Von Neumann entropy via the entropies of measurements, *Rep. Math. Phys.* 13 67–71.
309. Staszewski, P. (1993) *Quantum Mechanics of Continuously Observed Systems*, N. Copernicus Univ. Press, Toruń.
310. Steinhaus, H. (1923) Les probabilités dénombrables et leur rapport a la théorie de la mesure, *Fund. Math.* 4, 286–310.
311. Stinespring, W.F. (1955) Positive functions on C^*-algebras, *Proc. Am. Math. Soc.* 6, 211–216.
312. Størmer, E. (1963) Positive linear maps of operator algebras, *Acta Math.* 110,

233–278.

313. Størmer, E. (1980) Decomposition of positive projections on C^*-algebras, *Math. Ann.* **27**, 21–41.
314. Størmer, E. (1982) Decomposable positive maps on C^*-algebras, *Proc. Am. Math. Soc.* **86**, 402–404.
315. Szafnicki, B. (1969) On the entropy of systems of operators, *Bull. Acad. Polon. Sci.*, *Sèr. math.phys.astr.* **17**, 405–409.
316. Szent-Györgyi, A. (1951) *Chemistry of Muscular Contraction*, 2nd ed., Academic Press, New York.
317. Szilard, L. (1929) Über die Entropieverminderung in einem thermodynamischen System bei Eingreifen intelligenter Wesen, *Z. Phys.* **53**, 840–856 (Engl. transl. 1964, On the decrease of entropy in a thermodynamic system by the intervention of intelligent beings, *Behavioral Science* **9**, 301–310).
318. Titchmarsh, E. C. (1951) *The Theory of the Riemann zeta-function*, Clarendon Press, Oxford.
319. Toda, M. (1983) *Nonlinear Waves and Solitons* (in Japanese), Nihon Hyōron-sha, Tokyo.
320. Tverberg, H. (1958) A new derivation of the information function, *Math. Scand.* **6**, 297–298.
321. Uhlmann, A. (1970) On the Shannon entropy and related functionals on convex sets, *Rep. Math. Phys.* **1**, 147–159.
322. Uhlmann, A. (1977) Relative entropy and the Wigner-Yanase-Dyson-Lieb concavity in an interpolation theory, *Commun. Math. Phys.* **54**, 21–32.
323. Umegaki, H. (1959) Conditional expectations in an operator algebra III, *Kodai Math. Sem. Rep.* **11**, 51–64.
324. Umegaki, H. (1962) Conditional expectations in an operator algebra IV, *Kodai Math. Sem. Rep.* **14**, 59–85.
325. Umegaki, H. (1962) Concitional expectations in an operator algebra IV, *Kodai Math. Sem. Rep.* **14**, 59–85.
326. Urbanik, K. (1961) Joint probability distributions of observables in quantum mechanics, *Studia Math.* **21**, 317–323.
327. Urbanik, K. (1964) The principle of increase of entropy in quantum mechanics *Trans. 3rd Prague Conf. Information Theory, Statistical Decision Functions, Random Processes*, Prague, 743.
328. Urbanik, K. (1972) On the concept of information, *Bull. Acad. Polon. Sci.*, *Sér. masth, astr. phys.* **20**, 887–890.
329. Urbanik, K. (1973) On the definition of information, *Rep. Math. Phys.* **4**, 289–301.
330. Van Hove, L. (1955) Quantum-mechanical perturbations giving rise ro a statistical transport equation, *Physica* **21**, 517–540.
331. Van Hove, L. (1957) The approach to equilibrium in quantum statistics, *Physica* **23**, 441–480.
332. Varadarajan, V.S. (1970) *Geometry of Quantum Theory*, vol. 2, Van Nostrand, New York.
333. Von Neumann, J. (1932) *Mathematische Grundlagen der Quantenmechanik*, Springer, Berlin (Engl. transl. Princeton Univ. Press, Princeton 1955, Russ. transl. Nauka, Moscow 1964).
334. Von Wright, G.H. (1957) *Logical Studies*, Routledge and Kegan Paul, London.
335. Watanabe, N., (1991): "Efficiency of optical modulations for photon number states", Quantum Probability and Related Topics **6**, pp. 489–498.
336. Wehrl, A. (1977) Remarks of A-entropy, *Rep. Math. Phys.* **12**, 385–394.
337. Wehrl, A. (1978) General properties of entropy, *Rev. Mod. Phys.* **60**, 221–260.
338. Wehrl, A. (1991) The many facets of entropy, *Rep. Math. Phys.* **30**, 119–129.
339. Weiss, P. (1967) Der Satz von der Eindeutigkeit der Unbestimmtheit, Eim Beitrag zum Begriff "nützliche Information", *Kybernetik* **4**, 65–67.

340. Whittaker, E. T. (1915) On the functions which are represented by the expansions of the interpolatory theory, *Proc. Roy. Soc. Ediburgh* **35**, 181–194.

341. Wichmann, E. H. (1963) Density matrices arising from incomplete measurement, *J. Math. Phys.* **4**, 884–896.

342. Wiener, N. (1948) *Cybernetics or Control and Communication in the Animal and the Machine*, Hermann, Paris (2nd ed.: MIT Press, Cambridge, Mass. 1961).

343. Wilcox. R. M. (1967) Exponential operators and parameter differentiation in quantum physics, *J. Math. Phys.* **8**, 962–982.

344. Wilson, A. (1965) *Latin Dictionary*, Hodder and Stoughton, London.

345. Wolfowitz, J. (1964) *Coding Theorems in Information Theory*, Springer, Berlin.

346. Woodward, P.M. (1953) *Probability and Information Theory, with Applications to Radar*, McGraw-Hill, New York, 2nd ed. Pergamon Press, London 1955, Polish transl. PWN, Warszawa 1959.

347. Woronowicz, S.L. (1976) Positive maps of low-dimensional matrix algebras, *Rep. Math. Phys.*, **10**, 165–183.

348. Yockey, H. P. (1992) *Information Theory and Molecular Biology*, Cambridge University Press, Cambridge.

349. Yanagi, K. (1982) On some properties of Gaussian channels, J. Math. Anal. Appl. **88**, 364–377.

350. Yosida, K. (1971) *Functional Analysis*, 3rd ed., Springer, Berlin.

351. Zadeh, L.A. (1968): *Probability measures of fuzzy events*, J. Math. Anal. Appl., **23**, pp. 421-427.

352. Zvonkin, A.K., and Levin, L.A. (1970) Complexity of finite objects and foundation of the concepts of information and complexity by means of theory of algorithms' (in Russian), *Usp. Math. Nauk* **25** No.6, 85–127.

353. Zurek, W.H. (1989) Thermodynamic cost of computation, algorithmic complexity and the information metric, *Nature* **341**, 119–124.

354. Zurek, W.H. (1989) Algorithmic information and physical entropy, *Phys. Rev.* **A40**, 4731–4751.

355. Zwanzig, R. (1960) Ensemble method in theory of irreversibility, *J. Chem. Phys.* **33**, 1338–1341.

Index

Fundamental Theories of Physics

22. A.O. Barut and A. van der Merwe (eds.): *Selected Scientific Papers of Alfred Landé*. [*1888-1975*]. 1988   ISBN 90-277-2594-2
23. W.T. Grandy, Jr.: *Foundations of Statistical Mechanics*.
    Vol. II: *Nonequilibrium Phenomena*. 1988   ISBN 90-277-2649-3
24. E.I. Bitsakis and C.A. Nicolaides (eds.): *The Concept of Probability*. Proceedings of the Delphi Conference (Delphi, Greece, 1987). 1989   ISBN 90-277-2679-5
25. A. van der Merwe, F. Selleri and G. Tarozzi (eds.): *Microphysical Reality and Quantum Formalism, Vol. 1*. Proceedings of the International Conference (Urbino, Italy, 1985). 1988   ISBN 90-277-2683-3
26. A. van der Merwe, F. Selleri and G. Tarozzi (eds.): *Microphysical Reality and Quantum Formalism, Vol. 2*. Proceedings of the International Conference (Urbino, Italy, 1985). 1988   ISBN 90-277-2684-1
27. I.D. Novikov and V.P. Frolov: *Physics of Black Holes*. 1989   ISBN 90-277-2685-X
28. G. Tarozzi and A. van der Merwe (eds.): *The Nature of Quantum Paradoxes*. Italian Studies in the Foundations and Philosophy of Modern Physics. 1988
                                                                    ISBN 90-277-2703-1
29. B.R. Iyer, N. Mukunda and C.V. Vishveshwara (eds.): *Gravitation, Gauge Theories and the Early Universe*. 1989   ISBN 90-277-2710-4
30. H. Mark and L. Wood (eds.): *Energy in Physics, War and Peace*. A Festschrift celebrating Edward Teller's 80th Birthday. 1988   ISBN 90-277-2775-9
31. G.J. Erickson and C.R. Smith (eds.): *Maximum-Entropy and Bayesian Methods in Science and Engineering*.
    Vol. I: *Foundations*. 1988   ISBN 90-277-2793-7
32. G.J. Erickson and C.R. Smith (eds.): *Maximum-Entropy and Bayesian Methods in Science and Engineering*.
    Vol. II: *Applications*. 1988   ISBN 90-277-2794-5
33. M.E. Noz and Y.S. Kim (eds.): *Special Relativity and Quantum Theory*. A Collection of Papers on the Poincaré Group. 1988   ISBN 90-277-2799-6
34. I.Yu. Kobzarev and Yu.I. Manin: *Elementary Particles. Mathematics, Physics and Philosophy*. 1989   ISBN 0-7923-0098-X
35. F. Selleri: *Quantum Paradoxes and Physical Reality*. 1990   ISBN 0-7923-0253-2
36. J. Skilling (ed.): *Maximum-Entropy and Bayesian Methods*. Proceedings of the 8th International Workshop (Cambridge, UK, 1988). 1989   ISBN 0-7923-0224-9
37. M. Kafatos (ed.): *Bell's Theorem, Quantum Theory and Conceptions of the Universe*. 1989   ISBN 0-7923-0496-9
38. Yu.A. Izyumov and V.N. Syromyatnikov: *Phase Transitions and Crystal Symmetry*. 1990   ISBN 0-7923-0542-6
39. P.F. Fougère (ed.): *Maximum-Entropy and Bayesian Methods*. Proceedings of the 9th International Workshop (Dartmouth, Massachusetts, USA, 1989). 1990
                                                                    ISBN 0-7923-0928-6
40. L. de Broglie: *Heisenberg's Uncertainties and the Probabilistic Interpretation of Wave Mechanics*. With Critical Notes of the Author. 1990   ISBN 0-7923-0929-4
41. W.T. Grandy, Jr.: *Relativistic Quantum Mechanics of Leptons and Fields*. 1991
                                                                    ISBN 0-7923-1049-7
42. Yu.L. Klimontovich: *Turbulent Motion and the Structure of Chaos*. A New Approach to the Statistical Theory of Open Systems. 1991   ISBN 0-7923-1114-0

Fundamental Theories of Physics

43. W.T. Grandy, Jr. and L.H. Schick (eds.): *Maximum-Entropy and Bayesian Methods.* Proceedings of the 10th International Workshop (Laramie, Wyoming, USA, 1990). 1991   ISBN 0-7923-1140-X
44. P.Pták and S. Pulmannová: *Orthomodular Structures as Quantum Logics.* Intrinsic Properties, State Space and Probabilistic Topics. 1991   ISBN 0-7923-1207-4
45. D. Hestenes and A. Weingartshofer (eds.): *The Electron.* New Theory and Experiment. 1991   ISBN 0-7923-1356-9
46. P.P.J.M. Schram: *Kinetic Theory of Gases and Plasmas.* 1991   ISBN 0-7923-1392-5
47. A. Micali, R. Boudet and J. Helmstetter (eds.): *Clifford Algebras and their Applications in Mathematical Physics.* 1992   ISBN 0-7923-1623-1
48. E. Prugovečki: *Quantum Geometry.* A Framework for Quantum General Relativity. 1992   ISBN 0-7923-1640-1
49. M.H. Mac Gregor: *The Enigmatic Electron.* 1992   ISBN 0-7923-1982-6
50. C.R. Smith, G.J. Erickson and P.O. Neudorfer (eds.): *Maximum Entropy and Bayesian Methods.* Proceedings of the 11th International Workshop (Seattle, 1991). 1993
                                                ISBN 0-7923-2031-X
51. D.J. Hoekzema: *The Quantum Labyrinth.* 1993   ISBN 0-7923-2066-2
52. Z. Oziewicz, B. Jancewicz and A. Borowiec (eds.): *Spinors, Twistors, Clifford Algebras and Quantum Deformations.* Proceedings of the Second Max Born Symposium (Wrocław, Poland, 1992). 1993   ISBN 0-7923-2251-7
53. A. Mohammad-Djafari and G. Demoment (eds.): *Maximum Entropy and Bayesian Methods.* Proceedings of the 12th International Workshop (Paris, France, 1992). 1993
                                                ISBN 0-7923-2280-0
54. M. Riesz: *Clifford Numbers and Spinors* with Riesz' Private Lectures to E. Folke Bolinder and a Historical Review by Pertti Lounesto. E.F. Bolinder and P. Lounesto (eds.). 1993   ISBN 0-7923-2299-1
55. F. Brackx, R. Delanghe and H. Serras (eds.): *Clifford Algebras and their Applications in Mathematical Physics.* Proceedings of the Third Conference (Deinze, 1993) 1993
                                                ISBN 0-7923-2347-5
56. J.R. Fanchi: *Parametrized Relativistic Quantum Theory.* 1993   ISBN 0-7923-2376-9
57. A. Peres: *Quantum Theory: Concepts and Methods.* 1993   ISBN 0-7923-2549-4
58. P.L. Antonelli, R.S. Ingarden and M. Matsumoto: *The Theory of Sprays and Finsler Spaces with Applications in Physics and Biology.* 1993   ISBN 0-7923-2577-X
59. R. Miron and M. Anastasiei: *The Geometry of Lagrange Spaces: Theory and Applications.* 1994   ISBN 0-7923-2591-5
60. G. Adomian: *Solving Frontier Problems of Physics: The Decomposition Method.* 1994
                                                ISBN 0-7923-2644-X
61. B.S. Kerner and V.V. Osipov: *Autosolitons.* A New Approach to Problems of Self-Organization and Turbulence. 1994   ISBN 0-7923-2816-7
62. G.R. Heidbreder (ed.): *Maximum Entropy and Bayesian Methods.* Proceedings of the 13th International Workshop (Santa Barbara, USA, 1993) 1996   ISBN 0-7923-2851-5
63. J. Peřina, Z. Hradil and B. Jurčo: *Quantum Optics and Fundamentals of Physics.* 1994
                                                ISBN 0-7923-3000-5

Fundamental Theories of Physics

64. M. Evans and J.-P. Vigier: *The Enigmatic Photon*. Volume 1: The Field $B^{(3)}$. 1994
ISBN 0-7923-3049-

65. C.K. Raju: *Time: Towards a Constistent Theory*. 1994   ISBN 0-7923-3103-
66. A.K.T. Assis: *Weber's Electrodynamics*. 1994   ISBN 0-7923-3137-
67. Yu. L. Klimontovich: *Statistical Theory of Open Systems*. Volume 1: A Unifie
Approach to Kinetic Description of Processes in Active Systems. 1995
ISBN 0-7923-3199-0; Pb: ISBN 0-7923-3242-

68. M. Evans and J.-P. Vigier: *The Enigmatic Photon*. Volume 2: Non-Abelian Electro
dynamics. 1995   ISBN 0-7923-3288-

69. G. Esposito: *Complex General Relativity*. 1995   ISBN 0-7923-3340-
70. J. Skilling and S. Sibisi (eds.): *Maximum Entropy and Bayesian Methods*. Proceeding
of the Fourteenth International Workshop on Maximum Entropy and Bayesia
Methods. 1996   ISBN 0-7923-3452-

71. C. Garola and A. Rossi (eds.): *The Foundations of Quantum Mechanics – Historica
Analysis and Open Questions*. 1995   ISBN 0-7923-3480-

72. A. Peres: *Quantum Theory: Concepts and Methods*. 1995 (see for hardback editior
Vol. 57)   ISBN Pb 0-7923-3632-

73. M. Ferrero and A. van der Merwe (eds.): *Fundamental Problems in Quantum Physics
1995   ISBN 0-7923-3670-

74. F.E. Schroeck, Jr.: *Quantum Mechanics on Phase Space*. 1996   ISBN 0-7923-3794-
75. L. de la Peña and A.M. Cetto: *The Quantum Dice*. An Introduction to Stochasti
Electrodynamics. 1996   ISBN 0-7923-3818-

76. P.L. Antonelli and R. Miron (eds.): *Lagrange and Finsler Geometry*. Applications t
Physics and Biology. 1996   ISBN 0-7923-3873-

77. M.W. Evans, J.-P. Vigier, S. Roy and S. Jeffers: *The Enigmatic Photon*. Volume 3
Theory and Practice of the $B^{(3)}$ Field. 1996   ISBN 0-7923-4044-

78. W.G.V. Rosser: *Interpretation of Classical Electromagnetism*. 1996
ISBN 0-7923-4187-

79. K.M. Hanson and R.N. Silver (eds.): *Maximum Entropy and Bayesian Methods*. 1996
ISBN 0-7923-4311-

80. S. Jeffers, S. Roy, J.-P. Vigier and G. Hunter (eds.): *The Present Status of the Quantur
Theory of Light*. Proceedings of a Symposium in Honour of Jean-Pierre Vigier. 1997
ISBN 0-7923-4337-

81. *Still to be published*
82. R. Miron: *The Geometry of Higher-Order Lagrange Spaces*. Applications to Mechanic
and Physics. 1997   ISBN 0-7923-4393-

83. T. Hakioğlu and A.S. Shumovsky (eds.): *Quantum Optics and the Spectroscopy a
Solids*. Concepts and Advances. 1997   ISBN 0-7923-4414-

84. A. Sitenko and V. Tartakovskii: *Theory of Nucleus*. Nuclear Structure and Nuclea
Interaction. 1997   ISBM 0-7923-4423-

85. G. Esposito, A.Yu. Kamenshchik and G. Pollifrone: *Euclidean Quantum Gravity o
Manifolds with Boundary*. 1997   ISBN 0-7923-4472-

86. R.S. Ingarden, A. Kossakowski and M. Ohya: *Information Dynamics and Ope
Systems*. Classical and Quantum Approach. 1997   ISBN 0-7923-4473-

KLUWER ACADEMIC PUBLISHERS – DORDRECHT / BOSTON / LONDON